GLACIAL GEOLOGY AND GEOMORPHOLOGY

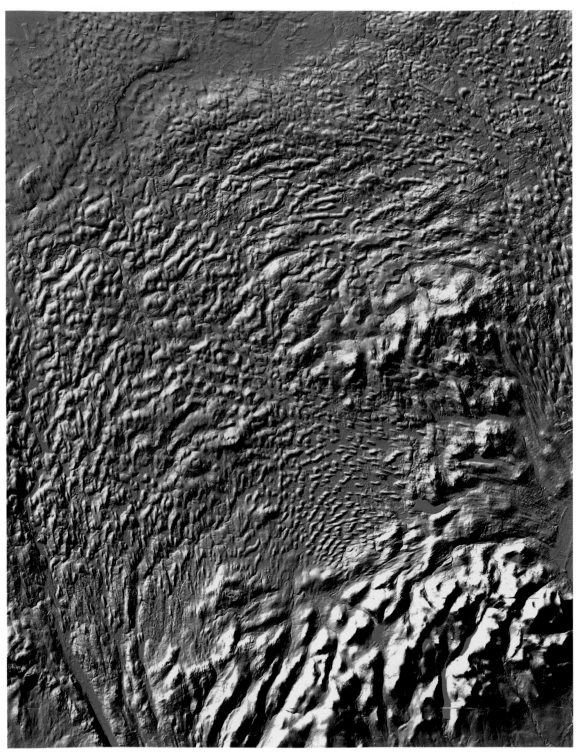

Subglacial flow-transverse, ribbed moraine overprinted by flow-parallel drumlins, south Ulster lowlands. View from Mourne Mountains looking north.

Glacial Geology and Geomorphology

The Landscapes of Ireland

Professor A. Marshall McCabe ScD, MRIA

Professor of Quaternary Science, University of Ulster

DUNEDIN

Published by
Dunedin Academic Press Ltd
Hudson House
8 Albany Street
Edinburgh EH1 3QB
Scotland

ISBN 978-1-903765-87-6

British Library Cataloguing in Publication Data
A catalogue record for this book is available from the British Library

Design and prepress production by Makar Publishing Production
Printed in Spain by Grafo SA

CONTENTS

LIST OF FIGURES

ACKNOWLEDGEMENTS

This book has benefited from generous support and encouragement from the Geological Survey of Northern Ireland, especially Garth Earls, Ian Mitchell and Alex Donald.

I thank my colleagues, including Jorie and Peter Clark, Kathy Delaney, Alex Donald, Paul Dunlop, Ian Enlander, Jim Hansom, Steve McCarron and Geoff Thomas, who have kindly supplied photographs and diagrams which have enhanced and improved the text. Kilian McDaid and Lisa Rodgers are also to be thanked for drawing the diagrams so expertly.

Finally, I must acknowledge the influence of J.R.L. Allen and D. Bowen who encouraged me to write this book.

1

Introduction

At first sight the glacial heritage of Ireland could be viewed simply as one of the many areas situated on mid-latitude continental margins that were repeatedly glaciated during the last two million years. However, in detail the glacigenic landscape is quite different to other neighbouring glaciated margins such as western Scotland and Scandinavia. Although Ireland sustained its own independent ice domes and centres like these other regions, its ice accumulation areas are characterised by a marked subglacial imprint of thick basal tills and one of the largest areas of continuous ribbed moraine in the world (Fig. 1.1). The island is not dominated by persistent areal and linear erosional patterns, overdeepened glacial troughs, rock basins and breached cols that are so typical of other mid-latitude glaciated continental margins. Causes of these general differences are difficult to quantify but are probably related to the fact that the glacial legacy of Ireland is dependent on growth and decay of independent, lowland ice sheets in an island setting (Fig. 1.2). The ice sheets on Scotland and Scandinavia were at least initially based in highland areas and responded to climate signals in slightly different modes and probably had different sensitivity thresholds of temperature and precipitation.

The glacial climate of Ireland was somewhat different to that in the remainder of the British Isles simply because of proximity to moisture from the Atlantic Ocean along the western seaboard combined with its extensive lowland glaciokarstic basins. Perhaps the combination of slight climatic differences and physiography has also increased the preservation potential of thick glacigenic deposits formed during deglaciation which generally offlap inland and northwards across the island. These marked offlap patterns across swathes of countryside record phases of widespread deglaciation on millennial timescales suggesting that the lowland ice sheets were more sensitive to climatic amelioration than highland ice masses. The widely held view that Scottish ice dominated the western glacial system especially in terms of patterns of isostatic depression is revisited in the text.

In general the book follows a progression from standard litho- and bio-stratigraphy to ice–ocean–climate–lithosphere interactions which characterise the dynamic Irish ice sheet system (Fig. 1.3a, b). The complexity of sediments and landforms formed around and below the last ice sheet in Ireland is clearly related to its history and character including major shifts in centres of ice dispersal, changes in ice-marginal configuration and the geomorphic work effected at the ice-rock interface (Fig. 1.4). These changes are mainly the ice sheet response to millennial timescale climate changes in the North Atlantic. The resulting hypotheses generated from field observations of landform patterns, stratigraphy and sedimentary variability in Ireland for well over one hundred and fifty years has contributed to the development of the subject area outside the island especially the contributions of Francis Synge and Frank Mitchell.

Ireland's position in the path of eastward moving climate systems during the Quaternary has created a potential reservoir for many of the climate signals related to changes in Atlantic sea surface temperatures. Other North Atlantic climate data in ice sheet cores and deep sea cores show that these climate signals were operating over centennial, millennial and much longer timescales (Bond et al., 1999). Because changes in climate directly affect the dominant processes operating at the earth's surface, landforms mosaics and subsurface deposits must record some of these signals which strongly influenced the main depositional and erosional systems that operated over the last two million years. Terrestrial records also testify to the sensitivity of the island to climate change which influenced repeated growth and decay of ice sheets across the central lowlands, highland boundary rim and continental shelf. Ice sheets were sensitive not only to land chilling by orbital forcing but to changes in the strength of the Gulf Stream and

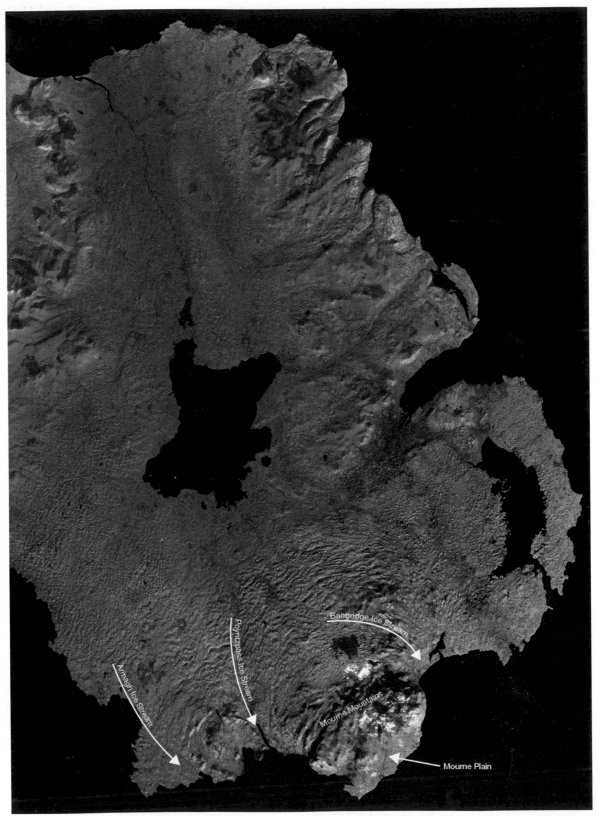

Fig. 1.1 Satellite image (Landsat TM5) showing widespread subglacial transverse bedforms (ribbed moraine) in the north Irish lowlands. At least three major ice streams from Armagh, Lough Neagh and Banbridge crosscut and form drumlins across the field of transverse ridges before ending at tidewater (BGS image).

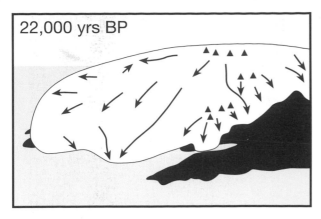

22,000 yrs BP

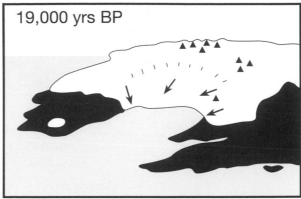

19,000 yrs BP

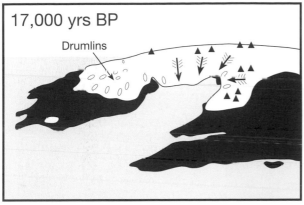

17,000 yrs BP

Drumlins

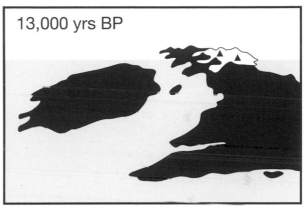

13,000 yrs BP

Fig. 1.2 Cartoons illustrating ice extent before 22 cal ka BP and general configuration during ice decay in the British Isles. Note that the thickest ice sheets were located along the western seaboard and on the Irish lowlands. During deglaciation widespread ice sheet readvance drumlinised the northern Irish lowlands and deposited morainal banks at tidewater margins. From Eyles and McCabe, 1989a.

Fig. 1.3a Tidewater glacier, eastern Spitzbergen. Gilbert-type deltas and dated marine deposits which are interbedded with terminal outwash show that the margins of the Irish ice sheet commonly ended either at tidewater or on the continental shelf and provide an analogy with contemporary tidewater ice margins.

Fig. 1.3b Basal glacial till from Rathcor, Dundalk Bay, consisting of dispersed, ice-bevelled clasts set in a fine-grained matrix derived from marine muds cannibalised and homogenised during ice sheet readvance.

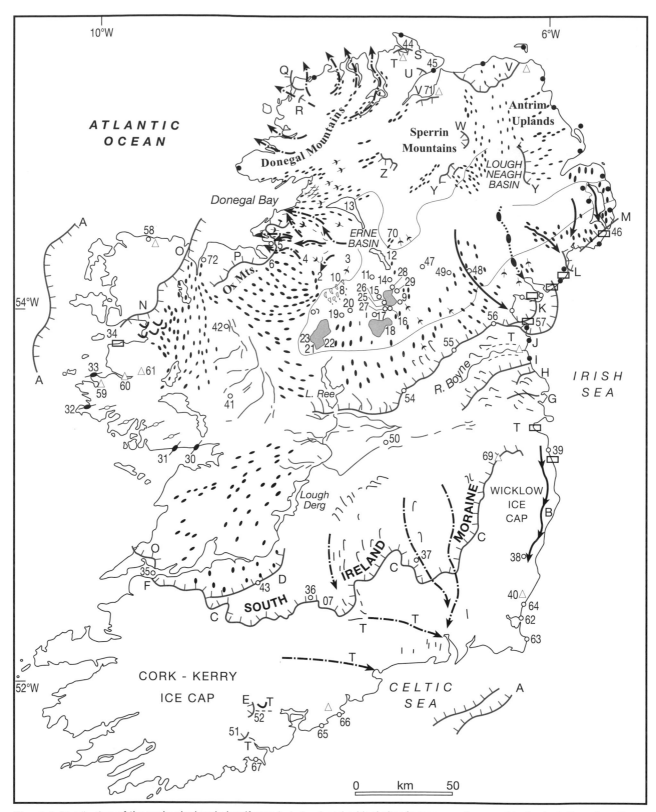

Fig. 1.4 Distribution of the main glacigenic landforms in Ireland and critical sites/exposures discussed in the text. Based on Synge (1970a) and published records of Charlesworth, Colhoun, Coxon, Farrington, McCabe, Sollas and Stephens. *See Keys on opposite page.*

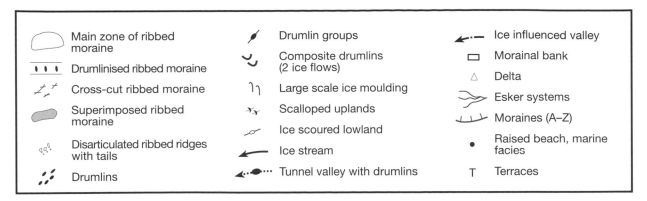

⌒	Main zone of ribbed moraine	✎	Drumlin groups
ı ı ı	Drumlinised ribbed moraine	⌣	Composite drumlins (2 ice flows)
⌁	Cross-cut ribbed moraine	⑂	Large scale ice moulding
⬭	Superimposed ribbed moraine	⤳	Scalloped uplands
⁎	Disarticulated ribbed ridges with tails	⌒	Ice scoured lowland
⁚⁚	Drumlins	←	Ice stream
		◀••••	Tunnel valley with drumlins

←•—	Ice influenced valley
▭	Morainal bank
△	Delta
⤳	Esker systems
⏖	Moraines (A–Z)
•	Raised beach, marine facies
T	Terraces

A	Moraine or subglacial incisions on continental shelf	1	Diffagher Valley	37	Kilkenny
B	Margin of Irish Sea glacier	2	Lough Allen	38	Gorey
C	SIEM	3	Cuilcagh Mountain	39	Bray
D	Fedamore	4	Belhavel Lough	40	Screen Hills
E	Killumney	5	Sligo	41	Tuam
F	Scattery Island	6	Ballysadare Bay	42	Kilkelly
G	Ward	7	Carrick on Shannon	43	Fedamore
H	Devlin	8	Ballinamore	44	Ballycrampsey
I	Ben Head	9	Cavan	45	Moville
J	Clogher Head - Kilkeel	10	SL. Anierin	46	Killard Point
K	Cranfield - Dunany - Athlone - Rathcor	11	Ballyconnell	47	Newbliss
L	Dunmore Head	12	Upper Lough Erne	48	Castleblaney
M	Killard Point	13	Lower Lough Erne	49	Ballybay
N	Furnace Lough	14	Belturbet	50	Tullamore
O	Killala	15	Killeshandra	51	Bandon
P	Ox Mountains	16	Crossdoney	52	Killumney
Q	Bloody Foreland	17	Arva	53	Athlone
R	Arduns	18	Lough Gowna	54	Mullingar
S	Ballycrampsey	19	Mohill	55	Kells
T	Carndonagh	20	Cloone	56	Ardee
U	Moville	21	Strokestown	57	Dunany Point
V	Armoy - Ballykelly - East Antrim	22	SL. Bawn	58	Glenulra
W	Tobermore	23	Kilglass Lough	59	Tullywee
Y	Lough Fea	24	Rivory	60	Killary Harbour
Z	Deerpark	25	Eonish	61	Strahlea
		26	Tullybrick	62	Tincone
		27	Tawlaght	63	St Helens
		28	Farnham	64	Blackwater Harbour
		29	Butlersbridge	65	Ballycotton
		30	Barna	66	Knockadoon Warren
		31	Spiddal	67	Courtmacsherry
		32	Ballyconnelly	68	Clonakilty Bay
		33	Kawrawer	69	Blessington
		34	Askillaun	70	Derryvree
		35	Scattery Island	71	Limavady
		36	Tipperary	72	Inishcrone

meridional overturning of the Atlantic Ocean (Ruddiman et al., 1980; Bowen, 1991). Therefore the glacial records in Ireland provide a testing ground for ideas on much larger climate-related events on both hemispheric and global scales (Clark et al., 2004). Therefore where possible reconstructions of entire regional and some larger ice sheet systems such as the Drumlin ice sheet are presented in the text if sufficient dating control is available.

The chapter organisation and purpose of this book is therefore closely linked to the idea that the glacial landscapes and glacial stratigraphies of Ireland are a product not only of mobile ice sheets but of repeated growth and decay cycles on Milankovitch timescales. A common theme is that the subglacial imprint of the last major ice sheet will be the most prominent landscape feature and that the preservation of older records will be fragmentary and buried. However, before the glacial records can be evaluated in depth it is essential to look at the different approaches that have been applied to landscape study through time, not least because of the contributions made by organisations such as The Geological Survey and numerous individuals. For example, it is tempting to suggest that the outdated ideas on catastrophism still have a role to play because interpretations of rapid environmental change during some glacigenic events occurred on centennial time-scales. The organisation of the last major glacial cycle certainly builds on the theme of rapid climatic, biological and glacial changes across the island. Perhaps the most prominent landscape of the island, traditionally known as the drumlin belt, provides detailed insights into how subglacial bedform patterns reflect extremely rapid changes in ice sheet configuration and conditions at the ice sediment interface. For over one hundred years records from The Irish Sea Glacier(s) has provided much needed inspiration on ideas surrounding the evolution of the British/Irish ice sheet (BIIS). These are still being tested today. It is argued on the basis of ice thickness and patterns of isostatic depression that the Irish ice masses played a dominant role in the BIIS system. The widely held view, based on erratic dispersal of Ailsa Craig microgranite, that Scottish ice dominated the last glacial system is not justified simply on the grounds that the centres of ice sheet dispersal on the Irish lowlands covered an area about twice that of Scotland.

Ice wastage after deglaciation of the continental shelf is separated into two chapters, one on terrestrial deglaciation and one specifically on eskers. The separation permits focus on esker ridges as the second most distinctive landform on the island but also on the nature, timing and pattern of deglaciation. The remaining chapters on ice sheet readvances and late-glacial sea levels are embedded in the history of the last ice sheet and provide accounts of the evidence used to reconstruct ice sheet systems. Highlighted are the links between ice activity, ice sheet loading of the lithosphere and crustal response, eustatic sea levels and the sedimentary or facies records. In some cases sediment exposures can relate to more than one event that are considered in separate chapters. For example, some sections from drumlins record tectonic events prior to final drumlinisation or shaping. Also certain sections in ice-marginal or glaciomarine sequences may provide evidence for a closely related ice sheet readvance.

A critical point to consider during environmental reconstruction is that expected facies sequences are rarely complete or preserved in the geological record (Fig. 1.5). The task of the geologist which is attempted in this book is therefore to piece together the available records within a framework that is not only age-constrained but is capable of being tested. In this respect detailed field maps, sedimentological logs and sketches and dating of events are basic tools used in initial field investigations and reconstructions of ice sheet history. However, a recurrent problem is that most ice sheet and geophysical models are unable to capture the current age-constrained field evidence.

In Ireland, unlike in the rest of the British Isles, there is a limited amount of stratigraphical and dating evidence for glacial and non-glacial events prior to the Midlandian cold stage. This term has been used to include all deposits that formed during the last major cold stage following the last interglacial (Mitchell et al., 1973). It follows the 1973 standard classification that accepted climatic fluctuations as the guiding principle for defining the stages of the British Quaternary (Bowen, 1999). At that time pollen analysis was considered to be a primary tool for zonation even though pollen biozone boundaries were diachronous and similar biozones could be expected to reoccur in response to climate at different times. More recently new dating methods including thermoluminescence (TL), electron spin resonance (ESR) and amino acid dating of bivalves has helped to support glacial geochronology. The application of amino acid dating of marine molluscs from diamicts around the Irish Sea Basin has for example demonstrated that the basin was flooded immediately prior to the traditional glacial maximum (Bowen et al., 2002). This discovery which questioned existing orthodoxy led the way for more detailed age calibration using AMS ^{14}C dating of microfaunas recording marine transgressions during the last glacial/deglacial cycle (Fig. 1.6). The combination of these dating methods with detailed facies analysis has now provided a powerful tool which has begun to document the complex series of events constituting the last glaciation and its termination (McCabe and Dunlop, 2006). The complexity and timing of these glacigenic events during the last glaciation has led to a complete revision of existing data sets. Patterns of nested lines of ice-contact moraines on the continental shelf off western Ireland record former ice limits but their significance is unknown through lack of dating control. However, their presence may be

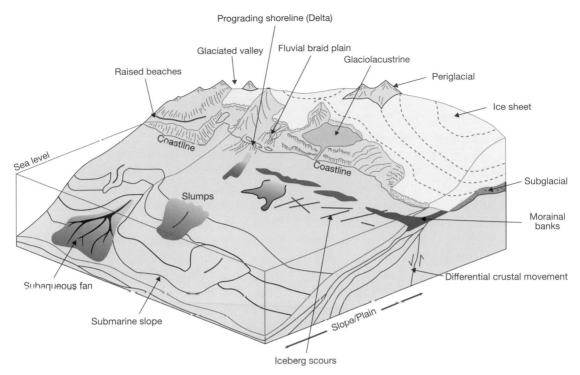

Fig. 1.5 Block diagram showing the main depositional environments found near an ice front. Sedimentary and geomorphic complexity will result when the ice front oscillates and decays from limits offshore. (After McCabe and Dunlop, 2006.) *By kind permission of GSNI.*

Fig. 1.6 Well-preserved species of foraminifera from marine muds deposited in the Irish Sea Basin (Dog Mills, Clogga, Ballyvaldon and Shellag). *Elphidium clavatum* (1–4) and *Haynesina orbiculare* (5–8). Monospecific samples of the former have been AMS ^{14}C dated to provide a deglacial stratigraphy for the western basin (McCabe et al., 2005, 2007b).

related to oscillating ice sheet margins that may even predate the traditional late glacial maximum (LGM, ~22 cal ka BP) and extend ice sheet history back into MIS Stage 3 (>30 cal ka BP). Many events during the last glacial termination (<22 cal ka BP) are now thought to occur on millennial timescales or less and some can be correlated with major events in the North Atlantic such as Heinrich events (McCabe et al., 2005, 2007b). Short summaries of most of the important lithostratigraphic units and formations used in this book can be found in Bowen's (1999) correlation of Quaternary Deposits in the British Isles. However, major interpretative problems, including the environmental significance and climatic signature of sedimentary successions and landforms, often result because localised studies of glacigenic sediments generally fail to consider the regional parameters controlling sedimentation in a basin or wider context.

If the location of Ireland in the North Atlantic is climatically significant then the gross physiography of the island provides an important anchor for patterns of glacier development. The island can be compared to a collapsed pie consisting of a flat to saucer-shaped central plain with prominent basins (Erne, Shannon, Lough Neagh) surrounded by a discontinuous rim of mountains. The central plain rarely rises above the 120 m contour and is developed across flat-lying beds of Carboniferous limestones pierced by some hilly inliers of older Devonian and volcanic rocks. Carboniferous sandstones form prominent outliers to the west of Lough Erne which channelled basal ice flows. The landscape could be termed a glaciokarst eroded by ice, scalloped into a range of ice-oriented lakes and draped by variable thicknesses of drifts. Generally the mountains on the coastal rim are compact and often fragmented allowing ice flows to move seaward from the lowlands. The steep-sided glens traversing the Cuilcagh Plateau country channelled sinuous basal ice flows towards Donegal Bay. Drumlin trends and other modified bedforms aligned along valley axes track these ice flows away from inland centres of ice dispersal. Occasionally isolated fjords such as Killary Harbour were formed during repeated ice advance onto the continental shelf, though clusters of similar troughs mark the position of ice streams on the former periphery of the Donegal Mountain ice cap (Fig. 1.7). Most of the peripheral mountain ranges contain corries, slopes, valleys, basins and streamlined features typical of mountain glaciations elsewhere (Fig. 1.7) (Farrington, 1938; Dury, 1957, 1958, 1959; Warren, 1979; Coxon and Browne, 1991). In mountains or uplands where there is strong structural and tectonic grain, the resulting weaknesses have been exploited and overdeepened reflecting a marked positive feedback between glacial erosion, linear ice flow and rock removal (Figs. 1.8, 1.9). Uncertainty exists, however, on the precise timings of erosion and the number of occasions that ice reoccupied mountain valley systems.

A significant aspect of glacial studies around different mountain blocks concerns the relationships between mountain (Local) and lowland (General) ice sheets. The model was based mainly on the works of Farrington (1934; 1942; 1944; 1949; 1953; 1954) which recognised expansions of mountain ice and their phased relationships with the general or lowland ice sheets. In some cases a relative stratigraphy between the ice masses can be recognised where there is a common depositional sink such as a glacially dammed lake bordering both ice margins. The scheme was developed to include correlations between greater and lesser Cork/Kerry glaciations with that elsewhere in Ireland (Farrington, 1954). These relationships have not been thoroughly tested by independent dating techniques and remain as lithofacies records and locations of gravelly ridges and spreads. Recent work (Ballantyne et al., 2006, 2007) has used exposure dating to determine the timing of deglaciation around some mountain groups such as the Wicklows. The results emphasise the time when the mountain ice and lowland ice separated, suggesting that some of the ice limits formerly ascribed to distinct ice sheet glaciations are in reality ice decay phenomena. This type of model has been embedded in the literature for about seventy years, strongly influencing contemporary thought and styles of publications. Current thought may place more emphasis on the millennial timescales now thought to dominate ice sheet history.

Like any other dominant geomorphic system ice sheets not only utilised older morphological lineaments such as structurally controlled valleys but also provided elements which strongly influenced postglacial systems. From a tectonic viewpoint, isostatic depression followed by rebound and structural readjustment may have resulted in significant neotectonics (Knight, 1999). More obvious effects such as pressure release following erosion around the margins of the Antrim uplands has contributed significantly to the development of multiple rotational slumps which now characterise the basalt scarp. To the north of Larne at Sallagh Braes, older landslips formed during early deglaciation were overridden and ice-moulded by North Channel ice during the subsequent Drumlin Readvance. The two generations of landslips illustrate the close coupling and cascading links between major glacial events and subsequent lithospheric responses (Wilson, 2004). More importantly glacigenic landforms provide the backcloth for biological diversity and drainage characteristics across the country. In a very general sense the intensity of this relationship increases from south to north and is well-marked within upland and mountain blocks. Locations of peatlands and raised bogs are closely related to local topographic contrasts which also give rise to rapid changes in soil type because moisture contrasts also mirror topographic changes. These patterns are best observed within the areas of ribbed moraine and drumlins which provide intricate though repeated patterns of slope-influenced drainage. The very sluggish water pathways through some areas of ribbed moraine especially in the inland

Fig. 1.7 Relief map of Ireland illustrating the position of peripheral mountains and extensive lowlands which sustained the main centres of ice dispersal.

basins of the Erne and Shannon are closely dependent on lake systems constrained by tortuous to curved outlines of the subjacent ribbed moraines and poorly-connected drainage ditches. In contrast the marked topographic NW–SE grain provided by drumlin orientation immediately north of the limiting moraine is drained efficiently by small streams from County Louth to County Longford. The efficiency of each system must impact on current levels of chemicals retained in both soil and water bodies. In many areas of the west efficient ice scouring during areal ice flow onto the continental shelf has resulted in subdued knock and lochan type topographies draped with postglacial bogs. Some of the best examples with local reliefs of 10–15 m developed

Fig. 1.8a Killary Harbour looking westwards. An example of a structurally controlled valley overdeepened by ice flow onto the continental shelf. Gilbert-type deltas at the head of the valley record high relative sea level during deglaciation.

Fig. 1.8b Glenmacnass, County Wicklow. An example of a deeply-eroded, linear U-shaped valley associated with a prominent mountain ice cap.

Fig. 1.9a Lough Acorrymore, Achill Island, western Ireland. A corrie basin consisting of at least two corrie basins and multiple boulder moraines situated immediately outside the outer corrie lip.

Fig. 1.9b Lough Nakeerogue, Achill Island. Possibly the lowest corrie basin in Ireland near present sea level. Currently the frontal moraine which consists of debris eroded and transported from the corrie basin is being eroded by marine action.

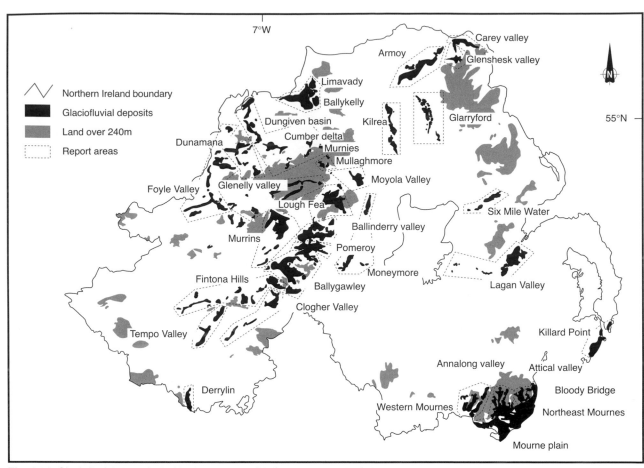

Fig. 1.10 Glaciofluvial deposits in Northern Ireland (McCarron et al., 1998). From a report on the scientific and aesthetic attributes of glaciofluvial landscapes in Northern Ireland, Countryside and Wildlife Branch of the Department of the Environment.

in response to spacings within joint systems. Perhaps one of the most important legacies of the ice age is the thick drifts which tend to mask or straighten coastal outlines between rocky headlands. Drift plugs occur at the heads of many bays along the east and south coasts. In a physiographic sense the Quaternary coastlines are subdued and the least spectacular of Ireland's coasts. Nevertheless, because they consist of unconsolidated sediments they are easily eroded and constitute a critical supply of sediment which maintains what is essentially a closed nearshore sediment system. Critical sediment loss to sediment sinks is therefore a current problem that cannot be solved easily because the major onshore flux of glacigenic sand finished during the final rise in postglacial sea level. Not only did the outwash from melting glaciers furnish debris for offshore bars, spits and spreads, but emergence of related nearshore forms provided a base and readily available sand supply to build extensive dune systems. In cases where an entire subglacial topography consisting of composite or single drumlins was submerged, local sediment cells and reorganisation of tombolas and spits add to the dynamic nature of postglacial and modern coastal setups. Finally, there is little doubt that sand and gravel landscapes are important aesthetically, providing diversity in drab, till-covered lowlands and spectacular hummocky relief in many glaciated valleys (McCabe, 1994). Across the island the rates of resource depletion are increasing to such an extent that conservation and preservation measures are essential if scientifically important landscapes are to survive (Fig. 1.10). Knight et al. (1999) concluded that many sand and gravel landscapes have a value well above that of their economic worth because of links to past cultures, aesthetic values, scientific values and natural landscapes. Rather than attempting to address these issues, this book is an exercise in historical geology focusing on description and explanation of glacigenic landscapes and deposits in Ireland.

2

Themes

Early observations

In the past our interpretations of landscape origin, form and evolution were piecemeal, lacking support from observational evidence and interpretation from North Atlantic records. However, many of the ideas and problems which still pervade the current literature were first identified by Victorian scientists and others well over 150 years ago. Prior to the glacial theory the 'contorted character and tumultuous arrangement of our glacial drifts' (Wright, 1937) and the seemingly chaotic arrangement of hills, lakes and hollows were difficult to interpret. Interpretations were made more difficult because of the very limited timescale of 6000 years contained in Archbishop Ussher's Received Chronology (Davies, 1964). The evolutionary view of landscapes therefore hardly existed because catastrophic concepts were at first sight more in tune with the existing religious orthodoxy. During the 18th and early 19th century scholars such as Kirwin linked glacial drifts with the term 'diluvium' and the earliest written accounts of a major event often termed The Noachian Deluge. Such catastrophic floods were perhaps one of the first unifying attempts to explain landscape patterns including striations, rock crushing, meltwater channels, overdeepened valleys, erratic carriage and drumlins. Weaver in 1819 saw the Wicklow landscape as the work of the raging torrents of a vast flood (Davies, 1964).

Later, cataclysmal diluvium mellowed and evolved into the Drift Theory which envisaged local glaciers releasing floating ice during a great submergence when most of the shelly gravels and so-called boulder clays formed. By 1843 Portlock could not decide whether or not the gravelly deposits of the north of Ireland were related to falling sea levels or products of major floods. Indeed the widespread shelly muds and boulder clays of County Londonderry were considered as calcareous Tertiary clays and distinct from superficial detritus.

Portlock (1843) considered that the widespread distribution of large granite erratics in County Londonderry was 'unavoidably connected with the action of floating ice'. Nevertheless Portlock (1843) recognised that it was difficult to reconcile rhythmically-bedded and delicate structures in sand and gravel beds with one violent rush of water. Portlock (1843) also recognised for the first time in Ireland that bimodal deposits may represent the simultaneous operation of two discrete processes, a marine current forming a bank and icebergs dumping a coarse fraction resulting in a fine gravel with boulders. These analogies between catastrophic events and the apparently chaotic aspects of glacial drifts were not only a manifestation of the prevailing intellectual environment but they predated significant Arctic research and observations on the geomorphic work of active ice sheets and glacial landscapes. Although modern work may show that the glacial agencies which eroded rocks and deposited debris throughout geological time are uniformitarian in context, many of our best models for explanation of glacial events require very dynamic ice sheet systems operating over short millennial to centennial timescales. In geological terms these timescales are catastrophic.

Up to the 1860s it was still common that notable authorities still regarded the widespread mantle of Drift as resulting from total submergence of Ireland. Apjohn's (1841) great NW to SE waves and Scouler's shells (1838) at 1000m in the Dublin Mountains 'confirmed' the depth of submergence. Well-marked glacial moraine ridges in Glenmalure, County Wicklow, were regarded by Oldham (1846) as wave generated and Kinahan (1862) pigeonholed eskers into barrier, shoal and fringe types based on marine analogies. Gradually throughout the late 19th century the debates and contests were between the supporters of the Glacial Theory and members of the Drift Lobby. The ideas and observations of workers such as Carvill-Lewis (moraines and no great submergence),

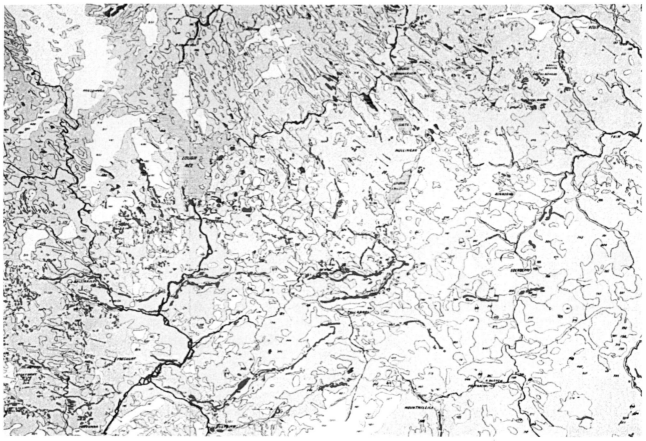

Fig. 2.1. Sollas's (1896) map of the esker ridges across the central midlands of Ireland. Eskers are in red lines and bogs/alluvium in yellow.

Close (drumlin distributions), Agassiz (ice-centres and moraines) and Sollas (eskers) provided a clear foundation for general acceptance that glaciers were the prime geomorphic agent which fashioned our landscapes (Fig. 2.1). Therefore the concepts of ice sheds, ice flow patterns, erratic distribution, ice frontal positions and especially major ice sheet systems came into vogue and are still central to mapping by field geologists (Figs. 2.2, 2.3).

Regional ice systems

Widespread observations by the Geological Survey continued to add to the glacial records at a variety of scales from individual erosional forms to landform patterns. Hallissy's (1914) early records of two different ridge orientations in the Rosslea–Monaghan area were compared to Close's (1866) 'curious mounds of boulder clay' and the presence or absence of blunt ends on upstream or downstream parts of the ridges. The main set of striae were parallel to NW–SE drumlins while Kilroe suggested to Hallissy that the NE–SW 'drumlin ridges' were asso-

ciated with an incursion by Scottish ice. Clearly, at this time the concept of flow-parallel and flow-transverse forms did not exist. Using over 600 sets of striae Kilroe (1888) began to recognise patterns of different ice sheet flows especially across the northern part of the island where The Scottish Glacial System produced ice flows from the northeast across the entire country and onto the Atlantic continental shelf. The Irish Glacial System involved a central snowfield stretching from northern Galway towards Lough Neagh and Belfast Lough (Fig. 1.7). Ice advanced north and south from this axis in a pattern similar to that of Hull's ice flow map. Kilroe considered that both ice systems were contemporaneous as the ice masses accumulated, but the Scottish System eventually dominated and moved westward uninterruptedly. We now have added the topics of stratigraphy, ice sheet limits and centres of ice dispersal to the literature together with the critical concepts of the extent of Irish and extraneous ice along the margins of the Irish Sea Basin. South of the Galway Bay-Strangford Lough line (limit of Scottish Ice System) Kilroe maintained that

Fig. 2.2 Compilation of lantern slides dating from the late 19th and early 20th century illustrating the interests of geologists and naturalists in landscape features. A) Large glacial erratic known as the Cloughmore stone at Rostrevor, County Down. B) The Scalp meltwater channel south County Dublin looking south. C) The Gap of Mamore, County Donegal cut by ice and meltwater. D). Steep ice-contact slopes of a moraine and adjacent meltwater channel near Gortin, County Tyrone. E) Summit surfaces and ice-contact faces of the delta moraines at Gortin, County Tyrone looking northwards. F) Field party examining the esker sediments in a tortuous ridge at Creggan, County Tyrone (Wright, 1937). *Compilation courtesy of Nigel McDowell*

ice moved southeastwards throughout the glacial epoch until the 'mer de glace' gave place to independent local glaciers and moraines marking the decline of glacial conditions on the island.

Ice limits, older and newer drift

Both Charlesworth (1924, 1939) and Dwerryhouse (1923) believed the Scottish and Irish ice systems to be contemporaneous though they argued that western Irish ice from Donegal crossed much of Ulster. Ice flows were in part reconstructed from erratic fans. Charlesworth's deglaciation models followed a similar pattern over much of Ireland and emphasised the role played by ice-dammed lakes and related overflow channels. Perhaps one of the most significant advances at this time was the correlation between the Carlingford Readvance, the Solway readvances and the Bride moraine of the Isle of Man (Charlesworth, 1939). Charlesworth (1928) also rediscovered the Southern Irish End Moraine (SIEM) which had been already identified by H. Carvill Lewis though it is not clear if Charlesworth (1928) regarded the moraine as the glacial maximum or more probably as the end-moraine of the Newer Drift. He observed that

Fig. 2.3 Compilation of lantern slides dating from the late 19th and early 20th century illustrating the interests of geologists and naturalists in landforms, sediments, sedimentary sequences on 'preglacial' marine platforms and megafaunal remains. A) Cross valley moraine in the Silent Valley, Mourne Mountains. B) The Happy Valley, Mourne Mountains with moraines, blockfields and a valley train. C) Rippled sand from the Greenhills esker, County Dublin. D) The marine rock-cut platform overlain by 'preglacial beach' gravel and head at Howe's Strand, County Cork (Wright and Muff, 1904). Note the fossil stack on the glaciated platform consisting mainly of periglacial head and the large, shore-parallel boulder (~2 m long) emerging from angular head at the base of the cliff. E) Map showing distribution of arcuate moraines blocking Lough Caragh, County Kerry, an early example of field mapping (e.g. Wright, 1937). F) Mammoth molars with their jaw bone from the Castlepook cave fauna, close to Doneraile, County Cork (Mitchell, 1976). *Compilation courtesy of Nigel McDowell.*

moraines were strikingly absent in the country outside the large end moraines. This observation was reinforced by comments that marginal drainage patterns were absent outside major ice sheet limits in the area of Older Drift where drift is less continuous, deeply weathered, boulders are more decomposed, drift topography is more smoothed due to later erosion and glacial mammalian faunas are recorded. Correlations were made between independent ice centres in the Wicklow Mountains and the Ivernian or Irish ice sheets suggesting contemporaneity. On a much broader scale Charlesworth went on to present a map of the entire Central Plain of Ireland depicting the mode of ice recession back toward the main centres of ice sheet dispersal in the north and west. Halts in the ice fronts were marked by confluent lobes and festooning of moraines which were thought to be continuous from coast to coast recording a rhythm superimposed on the main recessional conditions.

Correlations between the SIEM and moraines in Germany, the Baltic and Wisconsin limits in the USA followed (Charlesworth, 1928). A more cautious observer might ask how the relatively small mountain groups sited on the periphery of the island nourished the large ice sheets across the flat Carboniferous lowlands of central Ireland.

As in the rest of the British Isles the first half of the 20th century was dominated by discussions on the distribution of Older and Newer Drifts. Both Charlesworth (1928) and Wright (1937) accepted the division of drifts into fresh landforms north of the SIEM and more subdued forms to the south. However, Wright (1937) remarked that the apparent contrasts in topographic expression may be due to differences in the styles of retreat of earlier and later ice sheets rather than differences in the degree of weathering. This statement is one of the first to recognise that different topographies may simply result from variation in glaciological ice sheet parameters rather than weathering contrasts or age differences.

General glaciations and mountain glaciations

Farrington (1942, 1944, 1949, 1953, 1957, 1965) in a series of papers from the 1930s began to use detailed field evidence to reconstruct an Eastern General and a Midland General Glaciation together with independent centres of ice dispersal located on peripheral mountain groups. In the Wicklow Mountains for example Farrington (1944, 1949, 1954) established possible relations between the two general glaciations and local ice advances from the mountains. This scheme was based on the relative positions of moraine ridges in mountain valleys and along the margins of the mountain mass. In some cases erratic carriage was used to support specific ice sheet events or limits bordering the mountainous area of granite. To a large extent Synge substantiated Farrington's work on Wicklow, and work on basic ideas of General Glaciations was repeated across the island (e.g. 1968, 1973). Traditionally, older events were assigned a Munsterian age (penultimate cold stage) and younger ones were placed in the Midlandian Cold Stage (last cold stage). Reliance on using morphostratigraphic units of this type for stratigraphic and age purposes was a useful advance in the interpretation of field data especially lithofacies variability. However, this approach failed to take account of the dynamic character of ice sheets, ice sheet oscillations and depositional patterns on the margins of topographically diverse mountains.

Francis Synge

Synge's main publications are based on detailed field observations from specific field areas such as the seminal paper on the Trim eskers and related moraines (Synge, 1950). Perhaps his greatest contributions were on the distribution of drumlins and the fact that conspicuous moraines bordered the drumlin fields across the island. Synge argued that this association resulted from a major readvance of ice towards the end of the last cold stage. This set the scene for later workers to re-examine the case for active rejuvenation of the ice sheet during deglaciation rather than monotonic retreat. Synge (1979a) reinforced the traditional view of fresh drift within SIEM limits (Midlandian) and older drift outside these limits which he termed Munsterian because their main lithological characteristics were typical of this type area. The two new terms were supported by extensive mapping by Synge summarised by excellent distribution maps of glacigenic deposits and landform patterns across Ireland. In a series of papers spanning about 33 years Synge published work on many coastal sections which have formed the basis for countless investigations by later authors. Sections generally were drawn to show relative stratigraphy and lithofacies variability though genetic terms such as Ballycroneen Till or Irish Shelly Till were adopted to add a sense of provenance (Synge, 1979a).

The Irish Sea Glacier

The Irish Sea Glacier was one of the main arteries receiving vast quantities of ice from the British ice sheet and transporting shelly drift onto coastal lowlands and onto the northern slopes of the Wicklow Mountains. Sedimentologically complex sequences of diamict, muddy diamict, mud, silt, sand and gravel found in near-coastal locations in the Irish Sea Basin are termed Irish Sea Drift. This umbrella term included deposits ploughed out from unconsolidated shelly sediments on the sea bed together with a remarkable erratic carriage of microgranite from Ailsa Craig in the Firth of Clyde as the ice moved southwards (Wright, 1937). Oscillations of the Irish Sea Glacier were identified from deposition of shelly tills onland before Irish ice moved offshore. The concept of local and extraneous ice sheets re-emerged when sections containing multiple lithofacies or sedimentary units were used to reconstruct glacial stratigraphy and multiple ice flows (Mitchell, 1960, 1972; Synge, 1964, 1979b). This methodology based on bed for bed correlations away from a reference area was eventually used by Mitchell et al., (1973) for their Correlation of Quaternary Deposits in the British Isles. The tenuous nature of this approach and the correlations basin-wide was outlined in a paper by Eyles and McCabe (1989a) which used a depositional systems approach to identify an event stratigraphy within the basin. The event stratigraphy recorded the entry of marine waters into a glacioeustatically depressed basin and the rapid retreat of the Irish Sea Glacier as a tidewater margin (Fig. 1.2). The 1973 correlations also recognised that the principles governing subdivision are actually the great climatic fluctuations or warm and cold stages of the Quaternary which are identified from type sites or type areas. Lists of significant deposits such as sub-till peat and especially morainic landforms were related to specific major stage events (Mitchell et al., 1973) though the problem remained that these stages were superimposed from preconceived British regional stages.

Sedimentology, dating and correlation

Towards the final quarter of the 20th century it became obvious that Quaternary Science in Ireland and indeed in the British Isles as a whole lagged behind sedimentological and climate research in North America.

Traditional methods using bed for bed correlations, the lack of reliable dating programmes and the widespread use of genetic terms in sediment descriptions prevented progress. Fresh approaches were necessary in order to record evidence of climate change, timing of climate shift, ice sheet variability and sensitivity of the island to changes elsewhere in the amphi-North Atlantic. Three themes have emerged from the literature:

First, the sedimentological descriptors and methodology used to investigate complex sections were not sophisticated enough for progress and environmental reconstruction. The use of genetic descriptors meant that the likely depositional environment could not be reliably constructed from complex exposures and it was therefore impossible to compare a local site with sedimentological models in the international literature (Walker, 1984). In a series of papers from 1984 onwards McCabe, Dardis and Hanvey attempted to reconstruct deglacial stratigraphy from glaciomarine sequences on the margins of the Irish Sea Basin. This work developed a facies analysis approach to the description and interpretation of glacigenic deposits and tried to compare facies and facies associations with both modern analogues and process geology (Fig. 2.4). The success of this approach culminated in a commissioned paper for Quaternary Science reviews by Eyles and McCabe

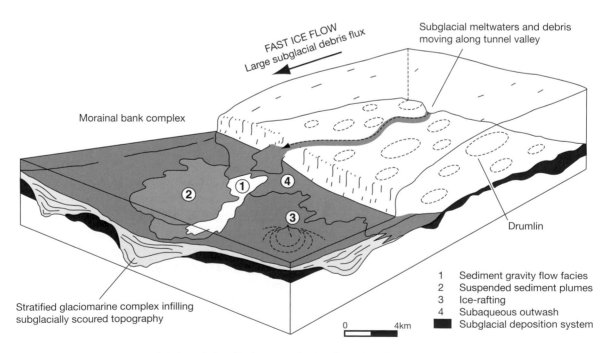

Fig. 2.4 Depositional model showing the relationship between fast ice flow, drumlinisation and subglacial sediment transfer towards tidewater ice margins where large morainal banks develop (after Eyles and McCabe, 1989a).

(1989a) on the glacio-isostatic event stratigraphy of the Irish Sea Basin. For the first time used a facies approach was used to reconstruct the detailed history of deglacial events in the basin based on widespread sedimentary records. Crucially this work identified that the entry of marine waters into the basin influenced the glaciology of adjacent ice sheets, the stratigraphic architecture of ice-marginal sediment wedges and the rates of ice recession (Fig. 1.2). From this time on facies logs, facies codes/descriptions and sedimentary models became the norm for descriptions and interpretations. Now it is possible where descriptions are detailed and non-genetic to change interpretations as new evidence or knowledge arrives. Current literature now considers detailed descriptions of glacigenic sedimentation from a wide range of extra glacial, marginal and subglacial settings based on specific well-exposed sections (e.g. Thomas and Chiverrell, 2006).

Second, dating, correlation and comparison are essential research tools if events identified from Ireland are to be assessed either in a wider climatic or ice sheet context. Coxon (1993) and his co-workers have restructured our understanding of Irish Pleistocene biostratigraphy in a series of papers on new interglacial sites which are in some cases supported by U-series dating and critical evaluations of temperate stages in Ireland extending the foundations laid by Jessen et al., (1959), Mitchell (1970, 1976,1981) and Watts (1959, 1977, 1985) (Figs. 2.5, 2.6). Because many of the peats have very similar floristic records it is difficult to decide how many interglacials are indeed preserved in our records (Dowling and Coxon, 2001). One unresolved problem is the apparent absence of a last interglacial site.

Changing climates during the early and middle parts of the last cold stage are identified from plants and animal remains from four sites at Aghnadarragh (McCabe et al., 1987), Derryvree (Colhoun et al., 1972), Greenagho (Dardis et al., 1985) and Hollymount (McCabe et al., 1978) in the northern part of the island. A wide range of mammoth remains were recovered from the cold climate deposits overlying the tectonised Tertiary lignites at Aghnadarragh, County Antrim. These beds are difficult to date with precision but occur between two regional till sheets formed at the start and end of the last cold stage complex. Rapid climatic change during the last glacial termination has now been determined by isotope dating. Bowen et al. (2002) provided a new approach using cosmogenic nuclide surface-exposure dating which began to improve the previously poor geo-

chronological control. Clusters of ^{36}Cl ages were identified suggesting deglacial events related to a dynamic ice sheet between 40 and 25 ka BP. The LGM occurred around >22 ka BP and was followed by an early deglaciation around 21 ka BP.

Third, reconstructions based on field data now show that the history of ice sheet growth and decay during the last cycle were related to millennial timescale events and climates. For example, during the latter parts of the 20th century before basic sedimentological principles were applied to glacigenic sequences all muds exposed around the coastline were thought to be some type of till even though they were interbedded with stratified facies. AMS ^{14}C radiocarbon dating of *in situ* microfaunas from these marine muds have age-constrained a wide range of events which characterise the last glaciation (Fig. 2.7). These include an early build up of ice in western areas (McCabe et al., 2007a), deep-isostatic depression and ice sheet oscillations prior to the LGM, early deglaciation (Bowen et al., 2002; McCabe et al., 2005), records of a global meltwater pulse ~ 19,000 years ago (Clark et al., 2004), at least two major ice sheet readvances around Heinrich 1 times followed by stagnation zone retreat (McCabe et al., 2007b). This deglacial history allows events identified from the deglacial stratigraphy from a relatively small ice mass adjacent to the climatically sensitive North Atlantic to be compared with millennial timescale events in the amphi-North Atlantic (McCabe and Clark, 1998, 2003; Clark et al., 2004; McCabe et al., 2005, 2007a, b). Furthermore, for the first time dated ice sheet fluctuations, ice sheet limits and ice-marginal marine deposits can be compared with confidence to subglacial imprints and bedform patterns of the terrestrial parts of the ice sheet. These ice sheet subsystems consisting of erosion during bedform formation, subglacial sediment transfer and sediment output into glaciomarine settings are closely linked and are part of the same readvance event (McCabe and Dunlop, 2006).

These deglaciation records also show that the ice sheet lost about two thirds of its mass shortly after the LGM when ice withdrew from the inner Celtic Sea, the Irish Sea and North Atlantic fringes onto land. After this the extensive subglacial transverse ridges across north central Ireland were finally formed prior to drumlinisation (McCabe and Dunlop, 2006). Final ice sheet decay from readvance limits across the north Irish Midlands and western seaboards occurred after 12.7–13.1 ^{14}C kyr BP when northeastern Ireland emerged from the last marine transgression (Fig. 2.7). In many places the

Series	Stage	Age	Substage	Comments
Holocene	Littletonian	10,000	Nahanagan Stadial	Named after glacier activity at Lough Nahanagan in the Wicklow Mountains (Colhoun & Synge 1980. Extensive glaciation has not been recognised in Ireland but many periglacial features and the evidence of small glaciers are found (Coxon 1988, Gray and Coxon 1991, Wilson 1990a and b and Walker *et al.* 1994).
Pleistocene	Midlandian — Late	Late-glacial — 11,000	Woodgrange Interstadial	This complex interstadial (with an early phase of climate amelioration and containing at least one period of erosion and climate deterioration) is recorded in many biogenic sequences from Irish Late-glacial sites (Watts 1977; 1985, Cwynar and Watts 1989, Walker *et al.* 1994).
		13,000 17,000	Drumlin Event	A distinct event (within the Drumlin Readvance Moraine of Synge 1969) producing drumlins. Evidence from north Mayo dates this event to around 17ka (McCabe *et al.* 1986) and the period is discussed in detail by McCabe and Dunlop, 2006.
		Glenavy Stadial	Main Event	The maximum ice advance of the last glaciation peaking by 20-24ka. Sequences of till and organic sediments from Aghnadarragh (McCabe, *et al.* 1987) allow this phase of glaciation to be put into context within the framework of the Midlandian cold stage.
	Midlandian — Middle	*c.* 25,000	Derryvree Cold Phase	Organic silts found between two tills at Derryvree (Colhoun *et al.* 1972) show a treeless, muskeg environment. The mammal remains from Castlepook Cave (Mitchell 1976, 1981; Stuart and van Wijngaarden-Bakker 1985) date from this period (34-35ka). Recent dates for mammal faunas from caves range from 32ka-20ka (Woodman and Monaghan 1993) indicating the possibility of ice free areas in Cork during the Glenavy Stadial.
		c. 40,000	Hollymount Cold Phase	Organic muds found at Hollymount (McCabe, Mitchell *et al.* 1978), Aghnadarragh (McCabe, *et al.* 1987) and Greenagho (Dardis *et al.* 1985). Fossils suggest cold, open treeless environments. Possibly a continental climate with high seasonality.
		> 48,000	Aghnadarragh Interstadial	Pollen and beetle evidence from Aghnadarragh (McCabe, *et al.* 1987) suggests cool temperate conditions with woodland, similar to that of Fennoscandia today. Dated to >48ka and tentatively correlated to the Chelford Interstadial (McCabe, 1987).
	Midlandian — Early		Fermanagh Stadial	Till pre-dating organic beds at Derryvree, Hollymount (McCabe *et al.*, 1978) and Aghnadarragh (McCabe *et al.*, 1987) are believed to have covered most of Ulster. Evidence (from the presence of certain tree taxa in the subsequent interstadial) suggests that the glaciation may have been short-lived (Gennard 1986; McCabe 1987).
		c. 115,000	'Kilfenora Interstadial'	UTD dates place cool temperate organic deposits at Fenit in Co.Kerry early within the Midlandian Glaciation (118,000 years BP –Heijnis 1992; Heijnis *et al.* 1993). The biogenic sediments represent cool conditions during Oxygen Isotope Stage 5a or (more likely) 5c.
	Last Interglacial	*c.* 120,000		The discovery of a reworked ball of organic sediment within the sands and gravels of the Screen Hills moraine (500m north of Blackwater Harbour within the Screen Member of Thomas and Summers 1983) gives hope of finding deposits of last interglacial age, as here for the first time in Ireland a *Carpinus* - rich pollen assemblage has been recorded (McCabe and Coxon, 1993). This material may represent Oxygen Isotope Stage 5e, a warm temperate stage, but the evidence to date is far from conclusive.
	Munsterian	*c.* 132,000		Widespread glacigenic sediments in the southern part of Ireland (Munster) have been long been regarded as belonging to an 'old' glaciation on the grounds that they show distinct assemblages of erratics, striae and glacial limits as well as exhibiting subdued relief, deep weathering profiles and a lack of 'fresh' glacial landforms (Mitchell *et al.* 1973; Synge 1968; Finch and Synge 1966; McCabe 1985; 1987). The lack of (any) stratigraphic control has meant that although the Munsterian deposits exhibit certain unique characteristics the relative age of the cold stage is unknown. A distinct possibility is that some 'Munsterian' deposits are in fact Midlandian (including Early Midlandian) in age, but this theory awaits further verification.
	Gortian	estimated minimum age *c.* 302,000	rapid termination — Gn IV / Gn IIIb / Gn IIIa / Gn II / Gn I / Pre-Gn 1-g	Eleven sites have been described from around Ireland that record part of a characteristic temperate stage deposit with a biostratigraphically identifiable record. The Gortian is represented by a unique record of vegetational succession and by a number of fossil assemblages that represent stages which have been described in a number of ways (e.g. by Mitchell 1981; Watts 1985; Coxon 1993). Particularly noticeable aspects of the Gortian are its sudden truncation (Coxon *et al.* 1994) and biogeographically intersting flora (Coxon and Waldren 1995). Opinion is divided as to the age of the Gortian (Watts 1985 and Warren 1985 give the basis of the two arguments). Biostratigraphically it resembles the Hoxnian of Britain and the Holsteinian of Europe. Amino-acid racemisation results on marine Gortian sediments from Cork Harbour (Scourse *et al.* 1992) confirm this suggestion and the interglacial may represent Oxygen Isotope Stage 9 or 11 (see Coxon 1993). The dates on this chart are tentative and based on Bowen *et al.* 1986.
	Pre-Gortian	estimated maximum age *c.* 428,000		Prior to the Gortian are sediments of late-glacial aspect, suggesting the temperate stage was preceded by a cool/cold stage. This stage is not represented by long or datable sequences, and the age is unknown.
	Ballyline	age unknown possibly > 428,000		A deposit of laminated, lacustrine, clay over 25 metres thick was discovered in 1979 by the Geological Survey of Ireland filling a solution feature in Carboniferous Limestone below glacial sediments near Ballyline, Co. Kilkenny (Coxon & Flegg 1985). From the evidence available the pollen assemblages can be seen to be typical of Middle Pleistocene sequences in Europe, but a firm correlation to a particular stage is not possible.
Pliocene	Pollnahallia (Pliocene-Pleistocene boundary?)	*c.* 2.3Ma		A complex network of gorges and caves in Carboniferous Limestone at Pollnahallia, Co.Galway, contains lignite deposits — now covered by superficial material including wind-blown silica-rich sands (Tertiary weathering residues) and glacigenic deposits. Palynological results (Coxon & Flegg 1987; Coxon & Coxon 1997) suggest that the lignite infilling the base of the limestone gorge is Pliocene or Early Pleistocene in age. Since the study a further continuous core through the lignite has been taken.

Fig. 2.5 The subdivision of the Quaternary Period in Ireland (after Mitchell et al., 1973 and Coxon, 1993).

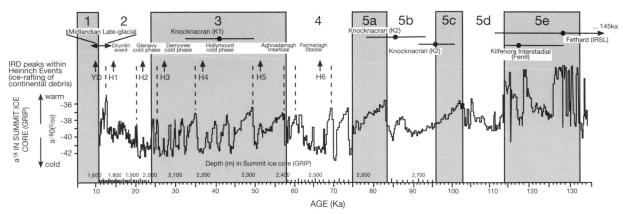

Fig. 2.6 The ranges of Irish Late Pleistocene dates in relation to the oxygen isotope record of the Summit ice core (GRIP) (after Dowling and Coxon, 2001).

general absence of well-defined moraines suggests that final deglaciation was characterised by stagnation zone retreat (McCabe and Clark, 1998). In some corries small ice masses reappeared during the Younger Dryas (11–10 [14]Ckyr BP) in the Wicklow Mountains (Colhoun and Synge, 1980).

Landform patterns

Knowledge on the distributions and origins of specific landforms and landform associations has expanded significantly from the essentially stratigraphic geomorphology of the 1980s. Many workers have attempted to describe and explain the significance of both the external and internal geometry of landforms especially the range of lithofacies present and the influence of glaciotectonics. Glacigenic and periglacial landforms can be pigeonholed into ten main categories which have received particular attention for a variety of different reasons. Each of these is particularly important because they contribute to knowledge of ice sheet history, environmental change and on the importance of comparison with landforms developed beneath and on the margins of other ice sheets.

1. It is only in the last few years that we recognised that the subglacial bedforms so prominent in the northern half of the island are dominated by subglacial transverse moraine (Ribbed or Rogen moraine) and not by flow-parallel drumlins. Dardis and McCabe (1983) first recognised that ribbed moraine morphology was widespread in County Armagh. Later the distribution and stratigraphic position of the transverse ridges was discussed in detail

(McCabe and Clark, 1998) and how later ice streams modified the earlier topography of transverse ridges (McCabe et al., 1999; Clark and Meehan, 2001). Subglacial sediment transfer towards tidewater ice sheet margins during drumlinisation resulted in the formation of morainal banks or spreads (Fig. 2.4). More recently Dunlop and Clark (2005) have discussed the morphological characteristics of the transverse ridges on a global scale.

2. Drumlins which are flow-parallel landforms overprinting earlier flow-transverse ridges are widespread in the north and west of the island. Drumlin stratigraphy was described sedimentologically by Dardis (1985a) in order to shed light on ice sheet history. Seven major facies associations within drumlins were related to distinct events during the course of the last glacial cycle (Dardis and McCabe, 1987; Dardis et al., 1984; McCabe 1993). Two major aspects were distinguished: drumlin formation includes all of the processes which have contributed to creation of the final drumlin form, including till agglomeration to streamlining and drumlinisation which is the processes of ice streamlining of sediment and rock near the base of the ice sheet.

3. Eskers are linear ridges composed mainly of stratified deposits deposited by meltwaters within ice walls. The west–east esker system across the central lowlands and the south–north system on the Plains of Mayo (Sollas, 1896; McCabe, 1985) help in the reconstruction of the final patterns and the nature of deglaciation. Although the polygenetic nature of esker ridges was recognised by Warren and Ashley

^{14}C	error	Mean cal.	cal age error	Lab Number	Location
12295	50	14152	94	SRR-6427	SLG-1
12430	85	14440	224	AA-34265	SLG-1
12335	80	14268	194	Beta - 48580	Rathlin Island
12740	95	15041.5	160.5	AA21822	Rough Island
13125	70	15502.5	179.5	AA68680	Rough Island
13785	115	16406.5	234.5	AA22820	Killard Point
13995	105	16676.5	229.5	AA22821	Killard Point
14045	100	16740.5	226.5	AA32315	Corvish
14157	69	16882	209	AA56700	Linns
14250	130	17039.5	284.5	AA56701	Rathcor Bay
14705	130	17776	272	AA21818	Cranfield Point
15020	110	18247	140	AA17693	Cooley Point
15025	95	18246.5	132.5	AA33831	Corvish
15190	150	18613	139	AA45967	Corvish
15190	85	18620	101	AA68976	Clogher Head
15300	90	18688	74	AA68975	Clogher Head
15390	110	18743.5	80.5	AA17694	Cooley Point
15400	140	18749	99	AA17695	Cooley Point
15450	45	18781	52	CAMS105063	Clogher Head
15605	140	18868.5	93.5	AA21819	Cranfield Point
15720	160	18941	125	AA45968	Corvish
15928	67	19087	90	AA56707	Belderg
15989	74	19147.5	105.5	AA56706	Belderg
16040	550	19220	416	CAMS105064	Clogher Head
16060	430	19221.5	348.5	AA45966	Corvish
16227	83	19393.5	85.5	AA56703	Belderg
16430	130	19532	71	AA56704	Belderg
16540	70	19724	78	CAMS89687	Kilkeel Steps
16540	120	19679.5	129.5	SSR-2713	Belderg
16580	120	19647	84	AA53589	Belderg
16640	70	19831	52	CAMS89686	Kilkeel Steps
16750	160	19926.5	141.5	AA22352	Kilkeel Steps
16760	130	19932.5	131.5	AA22351	Kilkeel Steps
16970	100	20085	120	SRR-2714	Fiddauntawnanoneen
16970	190	20094	190	CAMS89688	Kilkeel Steps
17140	110	20255.5	135.5	AA33832	Corvish
18275	99	21777.5	224.5	AA56705	Belderg
21130	220	25360	310	AA56702	Glenura
21920	90	26375	115	CAMS115268	Glenura
23400	110	27940	145	CAMS111594	Glenura
23530	110	28075	145	CAMS111595	Glenura
23630	90	28180	130	CAMS105065	Glenura
23740	110	28295	145	CAMS105068	Glenura
24380	120	28970	150	CAMS115273	Glenura
24630	130	29270	235	CAMS115271	Glenura
26930	160	31870	150	CAMS115272	Glenura
27200	130	32085	130	CAMS115066	Glenura
27380	170	32230	155	CAMS115266	Glenura
29660	220	34850	255	CAMS115274	Glenura
33550	340	38440	670	CAMS115267	Glenura
34470	390	39800	625	CAMS115269	Glenura
39540	490	43890	520	CAMS105067	Glenura

Fig. 2.7 Radiocarbon and calibrated ages from marine and glaciomarine muds described in the text.

(1994) they assumed that the features were related to one deglaciation of an ice sheet into discrete domes. Delaney (2002) clearly demonstrated that the eskers in the Lough Ree area, central Ireland, belonged to two distinct deglaciation phases. This evidence shows that the distribution of the two major esker systems in Ireland cannot be used to separate and identify a northern ice dome and a central ice dome model ~18,000 years BP. The Midland esker system across the Central Plain formed immediately after early deglaciation (~21 kyr BP) whereas the Dunmore/Ballyhaunas system formed after the Killard Point Stadial (~16 kyr BP) (McCabe et al., 1998).

4. Subglacial lakes and channels have been identified from the Enniskerry basin (McCabe and O'Cofaigh, 1994). Field relationships between stratified fills and rock-cut channel patterns along the eastern flanks of the Wicklow Mountains record ice-directed drainage along the western margin of the Irish Sea Glacier. The subglacial drainage system consisting of deep rock-cut channels separated by broad basins can be traced as a landscape unit for about 30 km from the Scalp channel south to beyond Wicklow Head. Formerly complex infills of this type have been used for stratigraphic purposes (Farrington, 1944) whereas recent detailed investigations between basin infills and adjacent channels show that there are close links between ice-directed erosional and depositional landscape features.

5. Morainal banks are either spreads or hummocky areas of sediment formed at tidewater when major glacial effluxes vent into the sea. Generally they are sedimentologically complex, characterised by rapid facies changes and composed of interbedded ice-contact facies contained within shallow, multi-storied channels (Eyles and McCabe, 1989a, 1989b). Good examples occur at Killard Point (McCabe et al., 1984) and Greystones where the spread occurs to the lee side of Bray Head which acted as a pinning point during ice-marginal retreat (McCabe and O'Cofaigh, 1995).

6. Cross-valley moraine ridges form during deglaciation usually where topographic constraints influence the ice-marginal configuration. In central Ulster Dardis (1982, 1985b) identified four main types of ridge based on detailed field mapping, palaeocurrent analysis, glaciotectonic structures, internal geometry and landform associations. The deltaic nature of sedimentation within these ridges and sediments along valley floors shows that lacustrine sedimentation was the norm because glacial lakes were impounded between the receding ice margin and the deglaciated Sperrin Mountains. The sedimentological work of Dardis (1982, 1985a,b) was a major step forward from the morphological approach of Colhoun (1970, 1972) which favoured a purely fluvioglacial origin for stratified deposits around the Sperrin Mountains. Cohen (1979) and Philcox (2000) provided particularly valuable work on the specific subaqueous mass transport mechanisms that operated close to high energy and low energy delta inputs into Glacial Lake Blessington, County Wicklow. Synge's (1979) detailed work on Glacial Lake Blessington focused on the complex deglacial history of the lake and possible interrelationships between Midland and local ice masses. Over many areas of central and western Ireland hummocky moraines and kame and kettle moraines are commonly described, though very few good sedimentological descriptions exist of their internal structure (McCabe, 1985; Coxon and Browne, 1991). However, until more details on their distribution and origins emerge it is premature to describe arcs of recessional moraines in relation to ice lobe recession (e.g. Charlesworth, 1928). Therefore moraines need to be accurately mapped and constrained possibly by exposure dating if they are to be used either as ice sheet limits or as elements for ice sheet models.

7. Although the use of moraine and outwash couplets has been used to reconstruct patterns of deglaciation in areas such as the Boyne valley and eastern County Meath (McCabe, 1979) and Sperrin Mountains (Colhoun, 1970) the sedimentology of the outwash is rarely described. For example, it is essential to know if a terrace is really a delta-terrace, braided outwash or erosional surfaces before local depositional environments and events can be assessed. A fact rarely commented upon is the absence of frontal moraines in most areas of the island. Reasons may be numerous but if no moraines occur within the limits of a major readvance limit this may signal that stagnation zone retreat has taken place. Phases of

very rapid ice recession have been recognised by the absence of moraines and outwash from large tracts of land, especially where well developed bedforms are present.

8. The well-defined topographies of erosional and depositional elements of mountain glaciers abound in the literature and are present within all our peripheral mountain groups which sustained independent ice caps (Farrington, 1934, 1938, 1949, 1953, 1966; Coxon and Browne, 1991; Synge, 1968). Progress in dating multiple moraines bordering corries is slow following identification of a Younger Dryas age for one moraine at Lough Nahanagan, County Wicklow (Colhoun and Synge, 1980). Various age estimates for corrie glaciations using cosmogenic nuclide surface-exposure dating on moraines bordering corrie basins have been proposed by Bowen et al., (2002) and range from 18 to 12.7 cal ka BP.

9. Descriptions of near-surface patterns in glacigenic sediments attributable to periglacial conditions are common in the literature though few authors dwell on the formative mechanisms. Ice wedge pseudomorphs are fairly common penetrating stratified outwash and concentric to festoon patterns are recognised from pebble realignments. More recently large-scale debris landforms including protalus ramparts and relict protalus rock glaciers have been described by Wilson (1990a, b, 1993, 2004) from the Donegal Mountains. Detailed sedimentological and morphological assessments of these landforms and their relationships with adjacent terrain suggest that slope failures following deglaciation are closely related to subsequent downslope movement of coarse debris (Wilson, 2004).

10. Understanding the history of the last ice sheet has long been hindered by the lack of dates on sea level changes during glaciation, the general reluctance to accept that isostatic depression was a significant factor influencing ice sheets and the absence of an accurate sea level curve following ice retreat. Fundamental work in a series of papers by Synge and Stephens in northern and north eastern Ireland provided very early insights into the significance of the extent of isostatic depression and position of ice sheet margins during deglaciation when global eustatic sea levels were thought to be low, possibly at −140 m (Yokohama et al., 2000). Raised clastic beaches were recognised to be largely contemporaneous with ice limits in northern Donegal and in eastern Ireland (Stephens and Synge, 1965, 1966a,b; Synge, 1977). Later, Eyles and McCabe (1989a) used sedimentological evidence from the Irish Sea Basin to reconstruct high relative sea levels during deglaciation and the impact of marine downdraw on ice streams within adjacent terrestrial ice sheets. This glacio-isostatic event stratigraphy has now been underpinned by AMS ^{14}C dates from marine muds within deglacial sediments (McCabe and Clark, 1998, 2003; Clark et al., 2004). Evidence for significant changes in relative sea level undoubtedly occur within the deglacial stratigraphy and can now be summarised into a sea level curve for northeastern Ireland which questions the parameters used in current geophysical models (McCabe et al. 2007c). Sea levels for earlier parts of the last cold stage are now available from western Ireland recording deep isostatic depression which greatly exceeds eustatic sea level lowering (McCabe et al., 2005, 2007a, b, c; Thomas and Chiverrell, 2006). These records of high relative sea level in western, northern and eastern Ireland suggest that high relative sea levels also existed along the south coast of the island during the last cold stage. The classic study of Wright and Muff (1904) on pre-glacial raised beach gravel on the south coast of Ireland requires serious revision because the 'preglacial' beach directly overlies a glaciated marine rock platform. High level raised Gilbert-type deltas immediately inland from this coast face directly into the immediately adjacent Celtic Sea and are difficult to explain as ice-dammed glaciolacustrine deposits unless ice advanced north onto the coast from the continental shelf.

Overview

The historical perspectives not only show the progress of geological thought but also how local evidence was generally interpreted within the context of fairly static models and ideas from elsewhere in Europe. Nevertheless the geothemes identified by early workers are still pertinent today and indeed some critical themes such as the degree and impact of isostatic depression and high relative sea levels on adjacent terrestrial ice sheets is now only being accepted (e.g. Thomas and Chiverrell, 2006). Our hypotheses and ideas on ice sheet activity are

Fig. 2.8 Glacially-transported Cretaceous flint core found within a drumlin lee-side deposit, Newtownards, County Down. It is possible that the artefact dates from the Middle Palaeolithic. Scale in cm. *By kind permission of Mr. Jon Stirland.*

far from complete because the evidence we are left with after successive phases of ice sheet scouring and meltwater erosion is fragmentary and individual pieces of evidence have vastly different degrees of resolution. For example, our interglacial records are patchy and are only recognised in very broad terms, though more detailed biological and stratigraphical evidence is available from the climatic vagaries of the last major glacial cycle. As expected, the range of evidence from the last (Late Midlandian) ice sheet is fairly well preserved and constrained by landform patterns and radiocarbon dating. While many uncertainties still exist on the nature of glacial erosion and transport there is no doubt that these processes operated across the whole of the island. One task therefore is to interpret these processes and landscape phenomena in the most practical way and within the constraints of known physical processes. It would be a mistake to over-interpret landscape evidence beyond

its natural limits. Prior to 1998 ice sheet models were based on an amalgam of undated flow lines, assumed LGM ages and ice sheet limits and little geologic data on ice sheet build up or decay (e.g. Boulton et al., 1977). Workers failed to recognise the dynamic character of the last ice sheet and the fact that the entire ice sheet system was responding to climate signals on a millennial time-scale or less (McCabe and Clark, 1998). The resulting geophysical models were therefore flawed and did not fit the known field evidence. Recent work has begun to address these issues by attempting to age-constrain the major fluctuations and evolution of the last ice sheet by radiocarbon dating marine beds and the sea level changes which are intimately associated with ice sheet history (Clark et al., 2004; McCabe et al., 2005). Dating of ice age events has also proved useful in assessing the provenance of reworked flints found in glacigenic deposits (Fig. 2.8).

3

Interglacials and Biostratigraphy

Introduction

The record of events before the last glacial cycle in Ireland is sparse and what exists cannot be correlated or provided with any reliable age estimates save perhaps, for the last 'interglacial' (Fig. 3.1). This arises from the erosional effects of not only the glaciers and ice sheets of the last glacial cycle but probably earlier ones too (Bowen, 1978). Preservation of earlier evidence is rare and relies to a large extent on sediment traps within solution hollows in Carboniferous limestone. Some are covered by glacial deposits which further enhanced their preservation and there is little doubt that many more remain unexposed. Deep-sea sediments preserve a record of interglacials during the 2.6 million years of the Pleistocene but only a handful may occur in Ireland and possibly represent the ten or so interglacials of the last million years.

One of the earliest records of interglacials in Ireland comes from Kinahan's notes (1865, 1878) on peaty deposits below drift near Gort, County Galway. The pioneering work of Jessen et al., 1959 on the biogenic sediments along the Boleyneendorrish river at Gort provided a basis for future studies on Irish biostratigraphy. However, the original studies focused on plant macrofossils and recently more detailed pollen analysis from bog and lake sediments has provided additional evidence for climate change (Watts, 1985). Following work in the Netherlands and eastern England it was customary to interpret interglacial vegetational history from pollen diagrams according to a simple model that reflected the onset, peak and closing phases of an interglacial. Thus there was a pre-temperate stage, early-temperate stage, late-temperate stage and a post-temperate stage during which the vegetation expanded from boreal forest characteristics to full mixed oak deciduous forest woodland, then subsequently to boreal conditions prior to the next ice age. While in general this succession reflects and is controlled by orbital

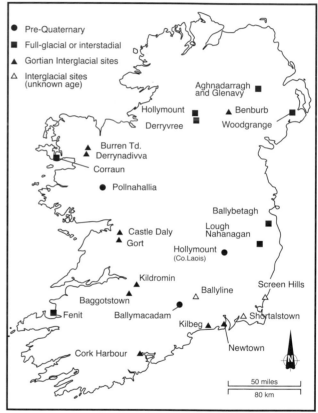

Fig. 3.1 Location map of sites of biostratigraphic importance in Ireland (after Coxon, 1993).

forcing of climate it is now clear that swift changes in vegetation were controlled by sub-orbital, millennial and centennial forcing in both the Pleistocene (Tzedakis, 2006) and Holocene. Pre-existing sampling intervals for pollen analysis would not have been able to pick up such changes and it is clear that millennial pacings run throughout the Quaternary although they are more accentuated during the extreme conditions of the ice ages. Nevertheless, plant fossils including pollen are used to construct climatic inferences from the known collective tolerances of the assemblage of species and their geographic ranges (Watts, 1985; 1988). Using the existing classifications of pollen biozones, Coxon (1993, 1996) has extended, compared and developed the frameworks of Mitchell (1981) and Watts (1985) into a system of pollen assemblage zones (Fig. 2.5). An important aspect of this approach is that interglacial data can now be presented in the context of a biostratigraphic framework (Fig. 2.6) (Coxon, 1996). It is clear that vegetational changes would have reflected climate as well as soil changes but until sites are revisited and sampling intervals for pollen greatly reduced the concealed millennial variability, especially rapid changes in vegetation, will remain undiscovered. Coxon (1996) argues that although many temperate records are incomplete, comparisons between long and shorter records allows correlations to be made between sites and therefore discrete parts of interglacial vegetational successions (Fig. 2.5).

Pollnahallia

Distinctive white sands and associated lignites occur within gorges and depressions developed on the karstic surfaces of Pollnahallia and Kilwullaun townlands, County Galway (Coxon and Flegg, 1987; Coxon et al., 2006). Although the local karstic topography is variable in form and scale Coxon et al., (2006) have been able to reconstruct a schematic cross section of the Pollnahallia deposit based on boreholes and resistivity lows (Fig. 3.2). An informal lithostratigraphy consisting of three main units is recognised (Coxon and Flegg, 1987). The

Pollnahallia organic silt and clay overlies limestone and ranges from thick (<1 m) massive and laminated lignites to oxidised organic material which accumulated either in shallow pools or as soils. Pure white sand known as the Pollnahallia sand overlies the organic sediments and is sometimes organised into large-scale cross beds suggestive of wind transport. The source of the silica sand is not known though deeply-weathered quartzites occur at localities in Connemara, 30 km to the west. A glacigenic unit termed the Headford till caps the sequence and seems to have disturbed the underlying facies by downward injections of water-saturated debris and brittle fractures including shear planes. Large rafts of limestone form an integral part of this till facies. The ice movement that formed the till came from the north though the number of erratics with a provenance to the west and south suggest that reworking of earlier sediments occurred (Coxon and Coxon, 1997). If this is the case then the Headford till predates the last ice sheet movement which generally moved northwards in this area.

Based on borehole 87/1 a pollen diagram consisting of five pollen assemblage biozones is recognised (Fig. 3.3). The data indicate a Pliocene age with a vegetation cover dominated by swamp cypress, ericaceous, cupressaceous and coniferous trees. Coxon et al., (2006) suggest that the climate was warm and wet similar to that of the swamps of eastern North America. They note that the disappearance of a number of thermophilous taxa and rising values of Ericaceae and *Juniperus* together with a facies change from clays to sands towards the top of the core probably records climatic deterioration at the end of the Pliocene. The exotic nature of some of the flora is important because many are now not native to Europe. A group of eight taxa (*Tsuga, Sciadopitys, Sequoia,*

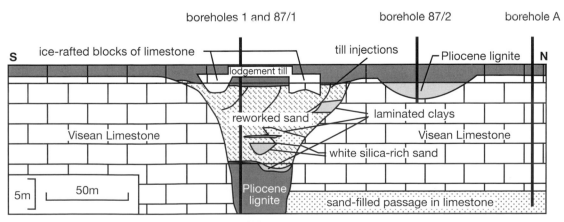

Fig. 3.2 A schematic cross-section of the deposit at Pollnahallia based on original surveys (Coxon and Flegg, 1987) and on unpublished boreholes carried out late in 1987 with funding from The Royal Society (after Coxon, 1993).

pab	depth (cm)	assemblage
87/1 -5	1050-1100	*Pinus* – *Betula* – Larix – Gramineae
87/1 -4	1100-1275	*Pinus* – *Corylus* – Ericaceae – *Carya*
87/1 -3	1275-1575	*Pinus* – *Taxus* – Taxodium – Ericaceae
87/1 -2	1575-1875	*Pinus* – *Corylus* – Ericaceae
87/1 -1	1875-2012	*Pinus* – *Taxodium* – Ericaceae – *Sequoia*

Fig. 3.3 *(left)* Preliminary pollen assemblage biozones from Pollnahallia (after Coxon and Flegg, 1987).

Fig. 3.4 *(below)* Percentage pollen diagram for the deposit at Ballyline (after Coxon and F legg, 1985 and Coxon, 1993).

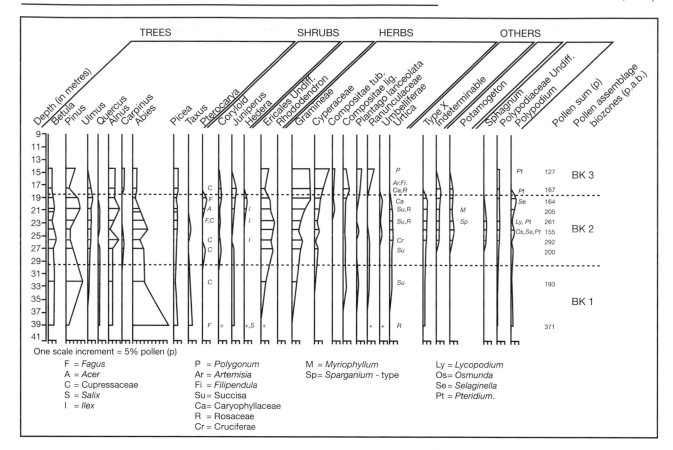

Taxodium, Taxodiaceae, Carya, Pterocarya and Vitis) indicates that the Pliocene flora was very different from that in most younger Irish Pleistocene sites.

Ballyline

The extensively karstified surface of the Irish lowlands provides sediment traps that are generally found during the course of mineral exploration (Coxon and Waldren, 1995). One such site of possible Middle Pleistocene age occurs in a solutional feature in Carboniferous limestone near Ballyline, County Kilkenny (Coxon and Flegg, 1985). The deposit consists of over 25 m of laminated clay overlain by 10 m of gravel and boulder clay. At the base, an *Abies–Pinus* p.a.b. dominated by *Abies* in a forest with open ground including *Taxus, Quercus,*

Ulmus, Betula, Picea, Juniperus, Graminae and Ericales. The overlying p.a.b. contains *Pinus* and *Abies* with peaks of *Pterocarya, Carpinus, Ulmus* and *Taxus*. Pollen of Type X and *Rhododendron* also occur. The upper Graminae–Cyperaceae–Pinus shows a decrease in most tree pollen and may represent climate deterioration with more open ground.

The pollen diagram (Fig. 3.4) is difficult to interpret because of the local site and the fact that the deposit probably contains much reworked material. Its age is difficult to determine but the general pollen taxa are similar to those recorded from a number of Middle Pleistocene interglacial stages in north-western Europe (Coxon and Flegg, 1985). Type X has been recorded from the Hoxnian and *Pterocarya* is uncommon in stages later than the Hoxnian. Coxon and Flegg (1985) stress that the Ballyline deposit is different from typical Gortian (later) sites because of the importance of *Abies* pollen, the presence of *Pterocarya* and isolated Fagus pollen, absence of *Buxus* and *Azolla*, high values of *Carpinus*, low values of *Rhododendron* and a continuous curve of *Ulmus*. These differences are significant enough to make palynological correlation with typical Gortian sites difficult. Nevertheless the fossils document a period of climatic conditions which allowed the development of a forest cover including *Abies, Picea, Ulmus, Alnus, Quercus, Pterocarya* and *Taxus*. It is also uncertain if the second *Carpinus* peak is diagnostic of a complex climatic oscillation and whether, and how much of an interglacial sequence is actually present at the site.

The Gortian

Since the discovery of temperate interglacial peats from the banks of the Boleyneendorrish river, near Gort County Galway, there has been debate on the age, stratigraphic position and possibly the number of interglacials embedded in the records (Jessen et al., 1959; Coxon, 1996). Eleven sites have been identified that contain biostratigraphical evidence associated with this temperate stage though sites differ in the range of material preserved, the probable parts of the interglacial recorded and the local depositional settings. A key element recognised by Coxon (1996) is that a well-balanced vegetational succession follows a pattern that is a logical outcome of a single interglacial progression (Fig. 3.5). The length of records from individual sites is variable but sites such as Gort, Baggotstown and Cork Harbour contain much of the interglacial and may serve as useful anchor points (Fig. 3.6).

Vegetational records can differ between sites and some of this variability can be attributed to the presence or absence of different substages. *Pinus* and *Betula* are present throughout the profiles and tend to mask other diverse woodland species. In some profiles charcoal is present suggesting that fire may have influenced the local vegetation especially *Pinus* and other less fire-resistant taxa (Coxon and Hannon, 1991). Abies expands in Zone IIIa and heath communities including *Rhododendron ponticum* with Ericaceae late in the interglacial. One characteristic is that the interglacial records seem to end abruptly in the Atlantic-climate forest stage (telocratic) termed the Gn IV substage. Although the precise meaning of this termination is unclear one site at Derrynadivva suggests that sedimentation continued

throughout this period because earlier deposited temperate fossils were reworked.

The standard view of the Gortian interglacial is that it is Middle Pleistocene in age. At most sites the age of the overlying glacigenic facies is either unknown or not proven in a regional context. Therefore dating relies partially on biostratigraphic correlation based on the presence or absence of key taxa in assemblages from other temperate deposits around Europe (Coxon, 1996). Coxon argues that the Gortian can be linked to the Hoxnian of Britain and Holsteinian of Europe because: 1. Relic taxa such as *Tsuga* of the Early Pleistocene do persist into the Gortian; 2. *Hippophae* is present at sites with the Pre-Gortian late-glacial (P-Gn I-g); 3. Presence of *Abies* in Gn III; 4. Presence of abundant *Azolla filiculoides*; 5. Presence of Type X; 6. Record of *Pterocarya* which disappears from Europe after the Middle Pleistocene; 7. An associated lack of *Carpinus* pollen. The development of acid *Empetrum* heath would be expected in the west of Ireland characterised by moist Atlantic climates associated with acid soil development and leaching late in the telocratic part of the interglacial cycle. If the Gortian is correlated with the Hoxnian then it belongs to oxygen isotope stage 11 (MIS11) with an age of 352–428 ka (Bowen et al., 1986a). Interglacial material from boreholes in Cork Harbour suggests that the Gortian Group may contain floristically similar deposits of different ages (Dowling and Coxon, 2001). Dating of organic material by uranium/thorium disequilibrium (U-series) methods suggests the Cork Harbour deposits are at least MIS5 in age (Dowling et al., 1998) whereas measurement of the degree of amino acid isoleucine epimerisation of forams suggests the deposits are MIS7. Earlier attempts to date other Gortian sites provided ages of 191 and 180 ka from Burren Townland and 350 ka from Gort (Dowling and Coxon, 2001).

At present it is difficult to place the Gortian deposits with confidence in any stage because even the stratigraphic position of the Holsteinian (one basis of correlation) is in doubt. The argument therefore centres on the possible presence of more than one warm stage within the Gortian and/or Holsteinian sedimentary records. In addition the Gortian sites are stratigraphically incomplete, not identical where known and poorly-constrained stratigraphically. There seems little doubt that temperate stages may produce very similar vegetation records and this situation coupled with the incomplete biostratigraphical records of Gortian events means that correlations or grouping of similar interglacial

Substage	Pollen assemblage
Gn IV	Marked termination of record during *Abies-Picea* — Ericaceae assemblage with notable re-working of thermophilous taxa
Gn IIIb	*Pinus-Betula-* with *Alnus-Taxus-Abies-Picea*— Ericaceae (including *Rhododendron ponticum*) Assemblage
Gn IIIa	*Pinus-Betula-* with *Alnus-Taxus-Abies-Picea* assemblage
Gn II	*Pinus-Betula-Quercus* assemblage. *Taxus* very important at some sites, with *Hedera* and occasional tree taxa including *Fraxinus*, *Corylus* and *Ulmus*
Gn I	*Betula-Pinus* assemblage
P-Gn 1-g	*Salix-Juniperus-Hippophae* assemblage including a herbaceous component

Fig. 3.5 Summary of the vegetational succession from the Gortian (after Coxon, 1996).

Fig. 3.6 Pollen zone ranges for Gortian interglacial sites (after Coxon, 1993,1996).

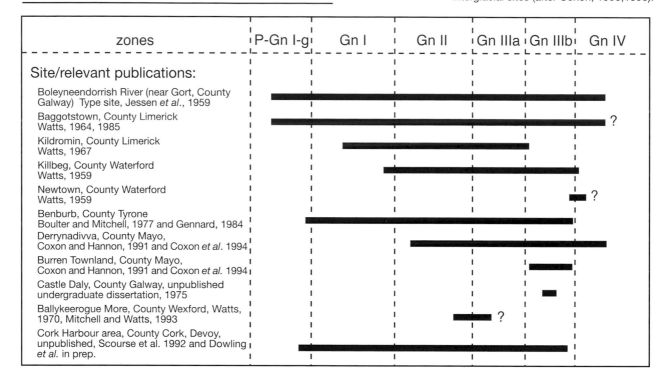

sequences is a flawed procedure. Some argue (Turner, 1998) that the apparent absences of some stages (MIS9, 11) is in some way due to the absence of accommodation space (Dowling and Coxon, 2001). However, a more pragmatic explanation is that palaeobotanists have lumped similar deposits of different ages into discrete pigeonholes simply on the basis that different interglacials must be floristically different. In this respect it has been argued that two temperate stages separated by a weak cold stage could have similar biotic signatures (Tzedakis and Bennett, 1995). For example, MIS10 was one of the less severe cold stages (Shackleton, 1969) and therefore could have provided the platform for progres-

sion of biological signatures from MIS11 to 9. Perhaps the vegetational records have been subject to over-interpretion and this can only be evaluated by absolute dating assessments.

The last interglacial stage

Deposits of the 'last interglacial' are well known in Europe where they are known as Eemian and in England as the Ipswichian. These are correlated with MIS5e or marine event 5.5 Its age remains debatable and it is likely that because of the diachronous response of vegetation to climate change it may have started earlier and ended later in some environments. Currently there is debate as to whether it lasted about ten or as long as twenty thousand years based on different evaluations such as orbital tuning of ocean isotope signals, high sea level events or long records from continental calcite deposits.

The apparent absence of a last interglacial sequence containing vegetative assemblages with affinities to the Ipswichian but distinct from the Gortian Group may simply be a result of the mobility, erosive and depositional patterns of most recent ice masses. In this respect the preservation of cool interstadial organic deposits of OI stages 3 or 4 beneath thick tills towards centres of ice sheet dispersal in the north of the island shows that earlier interglacial deposits are rarely preserved.

Mitchell (1976, 1981) suggested that the upper part of the sequence at Baggotstown (Watts, 1964) and the estuarine sand at Shortalstown, County Wexford (Colhoun and Mitchell, 1971) might form part of the last interglacial record. However, both of these records are poorly documented stratigraphically. The Baggotstown pollen diagram has been reconstructed from a spoil tip and the Shortalstown site has been disturbed by glaciotectonics. At Shortalstown pollen samples were taken from a 4 m section along a drain containing fossiliferous rafts of sand contained within a thrust complex of diamict and gravel. The relative pollen content from the sands was small but contained *Betula*, *Pinus* and *Quercus* and small amounts of *Ulmus*, occasional grains

of *Alnus*, *Corylus*, *Picea* and *Hedera* (Fig. 3.7). The proportions of pollen suggest that the countryside was forested and the climate was temperate but the precise depositional site is unknown. Colhoun and Mitchell (1971) argue that the deposit at Shortalstown resembles Zone e of the Ipswichian and is different floristically to the Gortian Group. However, it is not certain that relatively small differences in taxa content and the presence, absence or preponderance of marker taxa between far-removed sites can be used for correlation purposes. McCabe and Coxon (1993) have described a resedimented ball of organic sediment contained within the sandy foresets of the Screen moraine, 500 m north of Blackwater Harbour, County Wexford. The sample contained abundant Carpinus pollen (14%), *Betula*, *Pinus*, *Quercus*, *Picea*, *Salix* and *Ilex*. This assemblage strongly resembles Continental Eemian sequences and contrasts markedly with Late Gortian records (Coxon, 1993) (Fig. 3.5). However, an important part of the basis of correlation is that *Carpinus* was an important and distinctive component of both Irish and European Late Pleistocene flora.

At Knocknacran, County Monaghan, organic-rich deposits overlie mantled gypsum karst and occur below about 35 m of till in an opencast gypsum mine. The massive, overcompacted till forms part of the regional belt of drumlinised transverse ribbed moraine which developed shortly after early deglaciation around 20 kyr BP. The organics are therefore sealed by till of the last glaciation. Sinkholes are overlain by weathered lavas,

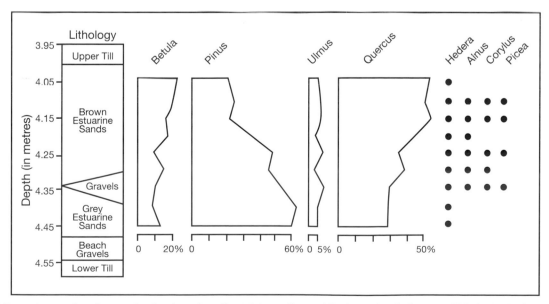

Fig. 3.7 Percentage pollen diagram for the deposit at Shortalstown, County Wexford (after Colhoun and Mitchell, 1971). *By permission of Elsevier.*

deformed green diamict with cobbles, cobble/pebble-rich organic debris with trunks of yew and oak, blue-grey diamict and massive brown till (Vaughan et al., 2004). In one section (K2) a nappe-like structure contains woody material oriented along the folds of the anticline. Pollen shows no zoning but is dominated by Poaceae (55–85%) with 5–15 percent *Alnus* and low frequencies of *Pinus*, *Betula*, *Taxus*, *Coryllus avellana* type and *Ilex* with occasional grains of *Larix*, *Juniperus*, Ericaceae and *Carpinus* (Vaughan et al., 2004). Herb pollen accounts for 90 percent of the pollen sum. A nearby section (K1) contains more tree pollen dominated by *Pinus* and *Betula* with small amounts of *Quercus*, *Corylus avellana*, *Ulmus*, *Alnus* and *Ilex* (Fig. 3.8). The fact that the organic debris are dispersed within gravelly diamict suggest erosion of an organic bank followed by mixing and redeposition by mass flow processes. It is probable that the organic-rich diamict was transported as a slurry or something akin to a bogburst which then lost its water content by percolation into sinkholes beneath. The deformation structures recorded are all near sinkholes which suggests that the folds are related to the position of these features. The scale of deformation is consistent with downslope movement towards structural lows rather than being driven by ice sheet loading. Uranium/thorium disequilibria (U-series) dating provided estimates of ca. 41 ka from K1 and 86 ka from K2. Vaughan et al., (2004) argue that K1 continued to take up uranium from groundwater long after K2 ceased to do so, suggesting that the deposits are older than 86 ka. The pollen suggests that the original deposit formed in a warmer climate than any post-Eemian event. The pollen assemblage has strong affinities with Eemian deposits though there is a lack of distinctive taxa which makes correlation difficult. The absence of *Picea* and *Abies* and other marker taxa makes correlation with earlier interglacial groupings such as the Gortian improbable.

Mitchell (1970) described peats below periglacial slope deposits between Fenit and Spa, County Kerry (Fig. 3.9). The organic sediments are interbedded with clay and silt beds which occur within sands and gravels of possible marine origin (Heijnis et al., 1993). The raised marine facies rest on a glaciated marine platform up to 6 m above sea level, vary from sand to pebbly gravel and

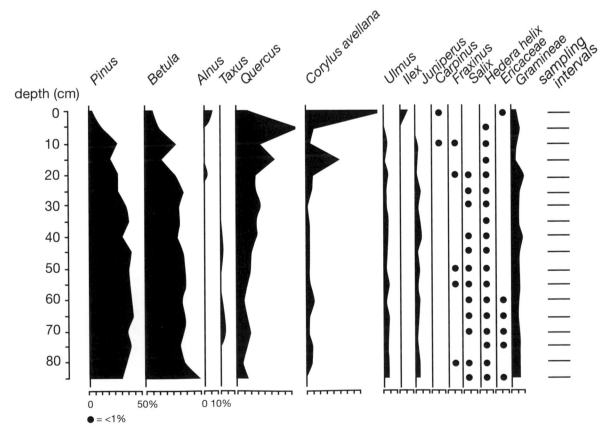

Fig. 3.8 Percentage pollen diagram of selected taxa from section K1 of the interglacial site at Knocknacran, County Monaghan (after Dowling and Coxon, 2001). *By permission of Elsevier*

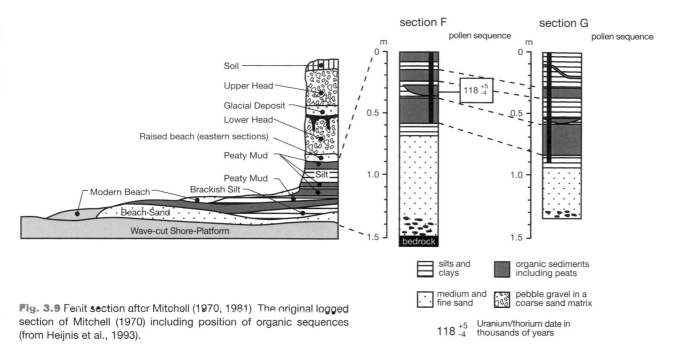

section F
pollen sequence

section G
pollen sequence

Fig. 3.9 Fenit section after Mitchell (1970, 1981) The original logged section of Mitchell (1970) including position of organic sequences (from Heijnis et al., 1993).

silts and clays	organic sediments including peats
medium and fine sand	pebble gravel in a coarse sand matrix

118^{+5}_{-4} Uranium/thorium date in thousands of years

contain angular shale clasts. Both Mitchell (1970) and Heijnis et al., (1993) record up to 10 m of poorly-sorted, mainly locally derived angular debris with a highly variable sandy matrix and sandstone erratics. This facies varies from a breccia to diamict and is clearly related to resedimentation by mass flow with clast fabrics aligned to local slopes. A prominent bed of sandstone boulders (<1 m long), cobbles and sand which occurs within the breccia has been attributed to ice rafting (Heijnis et al., 1993). Various types and patterns of soil structures including ice wedge casts and cryoturbation festoons have been noted from the section, but similar features occur widely where there are grain size, permeability and loading contrasts within a variable sediment pile. They have been interpreted as frost structures but are not diagnostic of permafrost conditions within this type of sedimentary succession which could be mainly related to mass flow processes. This interpretation is strengthened by the fact that Heijnis et al., (1993) recorded discontinuous lenses of peat mud interbedded with the breccia suggesting that the entire sequence is related to unstable slopes, erosion and remobilisation of a wide variety of inorganic and organic sediments from upslope reservoirs and resedimentation across the back of the rock platform. In addition the fact that the organic sediments are delicately interbedded with possible cobble/pebble/sand beach deposits deposited in a much higher energy regime is difficult to explain.

Similarly the preservation potential of such a depositional sequence is extremely low on a rock platform near sea level. If the general scenario of rapid slope failure and deposition on the platform is the case then a likely trigger would be basal slope erosion by a rising sea level sometime after the organic material formed. The presence of a prominent boulder bed within the sequence has been associated with outwash from an ice advance (Mitchell, 1970) and ice rafting (Heijnis et al., 1993). If either of these hypotheses are correct then it is likely that the sequence accumulated in shallow water.

The standard interpretation of the organic sediments consists of four local pollen assemblage zones (PAB) (Heijnis et al., 1993): Zone F1 Graminae-*Pinus* PAB with high herb pollen and traces of *Quercus*; Zone F2 Graminae-*Pinus*-Ericaceae PAB and *Calluna* and *Empetrum* heaths. Small percentages of Alnus, Corylus and *Ilex* occur; Zone F3 Graminae-Cyperaceae-*Quercus* PAB and other tree pollen is absent; Zone F4 Graminae-Ericaceae-*Pinus* PAB with other tree pollen (Fig. 3.10). The pollen data reflect periods of open pinewoods, heaths, increases in herbs and possibly a climate amelioration in Zone 4. The assemblages are distinct floristically from both the Gortian and interstadial records of the Early and Middle Midlandian but may record a time period towards the end of an interglacial. Results of U-series measurements suggest that the biogenics were deposited in cool temperate conditions between

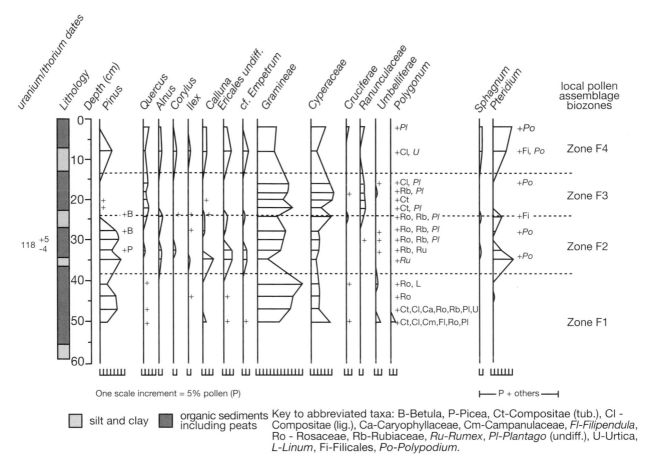

Fig. 3.10 Percentage pollen diagram from Fenit. The percentages are of the sum of terrestrial pollen types and occurrences of less than 1 percent are shown on the figure as a + symbol (after Heijnis et al., 1993).

115,000 and 120,000 years ago (Heijnis et al., 1993). This corrected age coincides with a period after the termination of the last interglacial at the end of 5e (5d, c) now termed the Kilfenora Interstadial.

Like many other sequences containing biogenic remains, the nature of the sedimentary sequence is critical in any interpretation of the fossil evidence. The biogenic material is plant detritus and as such records transport and redeposition in much the same way as the subjacent sand and gravel. The organic and inorganic sediments, because they are delicately and inextricably interbedded, must have been deposited at much the same time rather than recording separate events. It is therefore argued that all of the sediments at Fenit are essentially paraglacial sediments derived from earlier glacial, slope and organic deposits. To some extent this view is reinforced by the presence of disseminated lenses of detrital peat mud interbedded within the thick breccia/diamict beds towards the top of the section.

From a sedimentological stance it is also significant that locally derived shale clasts occur throughout the sequence and testify to slope instability. Although the organics may date from the close of the last interglacial it is a matter for debate when the organics were finally emplaced on the platform together with beach gravel. Mitchell's (1970) description and sketch of a synclinal hollow in the rock platform infilled sympathetically with conformable sand, temperate detrital peat mud, laminated sand and breccia (head) supports the slope model. Given the coastal position of the sequence it is likely that the solifluction processes which resulted in sedimentary variability within the sequence are associated with basal erosion of a former cliff line. A relative sea level similar or slightly higher to that of today could have triggered slope instability, cleaned the platform and then fallen. This scenario is consistent with initial erosion and resedimentation of surficial organic sediments followed by deeper erosion, removal of glacigenic clasts downslope and finally deeper slope stripping producing breccias. Timing of these events is unknown but likely phases are during cold periods which may have been prominent in southwestern Ireland during the Early and Middle Midlandian. The long (~20,000 years) period charcaterised by rapid environmental changes known to have occurred during the Late Midlandian must also be considered because there are prominent surficial paraglacial sediments in this area.

Overview

Dowling and Coxon (2001) consider that Ireland's Pleistocene record begins with the deterioration in climate following the deposition of the Late Tertiary biogenic sediments at Pollnahallia. Deposits preserved in karstic surfaces such as Ballyline are difficult to correlate floristically with other Middle Pleistocene temperate stages and highlight many of the problems surrounding the stratigraphic status of interglacial peats in Ireland. Although it is generally agreed that most of the Gortian Group of peaty deposits are similar in floristic composition, they may belong to different interglacials, possibly MIS9 or 11. A recurrent problem here is the difficulty involved in relating the stratigraphic position of the interglacials either to well-established regional till sheets or glacigenic beds. Some sequences are also influenced either by glacitectonics or other forms of deformation and it is difficult to construct a firm local stratigraphy. Therefore stratigraphic arguments based on simple counts from the top procedures are fatally flawed. Although there is no satisfactory site representing the last interglacial, the organic-rich deposits at Knocknacran are younger than the Gortian Group and occur immediately beneath tills deposited during the last glaciation. This deposit, because it is only preserved in a topographic low within gypsum karst, testifies to the efficiency of glacial erosion of most interglacial deposits during the oscillations and phases of fast ice flow which characterised the last glaciation. It also poses the question concerning the mechanisms responsible for generation of the very thick (>40 m locally) massive basal tills overlying the Knocknacran karstic surface.

4

The Last Major Glacial Cycle

Introduction

The last major glacial cycle can be defined on orbital timescales starting around 115,000 years ago and ending at 11,500 years ago at the opening of Postglacial climatic amelioration (Fig. 4.1). Terrestrial stratigraphies and evidence from climate proxies around the North Atlantic show that during this time the activity of the hydrological cycle and ice sheets varied in response to millennial timescale signals. The traditional stage term Midlandian which equates broadly with the last major glacial cycle comes from Farrington's term Midland General describing the deposits of the last major lowland ice sheet in Ireland (Mitchell et al., 1973). The four main sites containing organic beds in the north of Ireland contain partial but important records of climate and ice sheet variability at this time. The organic beds are stratigraphically important because they occur within a layer-cake stratigraphy, underlain and overlain by regional till sheets whose provenance and age is known (Fig. 4.2) (cf. Watts, 1985).

Simple lithostratigraphic approaches to stratigraphy (e.g. Warren, 1985) proposed for the Midlandian based on beach deposits, sites distant from critical glacigenic sequences and on undated sites containing no biozones cannot be used as standard. At present because of the relative absence of full records spanning the Midlandian it is best to use informal terms (Late, Middle and Late) which provide a working stratigraphic framework. In addition there are excellent biogenic records of climate change spanning parts of the Early and Middle Midlandian which must be considered in relation to wider climate events in the amphi-North Atlantic. The organic sediments preserved beneath Late Midlandian tills are important climatic indicators which could eventually be related or compared to the millennial timescale oscillations identified from ice cores in Greenland (Grootes et al., 1993; Clark et al., 2007). Recent work has shown

that MIS 3 (59–29 cal kyr BP) is an important interval characterised by climate oscillations on a millennial timescale (Voelker et al., 2002). Perhaps future dating of these terrestrial records will help determination of driving factors, lags and leads between various climate proxies in the North Atlantic region.

Location of the evidence

The four sections with interstadial and cool climate organic beds are located in the northern half of the island very near the largest centres of ice sheet dispersal. Three of the four deposits are present between thick, regional till sheets. The fossiliferous sediments are mainly *in situ* and have not been transported by ice away from their depositional site. In some cases the upper beds of sequences at Derryvree and Aghnadarragh show evidence of glaciotectonic shearing. These essentially cool to cold sequences have not been recorded from central and southern Ireland where glacial sediment transfer from land onto the continental shelf has generally been more efficient than in the north. Preservation of interstadial organic deposits occurs in the lowland basins of Lough Neagh and Lough Erne below 30 m OD where the regional upper tills are drumlinised. Clearly, the enhanced preservation potential of the organic beds is related not only to deposition of Late Midlandian tills but to the fact that the ice masses were unable to transfer subglacial detritus effectively out from central zones of the ice sheets. This suggests that the ice sheets simply moved inorganic detritus around core areas of ice sheet dispersal as ice centres changed. If this is the case then it was the mobility of the ice sheets during the Late Midlandian which enhanced preservation of interstadial beds within basinal zones. It is also significant that the subglacial ribbed moraines common in these areas document sticky ice conditions characterised by surficial drumlinisation and minimal sediment removal.

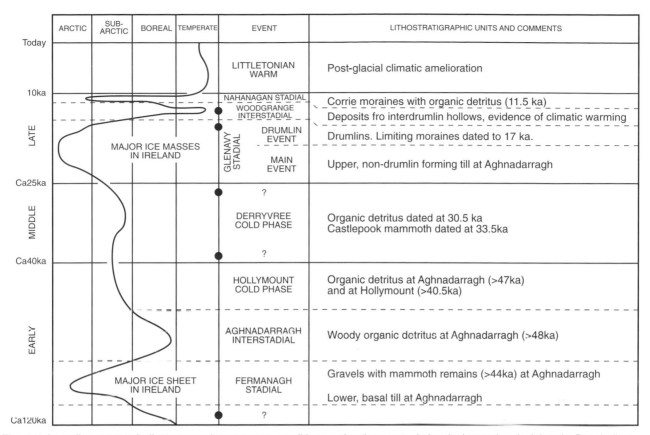

	ARCTIC	SUB-ARCTIC	BOREAL	TEMPERATE	EVENT	LITHOSTRATIGRAPHIC UNITS AND COMMENTS
Today					LITTLETONIAN WARM	Post-glacial climatic amelioration
10ka					NAHANAGAN STADIAL	Corrie moraines with organic detritus (11.5 ka)
					WOODGRANGE INTERSTADIAL	Deposits fro interdrumlin hollows, evidence of climatic warming
LATE		MAJOR ICE MASSES IN IRELAND			GLENAVY STADIAL — DRUMLIN EVENT	Drumlins. Limiting moraines dated to 17 ka.
					MAIN EVENT	Upper, non-drumlin forming till at Aghnadarragh
Ca25ka					?	
MIDDLE					DERRYVREE COLD PHASE	Organic detritus dated at 30.5 ka Castlepook mammoth dated at 33.5ka
Ca40ka					?	
EARLY					HOLLYMOUNT COLD PHASE	Organic detritus at Aghnadarragh (>47ka) and at Hollymount (>40.5ka)
					AGHNADARRAGH INTERSTADIAL	Woody organic detritus at Aghnadarragh (>48ka)
		MAJOR ICE SHEET IN IRELAND			FERMANAGH STADIAL	Gravels with mammoth remains (>44ka) at Aghnadarragh Lower, basal till at Aghnadarragh
Ca120ka					?	

Fig. 4.1 An outline curve to indicate general temperature conditions and major events during the last major glacial cycle. Dots indicate position of known unconformities. (modified after Mitchell, 1976).

The main events and possible temperature oscillations which occurred during the last major glacial cycle are summarised for comparative purposes (Fig. 4.1). Suggested correlations between the different sites in the north of the island are based not only on biostratigraphy but on the presence of the two regional till sheets which define the regional glaciations before and after deposition of the interstadial organic beds. Because of the incompleteness of the evidence it is not possible to establish formal upper and lower boundary definitions for stratotypes especially when non-sequences are recognised (Fig. 4.2). Five major events identified from direct glacial and organic beds are recognised from the four sites.

Fermanagh Stadial

In the Erne (Derryvree, Hollymount) and Lough Neagh (Aghnadarragh) basins thick (>5 m) till sheets underlie *in situ* fossiliferous organic deposits which date from the latter part of the Early and Middle Midlandian

(Fig. 4.2) (McCabe, 1987). Although these sites are 80 km apart they contain massive, overcompacted till that contains erratics derived from the Central Tyrone Igneous Complex. Local directional indicators in both areas are consistent with ice flow from Central Ulster and suggest that the far-travelled igneous erratics are in primary position.

The Boveagh Till (Colhoun, 1970, 1971) which is widely distributed in north County Londonderry (Portlock, 1843) occurs in a similar stratigraphical position to the lower tills in the Erne and Lough Neagh basins. It is overlain either by tills deposited by northward ice flow during the last glaciation or by the thick deposits of morainic gravel of the Ballykelly moraine. In many cases where the fine-grained till facies occur near the surface the texture of the sediment ranges from sandy to pebbly beds interbedded with muddy beds. Most fine-grained beds contain comminuted marine shell fragments together with intact shells of *Turritella*. The most complete exposures occur in river sections around Sistrakeel where up to 3 m of fossiliferous mud rests

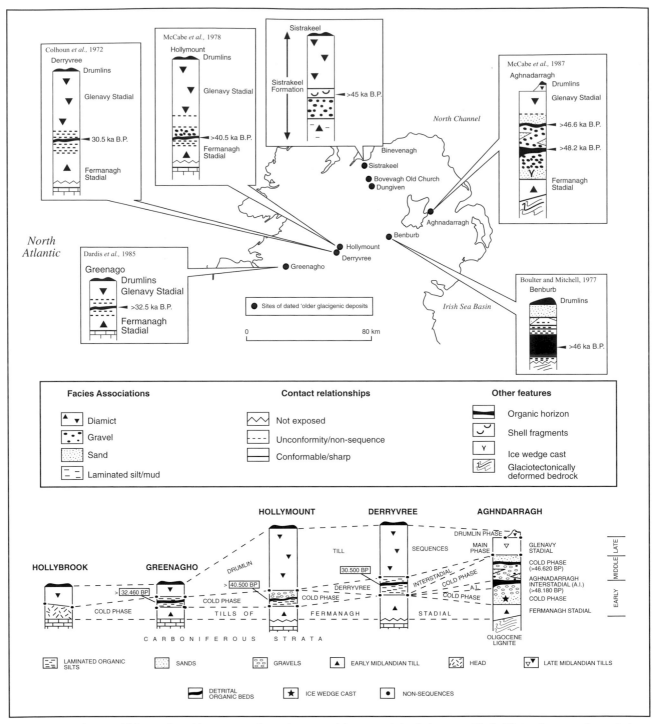

Fig. 4.2 Locations and lithostratigraphic summary from sites containing organic beds between tills in Ulster. Suggested correlations between the organic sediments and till sheets are suggested in the lower diagram (after McCabe, 1987).

on a sandy diamict and is overlain by a till with basalt erratics derived from Binevenagh to the northeast (Fig. 4.2). Because the sandy diamict is associated with an ice movement offshore it is likely that deposition of the overlying mud and till (Bovevagh Till) occurred when inland ice decayed. This phasing relationship between Scottish ice moving southwestwards and the slightly earlier advance of Irish ice northwards is difficult to explain with certainty. The major Scottish ice sheet advance into the Dungiven lowlands deposited shelly beds at Bovevagh Old Church, 40 km south of the north coastline (Fig. 4.2).

Although the muds at Sistrakeel are clearly deformed by ice many exposures around the Dungiven basin show conformable relations between mud, sand and gravel beds suggesting that minimal transportation occurred. In addition the vast amounts of mud recorded along river sections in the basin were hardly transported from the sea bed right across the Binevenagh ridge (384 m OD) without substantial mixing and deformation. It is suggested that the widespread distribution of shelly facies near the surface, conformable relations with other stratified sediments and the relative absence of intense sediment deformation suggests that Scottish ice simply overran a major marine embayment that was a precursor to the present day Lough Foyle. Although this model is based mainly on sediments and the fact that shells from Sistrakeel and Bovevagh Old Church are older than 41–55 ka BP, it suggests a very dynamic link between the two ice sheets. The widespread presence of marine sediments in the Dungiven lowlands must be related not only to deep isostatic depression but to the phasing relations between the two ice sheets. Isostatic depression must be due mainly to extensive and thick ice across the Irish lowlands rather than on restricted upland ice dispersal from the Scottish highlands. The latter areas are small in relation to the potential for ice growth across the entire north Irish lowlands. The stratigraphy at the Sistrakeel type site supports the phasing between the ice sheets. In general the microfaunal assemblages reflect shallow water, glaciomarine to marine conditions and a relatively warm palaeoclimate possibly approaching interstadial in character. However, most assemblages show evidence for resedimentation and thus reflect bottomwater processes and possibly meltwater discharge from the land. In these respects the microfaunas from muddy sediment may suggest a time lag between the decay of Irish ice and advance of Scottish ice, allowing some sediment reworking within the Dungiven basin.

Records from the later Late Midlandian ice sheet clearly show that ice sheet growth and decay occurred on millennial timescales and that ice centres were mobile during the glacial cycle in response to North Atlantic climate. It is therefore argued that development of large ice sheets during the Early Midlandian cannot have been confined to the northern parts of the island. Given the timeframes involved and the idea of rapid ice sheet growth and decay, there was sufficient time for Early Midlandian ice to cover most of Ireland in a manner similar

to that during the Late Midlandian. It is proposed that the deposits preserved at Tullyallen Quarry, near Drogheda are of the same general age as the Early Midlandian tills farther north. This hypothesis is supported by a number of considerations. First, the *in situ* marine shells (*Delectopectin greenlandicus*) within the glaciomarine part of the sequence at Tullyallen Quarry have AMS ^{14}C ages of 41,800±1500 and 45,200±2000 BP (Fig. 4.3). Second, the deposits are at the surface and could not have survived many glaciations and are stratigraphically distinct from old glacigenic deposits preserved in limestone hollows. In addition, the upper parts of the sequence are deformed by Late Midlandian ice (Colhoun and McCabe, 1973). Third, the stratigraphic duplet of basal till overlain by *in situ* glaciomarine silt and sand records a major glacial cycle in its own right. This comprises an ice movement into the Irish Sea Basin followed by ice sheet decay and deep isostatic depression similar to the pattern during the last glacial cycle. The Drogheda till records a major ice movement southeastwards into the Irish Sea Basin from centres of inland ice dispersal to the northwest in north central Ireland. This demonstrates that the Irish Sea Basin was essentially a sink being fed from adjacent land areas, a system pattern repeated during the Late Midlandian. Because the overlying glaciomarine deposits are *in situ* and occur between 29 and 40 m OD they record deep isostatic depression on the margin of the basin during deglaciation. Comparison between these Early Midlandian and Late Midlandian isostatic records in the northern Irish Sea area suggests that the earlier ice sheet was somewhat larger.

The lower till sheets identified as early Midlandian all occur beneath Late Midlandian tills or below fossiliferous interstadial deposits generally in basinal situations where some protection from later ice sheet erosion occurred. These areas in the north of the island provide evidence for sediment accumulation including till agglomeration, and lack the widespread erosion so typical of more western sectors of the ice sheet. The sites described show that the ice covered all of the northern part of the island, moved into the Irish Sea basin and deeply depressed the land both in northern County Londonderry and in east central Ireland at Drogheda. It can therefore be argued on the basis of the isostatic evidence alone that the ice sheet was greater and thicker than the later Late Midlandian maximum which ended on the continental shelf. If this is the case the available data point to an ice mass approaching one kilometre thick with the main centres of ice dispersal centred on

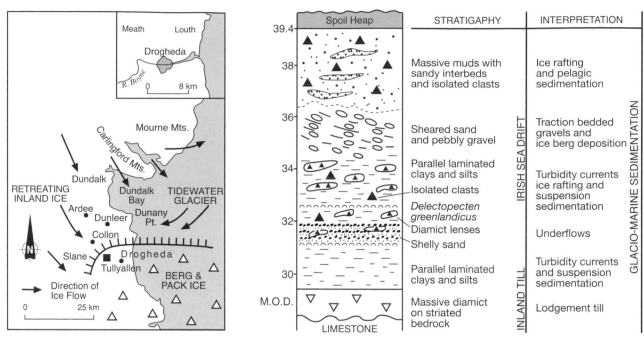

Fig. 4.3 Stratigraphy at Mell and Tullyallen townlands, Drogheda, County Louth and possible position of ice sheet margins during deposition of the marine silts and clays (after McCabe, 1987).

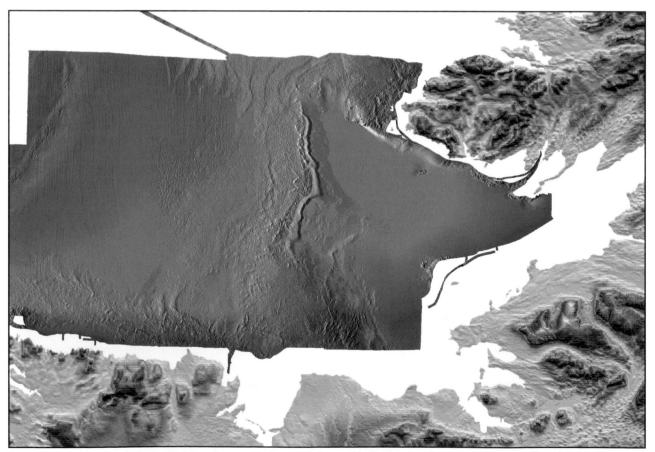

Fig. 4.4 Nested series of ice-contact moraines on the inner continental shelf off the north coast of Mayo. Multibeam bathymetry data acquired in 2002 onboard the R. V. *Celtic Voyager* under the Irish National Seabed Survey. *(Copyright, Geological Survey of Ireland, 2002)*

the Irish lowlands supplying ice streams which reached the continental shelf. Some of the moraines situated kilometres beyond the Late Midlandian advances mapped by King et al. (1998) on the western shelf may well belong to this event (Fig. 4.4).

There is little reason to believe that there were great differences between the main elements of the Early and Late Midlandian ice sheets except scale. The larger scale of the earlier glaciation may reflect a more active hydrological cycle and precipitation pattern following an interglacial and a relatively warm ocean. In contrast the Late Midlandian glaciation followed a relatively cold period and cooler sea surface temperatures in the North Atlantic. The terrestrial evidence, especially the widespread distribution of lower till facies and dating evidence from overlying organic beds, suggests similarities between the Early Midlandian and Late Midlandian ice sheets. First, the main event was less than 10 ka duration. Second, lowland ice centres were important sources of ice streams which ended on the continental shelf. Third, the Irish Sea Basin was not an ice dispersal area but an ice sink which became a marine seaway during deglaciation. Fourth, the raised glaciomarine deposits on the margins of the Irish Sea Basin testify to deep isostatic depression. In both cases this persisted well into the deglaciation phase. This relationship shows that there was delayed land uplift because substantial ice masses still remained on the Irish lowlands even though the Irish Sea Basin experienced an early deglaciation. This relationship during both glaciations stresses the roles played by deep isostatic depression during glacial terminations. Fifth, Scottish ice readvanced into the northern lowlands towards the end of both glaciations. This phasing of ice sheet movements when the Irish ice waned while the Scottish ice advanced on a significant scale is difficult to explain but it may mark the difference between a more climatically sensitive lowland ice sheet and a slightly more robust highland ice sheet with respect to different accumulation-area ratios.

At Aghnadarragh the lower till passes upwards into mud and laminated sand containing isolated clasts (Figs. 4.5, 4.6). These facies are thought to represent deglaciation and deposition in shallow water on top of unweathered till by suspension sedimentation followed by density currents. The ice wedge pseudomorph which truncates the bedding and penetrates downwards from the top of the sands is similar to examples described by Colhoun (1971). The structure of the wedge is dissimilar to that of pull-apart structures lacking their typical bed-sag continuity. The transitional junction between the sand and the overlying gravelly diamict indicates a change in process from current activity to mass flow. Dispersed large clasts in the chaotically organised, crudely-stratified diamict, which grades laterally into massive pebbly-gravel, suggests that flow types varied between cohesive debris flows with clast freighting and gravelly, high-density flows. It is thought that the 4 m sequence is associated with resedimentation from unstable slopes of a gravelly fan-delta in the vicinity of a small water body. This interpretation is supported by the presence of a sparse fauna of Arctic aspect together with mammoth remains. The population of *Mammuthus primigenius* consists of well-preserved teeth that were freighted into position in the gravelly diamict because many roots are almost intact and there is little evidence of surface exposure, splitting, erosion and damage (Fig. 4.7). In addition, the presence of tusks up to 1 m in length, fragmentary scapula and limb bones testify to an almost *in situ* biocoenosis. Clearly, the fine state of preservation of the fossil material is consistent with the hypothesis that the animals were living near the water body as unsorted and sorted debris was deposited and resedimented. The mammoth tooth enamel have uranium-series ages of between 65 and 95 ka BP pointing to an Early Midlandian age for the gravelly-diamict. Together the lithostratigraphy, biostratigraphy and geochronology from the basal facies at Aghnadarragh are consistent with a major deglaciation sometime during OIS 5d or 4. Elsewhere in Ireland there is no firm evidence for glacial activity even though multiple till sequences do exist. Similarly, the relative stratigraphy of repeated expansions of mountain ice has been proposed but the ages of the earlier events remains unknown. However, it is certain that conditions for ice cap development at this time must have been favourable because of the cold climate and supply of moisture for the lowland ice sheet.

Aghnadarragh Interstadial

This interstadial has only been recorded from the lignite pit at Aghnadarragh where an episode of climatic warming is recorded by peats overlying glacigenic deposits of the Fermanagh Stadial (Fig. 4.6). The woody peat detritus occurs in two basins separated by a ridge of lignite which was created by ice sheet compression when the upper till was formed. Radiocarbon ages from wood in the peat beds are 48.1 and 47.3 ka BP and are infinite.

Fig 4.5 Section on the northern face of the lignite pit at Aghnadarragh, County Antrim. Note the upper till (unit 10, Fig. 4.6) ending against a ridge of deformed lignite (upper left) and the dark organic deposits (unit 8, Fig. 4.6) below the light coloured sands (unit 9, Fig. 4.6).

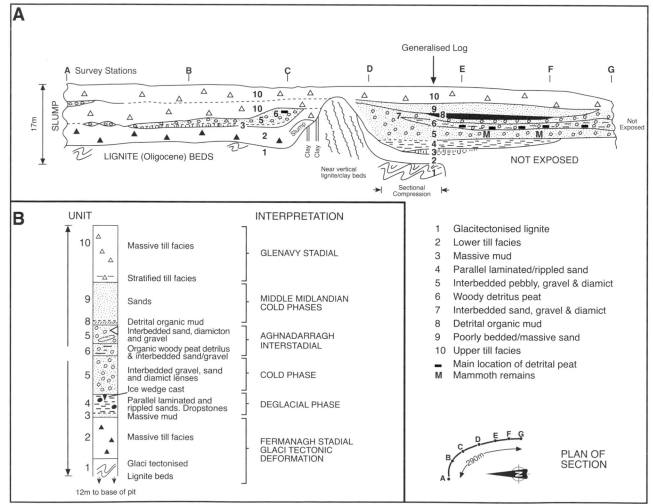

Fig. 4.6 Section at Aghnadarragh near Glenavy, County Antrim (after McCabe et al., 1987).

Fig. 4.7 Mammoth molar within gravelly debris flows (unit 5, Fig. 4.6), Aghnadarragh, County Antrim.

Luminescence dating of overlying sediments suggests that the interstadial occurred sometime before 100 kyr BP. The organic beds consist of compressed sheets of woody detritus characterised by abundant smoothed, flattened twigs and wood pebbles within sheets of black peaty material. Woody detritus is intimately bedded with laminated sand and occasionally pebble gravel. Most wood has little surviving identifiable structure but contains evidence of *Pinus*, *Picea* and a hardwood. Cones of *Pinus sylvestris*, fruits of *Betula pendula*, *Betula pubescens* and *Betula nana* were recorded. Megaspores of *Selaginella selaginoides* (L) Link, an open grassland indicator, are also found in the peats. Pollen and macro-remains of water plants including *Potamogeton natans* L., *Callitriche* sp., *Typha latifolia*, *Eleocharis* sp., and *Carex* all point to nearby marshy conditions. The main contributors to the pollen assemblage are *Pinus*, *Betula*, *Picea* and Coryloid pollen. However, the preservation of the Coryloid pollen was poor and specific identification was not secure, possibly suggesting that this pollen together with occasional grains of *Taxus* pollen were derived from an earlier deposit. Species composition of the pollen profile is fairly constant including derived Tertiary pollen.

An allochthonous origin for the woody detritus is indicated by the combined presence of pollen and macroremains of indicators of moving water, water-rolled wood and high levels of crumpled, broken and un-identifiable pollen (<26%). Because the detritus reflects transport the pollen profile does not record a vegetative succession. The pollen indicates that the immediate area was a grass and sedge marsh fed by streams transporting rolled wood. The presence of Tertiary material indicates that recycling was prominent, though the presence of both pollen and wood of *Picea* confirms it is not derived primarily from earlier deposits. Ericaceae pollen including *Calluna vulgaris* and *Empetrum* probably document the vegetation from the floor of the surrounding coniferous forests. Insect remains from the peat were broken up and there were marked concentrations of the more laminar fragments such as elytra and pronota suggesting that the fossil assemblage had undergone some transportation and sorting. However, the fact that the fauna as a whole provides an ecologically consistent picture means that there has not been reworking of an earlier deposit. The insects represent habitats not far removed from the depositional site. The assemblage is not Arctic because many recorded species reach as far north as the southern half of Fennoscandia at the present day. Using the Mutual Climatic Range Method (Atkinson et al., 1986) the thermal climate may be summarised as mean July temperature (+15°C to +18°) and mean January temperature (-11°C to +4°C). These temperatures are just within the range of values for the present day at Armagh.

The insect, pollen and macroscopic plant remains present an ecologically consistent picture of *Betula–Pinus–Picea* woodlands with adjacent areas of swamp. Rolled wood is consistent with prolonged to-and-fro rolling possibly on a local beach, though the presence of well-preserved cones and beetles indicates that parts of the deposit were simply flushed rapidly by floods into the depositional site. The overall evidence from pollen, macroscopic plant remains and insects shows that the assemblage is not a result or reworking from an earlier deposit because the hydrodynamic properties of the majority of fossils would have caused a greater degree of sorting than is actually the case. The woodland association could have developed towards the end of an interglacial or during a shorter interstadial. On stratigraphic grounds the peat occurs between cold stage deposits and this probably precludes development during a telocratic stage of an interglacial. Therefore the Aghnadarragh peat is interstadial in aspect and represents conditions that were warm enough for a sufficiently long period to allow migration of tree species eliminated by earlier glacial conditions (ie. lower till) at the site.

Hollymount Cold Phase

This phase groups together three sites situated in Ulster containing fossil assemblages of cold climate affinities recording deterioration of climate after the

Aghnadarragh Interstadial (Fig. 4.1). Organic fresh-water muds at Hollymount (McCabe et al., 1978), Aghnadarragh (McCabe et al., 1987) and Greenagho (Dardis et al., 1985) cannot be dated accurately but stratigraphically they are placed towards the end of the Early Midlandian (Fig. 4.2). The silts at Hollymount are radiocarbon dated to 41.5 ka BP, the upper detrital organic mud at Aghnadarragh to 46.62 ka BP and the Greenagho silts at 34,460±270 BP. The latter date is a minimum one and no recognisable pollen was identified from this site. Luminescence ages for the silts at Hollymount show much older estimates of 109±17 ka and 99±12 ka BP while a similar estimate of 97±13 ka BP was obtained from the upper detrital mud at Aghnadarragh. Palaeobotanical evidence from Hollymount indicates that the local environment was open countryside, sparsely occupied northern flora of open ground dominated by Graminae and Cyperaceae along with *Betula nana* and northern/montane beetles. Similar environmental conditions are also evident from the disseminated organic mud towards the top of the Aghnadarragh exposure. This herb biozone is dominated by a local treeless environment with a Cyperaceae–Graminae assemblage and *Calluna vulgaris* and *Betula nana*. Pollen of predominantly cold species such as *Artemisia*, *Thalictrum* and *Armeria* (type B) are present together with spores of *Selaginella selaginoides* and Filicales. In general the balance of aquatic to land species, in terms of both pollen and macro-remains, suggests that wet marshy conditions prevailed. The insect remains were in a fragmentary state but not eroded and it was possible to find complete sclerites and jointed pairs of elytra in the field. This evidence together with finely-disseminated organic remains within thin, discontinuous sandy streaks points to very quiet deposition in small ponds. All beetles record a cold treeless environment and the Carabid *Diachella arctica* is exclusively Arctic–Alpine. The thermal climate is summarised with average July temperatures between –11 to +13°C and the average January temperatures from –18 to –7°C. However, the presence of relatively warm summers is difficult to reconcile with the absence of trees. It is possible that treelessness in such a period covered by herb biozones is associated with low winter temperatures, soil instability, drying winds, waterlogged soils in the summer and perhaps persistent cool winds. The high albedo associated with snow cover and herb vegetation may have caused a shorter growing season in spite of the summer temperature, with the persistence of snow cover related to the cool Atlantic Ocean.

Environments of this general type occur, for example, in parts of Siberia where a cool ocean borders a continental mass (West, 1977).

Derryvree Cold Phase

At Derryvree, County Fermanagh, road cuts exposed fresh water *in situ* organic beds between two regional till sheets (Colhoun et al., 1972) (Fig. 4.8). The lower till facies is identical to that at Hollymount a few kilometres to the north and formed during the Fermanagh Stadial. The upper till forms a drumlin carapace over the lower facies and forms a regionally extensive till known as the Maguiresbridge Till. Along about 60 percent of the section both tills are separated by thrust zones, shear planes and sharp contacts. However, in the western part of the section two basins infilled with undisturbed sands and organic debris are preserved between the two till facies. Elsewhere rafts of lower or Derryvree Till and organic detritus have been sheared into the upper till from a northerly direction. The bedding of the laminated sands on cm-scale is subparallel to the basal outline of the depressions and shows no evidence for tectonic shearing and is considered to be *in situ*. A sample of moss-rich detritus mud from the largest basin gave a bulk radiocarbon age of 30,500±1170/1030 years BP showing for the first time that non-glacial conditions occurred between deposition of the two till sheets (Figs. 4.8, 4.9).

In all samples the pollen counts were low and in poor condition though the ten species of moss remains were the most numerous fossils and form the biological matrix which was *in situ* or almost so. This inference is based on the fact that the large tufts of *Dichodontium pellucidum*, which approaches var. *fagimontanum* (Brid.) Schimp, is the common growth form of this species found undamaged in the silts. In addition dissection of tufts of *Dichodontium* revealed quantities of *Philonotis* and *Drepanocladus* indicating that these moss species were growing together on a streamside or wet rocks. The local environment was treeless and muskeg in character with high pollen values (89%) of Gramineae and Cyperaceae. Lesser values (4%) of *Salix herbacea* were present and some long distance pollen. In summary grasses, sedges and mosses at least were conspicuous locally in a landscape which supported a varied vegetation with the aspect of tundra. Irish Late Midlandian deposits (Pollen zones 1–3) have yielded the great majority of the taxa at Derryvree including *Arenaria ciliate* agg., *Dryas octopetala*, *Silene acaulis, Saxifraga oppositifolia* and hypnoid saxifrages including seeds ascribed to *S. rosacea*. The first and last named forms are two of the most celebrated plants and are important phytogeographically. Their presence at Derryvree implies much wider ranges than at present for disjunct montane species before the last glaciation and may also point to periglacial survival in Ireland followed by increase in the final stages of the last glacial termination rather than immigration at that period. The cliffs forming the margins of massifs such as Benbulbin are important habitats for the rare Arctic–Alpine plant species in Ireland today and similar refuges may have occurred farther south during maximum glaciation.

The beetles, alone conjure up a picture of a sandy-bottomed pool, rich in plant detritus and surrounded predominantly by moss and associated plants forming a treeless muskeg environment. However, the omission of certain families which would be expected is strange. Most significant is the absence of Curculionidae (weevils) and dung beetles (*Geotrupes*, *Aphodius*) which

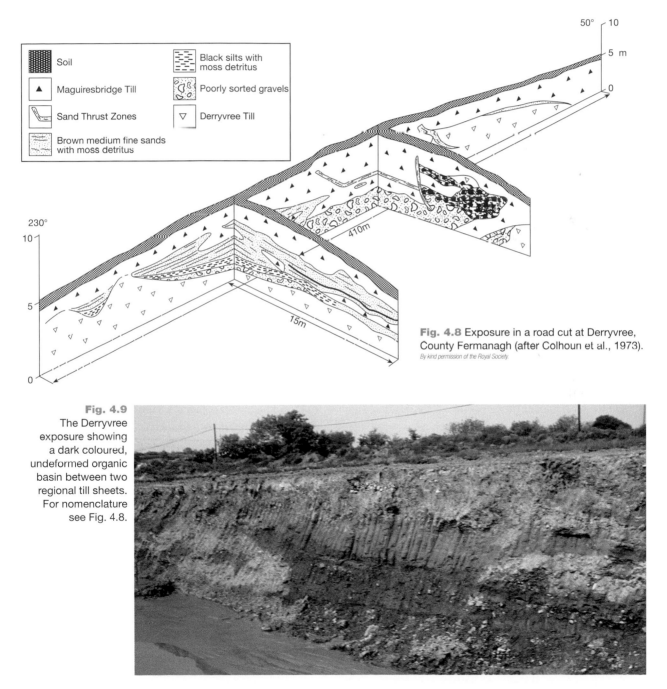

Fig. 4.8 Exposure in a road cut at Derryvree, County Fermanagh (after Colhoun et al., 1973).
By kind permission of the Royal Society.

Fig. 4.9 The Derryvree exposure showing a dark coloured, undeformed organic basin between two regional till sheets. For nomenclature see Fig. 4.8.

Legend:
- Soil
- Maguiresbridge Till
- Sand Thrust Zones
- Brown medium fine sands with moss detritus
- Black silts with moss detritus
- Poorly sorted gravels
- Derryvree Till

are known from similar deposits in England (Coope, 1968). It is argued that the absence of weevils is due to their normal lack of flying powers and the idea that they were eliminated by earlier ice and cold. If this is the case then a marine barrier existed between Ireland and the rest of the British Isles which prohibited the immigration of all but the fliers. Nearly all the specimens from Derryvree are fully macropterous and four are winged but not flying. Shotton (1962) has suggested that this marine barrier was in existence until well after the maximum of the Late Midlandian glaciation. This idea of a water-barrier forbidding colonisation is supported by the absence of dung beetles, which implies at least locally that mammals were not present and in fact did not return until a late-glacial marine lowstand around 14 ka BP.

In contrast, Mitchell (1976, 1981) has reviewed mammalian fauna from the Castlepook Cave, County Cork where a mammoth molar gave a date of 35,000±1200 BP and one from a spotted hyaena bone 34,300±1800 BP. If these dates are valid it is possible that mammals either did not reach the northern part of the country during Derryvree times or for ecological reasons were restricted to the south. However, there is little information on the taphonomy of the fossil assemblage, and its stratigraphic position is in question. Nevertheless, both sites provide the only unequivocal evidence for environmental conditions towards the end of the Middle Midlandian. The herb biozone clearly shows that no large ice sheets existed at this specific time in the north of the island. However, this comment must be viewed in terms of millennial timescale climate change in the North Atlantic and the biozone represents but a fraction of Middle Midlandian time. Ice sheets could have been growing on the western seaboard where precipitation was heavy and in mountains where mass balances were favourable. Field evidence from the south side of Donegal Bay at Glenulra confirms that ice sheets were waxing and waning before 30 ka BP, the prelude to more extensive glacierisation during the Late Midlandian (McCabe et al., 2007a).

Late Midlandian Events

It is to be expected that the deposits and landforms formed during the last glaciation will be better preserved than those formed during earlier times. The widespread distribution and apparent stratigraphic complexity of Late Midlandian drifts supports this observation.

However, like the earlier parts of the Midlandian, standard models of ice sheet events were simple. They depicted ice flow onto part of the continental shelf at the LGM followed largely by a deglaciation characterised by minor readvances during monolithic ice retreat towards centres of ice sheet dispersal. Ice sheet development occurred after the Derryvree cold phase followed by a LGM around 23 cal ka BP. Standard models with the notable exception of Synge (1968, 1969, 1970) failed to recognise the significance of major readvances during deglaciation. The terrestrial field evidence present along the emergent continental margins contains sedimentary and marine fossil records that can be related to events from the climatically sensitive northeastern Atlantic. New perspectives on the detailed evolution and scale of the last ice sheet can now be formulated because fossiliferous marine beds within the deglacial succession can now be dated by AMS ^{14}C and the time of deglaciation from terrestrial moraines can be dated from surface exposure dates using cosmogenic nuclides that have accumulated since ice melting (Ballantyne et al., 2006, 2007; Clark et al., 2007a,b).

Identification of discrete Late Midlandian events which can then be used to reconstruct ice sheet history during the last glacial/deglacial cycle (between 35 and 10 ka BP) is now possible because of four major research advances. Central to these is the fact that the relatively small mountain masses situated on the periphery of the island are much too small to provide enough ice to sustain lowland ice sheets, and simply were confluent with the large lowland ice sheet. First, subglacial imprints show that there were significant changes in the configuration and development of the glaciological mobile lowland ice sheet. Second, stratigraphies, striae and erratic patterns support the idea of a mobile ice sheet with shifting centres of ice sheet dispersal. Third, major controls on ice sheet mass and configuration seem to be linked to rapid changes in temperature and associated climate in the North Atlantic Ocean. Fourth, the ice thickness possibly over 800 m was always likely to result in deep isostatic depression which was greater than eustatic fall in sea level. A critical inference is that marine downdraw beginning at ice sheet margins, perpetuated ice flows deep within the heart of the ice sheet. This mechanism was a major factor in ice sheet decay from maximum position on the inner shelf. Field data in the form of raised marine features and emergent facies sequences from around the island testify to the efficiency of isostatic depression during glaciation.

Furthermore, even though many basal ice flow lines are delicately moulded to topography, the vast majority of directional indicators exit into major bays fringing the continental shelf. Some of these may be considered to be ice streams and directly responsible for multiple overprinting of subglacial bedforms and marked erosional overdeepening of glacial troughs within mountainous terrain.

Conventional notions of central snowfields and radial ice flow patterns onto the continental shelf, followed by monolithic ice sheet retreat back to major centres of ice sheet dispersal, tell us little about ice sheet development. However, over sixty radiocarbon dates now underpin a marine, glaciomarine and glacial stratigraphy which developed as glaciation progressed. At least eight age-constrained stratigraphic elements show that the anatomy of the last cold stage was complex, which strongly suggests that the sketchy events recorded from the earlier parts of the Midlandian are partial records only. The main stratigraphic elements include (Fig. 4.10):

1. **Advance onto the continental shelf.** Large linear features subparallel to the western coast of Ireland and iceberg furrows have been attributed to ice advance onto the continental shelf (King et al., 1998). Unfortunately, accurate descriptions of these features are not available and their size is much more than other moraines described from other shelf zones. The features are undated and may be associated with ice sheet oscillations onto the shelf from Early Midlandian times or even earlier. The fact that there are a series of nested lines of moraine may support the idea that ice sheet margins were oscillating on western coasts during much of the Midlandian (Fig. 4.4). This interpretation is consistent with the presence of interstadial and cool climate horizons between regional tills (MIS 4 and 3) within the adjacent terrestrial stratigraphy at sites such as Derryvree. In a general sense these records can also be interpreted in terms of rapidly changing North Atlantic climate events and a coupled ice–ocean–atmosphere system. Analysis of climate proxies suggest that a ~7-kyr oscillation occurred during MIS 3 and that it was dependent on Atlantic meriodional overturning circulation and attendant change in cross-equatorial ocean heat transport (Clark et al., 2007). It is also evident from these climate proxies and field records that the crude subdivision of the Midlandian into Early, Middle and Late is somewhat artificial because climatic oscillations across a range of scales were occurring throughout this time period. Ice sheet responses on land to climate change were probably cyclic and involved ice-marginal oscillations along the margins of the continental shelf.

2. **Early build-up of ice in the western sector.** The raised marine and ice-contact deposits on the south side of Donegal Bay at Glenulra Farm provide clear evidence for a thick ice mass over western Ireland following the Derryvree Cold Phase (McCabe et al., 2007a). AMS ^{14}C ages from the *in situ* marine microfauna at Glenulra Farm indicate that after the ice sheet retreated from the continental shelf, relative sea level remained high (>70 m above sea level) from 28.3 to 25.4 cal yr BP (McCabe et al., 2007a). Striated rock on this coast records ice advance onto the shelf at ca. 28 cal ka BP prior to the marine transgression. The events are the oldest known from the Late Midlandian, pre-date the conventional LGM and mark the onset of the Late Midlandian glacial cycle. They record the presence of a thick and persistent ice mass over western Ireland with major isostatic loading sufficient to maintain high relative sea levels, possibly for ten thousand years. The evidence from Donegal Bay is consistent with the presence of a marine ice margin off the western Scottish coast since 45 cal BP. and the delivery of IRD onto the continental shelf (Knutz et al., 2001).

The early growth of thick ice masses on the western seaboard is likely to be a function of high precipitation and a cool adjacent ocean. The presence of major erratic fans showing marked eastward dispersal for anything up to 180 km suggests that at this time ice from the west advanced into the Irish midlands. Three main erratic fans which have not undergone significant redistribution are linked to ice growth and movement from the west. First, Galway granite erratics are widely distributed generally southeastwards from their primary source on the northern margin of Galway Bay (Fig. 4.10) (Synge, 1979a). The carriage extends southeast to Slieve Bloom (110 km), south to Watergrass Hill near Cork Harbour (170 km) and into Offaly and Laois (130 km). There is little doubt that tills are associated with this ice advance (Gallagher et al., 1996) though satellite imagery clearly shows that these deposits have been overprinted by later southerly and south-

Fig. 4.10 Millennial timescale events, anatomy and development of the last ice sheet in Ireland. The model is based on age-constrained ice limits, coeval ice flow vectors, dated marine beds, prominent moraines, patterns of deglacial features and on probable values for regional isostatic depression.

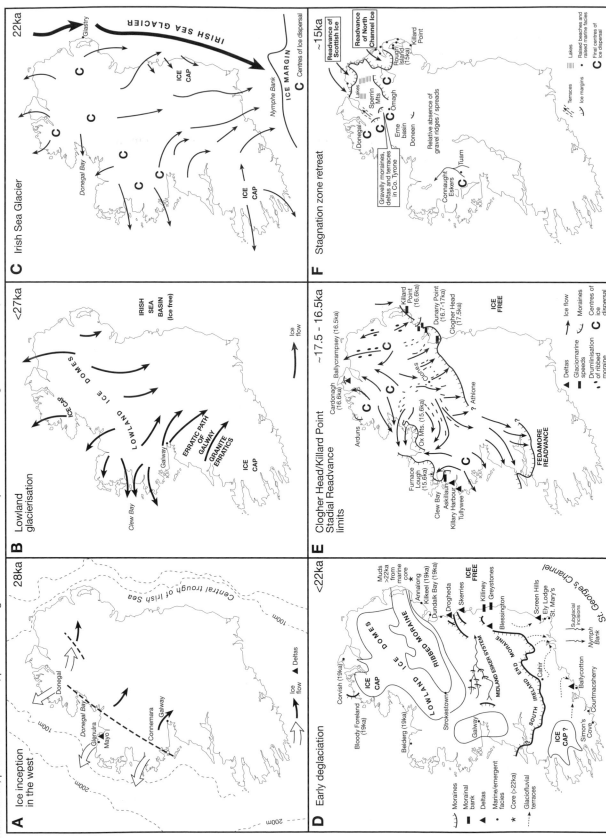

westerly ice flows. Second, Charlesworth (1924) and Colhoun (1971a) have shown that an erratic train of Donegal granites records a northeasterly ice low into the Sperrins and adjacent lowlands. Again, this phase of ice movement has been overprinted by later ice movements when ice from the great lowland dispersal basins of Lough Neagh and Omagh overtopped the Sperrins. Third, Dardis (1981) recognised from the erratics and facies variations within drumlin stratigraphies in south-central Ulster that the ice accumulation zone shifted from highland (Sperrins and Donegal) to lowland basins during the course of glaciation. A major implication is that drift accumulation occurred well before the final streamlining event (drumlinisation) at the end of the Late-Midlandian.

3. **Lowland glacierisation.** The next phase of glaciation must have involved the build-up of thick ice masses across the entire lowlands of the island. Although ice was probably building up across the lowlands at the same time as the ice buildup in the west, it reached a maximum when it began to reverse ice flow across the central lowlands from easterly to westerly directions. At this time the ice sheet anatomy seems to have consisted of large, thick domes of ice drained by ice streams which now began to gravity feed towards the major marine outlets in the west. Whether or not all of the western centres of ice dispersal waned is not known, though the axes of large streamlined rock ribs inland of Clew Bay record ice flow onto the inner shelf (Fig. 4.11). Although it is difficult to date this phase of glacierisation, the widespread presence of glacigenic drift and erosional marks right across the midlands and the later decay patterns of moraines and outwash testify to the presence of thick ice (>800 m). To a major extent thick ice at this time is also necessary to explain later isostatic rebound patterns together with surface exposure ages recorded by Ballantyne et al. (2006, 2007).

4. **The Irish Sea sink.** Ice flow along the Irish Sea Basin is often regarded as the product of a major ice stream that either drained the BIIS throughout much of the last glacial cycle or moved across adjacent land areas. However, the radiocarbon chronology and patterns of ice sheet growth and readvance from adjacent land areas paint quite a different picture. Most directional indicators, especially in the northern parts of the basin, show consistent centripetal ice sheet flow into the basin which therefore was acting as a kind of ice sink and conduit rather than an area of ice growth. Evidence from Glastry in the eastern Ards Peninsula, County Down, suggests that the northern basin was largely ice-free until after 24 ^{14}C kyr BP (~28 cal ka) (Hill and Prior, 1968). This inference is based on radiocarbon-dated marine shells which occur in the lower till in the Ards Peninsula. Therefore the marine shells must have been living in the vicinity before the ice advance southwards along the North Channel which deposited the lower till. This stratigraphy is also consistent with deposition of the upper till in the Ards Peninsula during the later ice sheet readvance across the area to the Killard Point moraine.

The fact that the Irish Sea Basin eventually became a major ice sheet conduit draining much of the BIIS ice sheet sometime after 28 and before 22 cal kyr BP shows that the basin was glaciated for much less than five thousand years, while the entire glacial cycle lasted for over twenty thousand years. A more important role for the basin was its position below the geographic centre of the BIIS. The southern maximum position of the ice lay somewhere in St. George's Channel and on the Nymphe Bank. There is no dating, sedimentological or geomorphic evidence for farther ice advance southward into the Celtic sea and onto the northern slopes of the Southwestern Peninsula of England. On practical grounds alone ice advance was stopped by the rapid western-facing expansion of the Celtic Sea Basin and by rising relative sea levels (Eyles and McCabe, 1989a). It is also extremely unlikely that the relatively narrow (>80 km) ice flow in St George's Channel could have sustained a bulbous shaped, piedmont glacier across the entire floor (>250 km) of the Celtic Sea at this time. A more realistic scenario is that the ice sheet margin simply lost momentum and halted when ice sheet flow lines were at their maximum. This interpretation is supported by the absence of any defining moraine and the presence of thick diamict sequences immediately to the north of St. George's Channel (Garrard, 1977).

To a large extent the concept of a late glacial maximum (LGM) has been used by many to depict ice sheet limits either on the continental shelf or on adjacent terrestrial margins. However, few if any sites have been dated with precision, and complications arise because in western ice sheet sectors

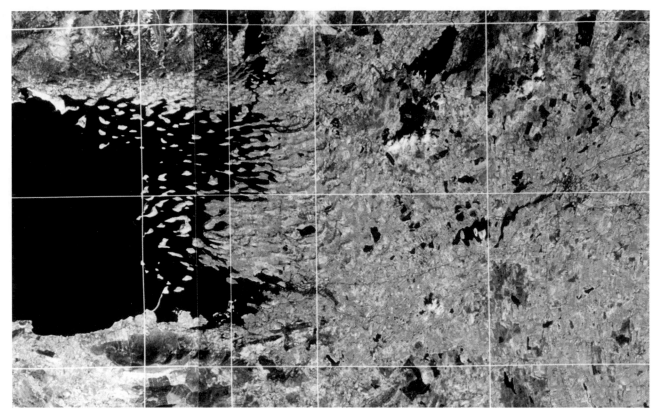

Fig. 4.11 Satellite image of inner Clew Bay, western Ireland, showing composite subglacial bedforms which are coloured orange to red. The main SW–NE axes of the ridges record ice flow into the bay and westward onto the continental shelf early in the last glacial cycle. Both ends of the ridges have been modified into northwesterly pointing horns, formed during later overprinting by ice moving northwestwards during the Killard Point Stadial (McCabe et al., 1998). Striae cut along the top of the rock-cored bedforms between Westport and Newport and the northwestern facing sediment horns are consistent with ice flow to the northwest which ended at moraines blocking the southern end of Furnace Lough. These moraines are coloured emerald blue and trend SW–NE on the northern margin of the bay.

pre-LGM ice sheets are more extensive than LGM limits (McCabe et al., 2007a). However, the approximate ice sheet limit in the Celtic Sea is often considered to represent the LGM largely because ice flows from the Irish lowlands, the Cork/Kerry ice cap and the Irish Sea Basin coalesced and moved south across the Nymphe Bank. Along the south coast of Ireland striae and erosional marks on the raised rock platform of Wright and Muff (1904), ice directed drainage channels across west–east bedrock ridges, subglacial bedform patterns and striae orientation right across the southern lowlands together provide a set of coherent ice flow indicators recording ice flow southwards into the Celtic Sea (McCabe and O'Cofaigh, 1996). Therefore the general picture of a LGM off the southern Irish coastline that was more extensive than ice sheet events earlier in the cycle has strong support from stratigraphic and ice flow data. In contrast ice sheet events in western ice

sheet sectors are more complicated and in fact it is difficult to locate the standard LGM which occurred around 19–23 ka BP.

5. **Early deglaciation.** Early deglaciation of the southern Irish coast and ice-marginal withdrawal from the Celtic Sea is recorded by a large system of subglacial tunnel valleys on the Nymphe Bank (Blundell et al., 1971) and the emergent marine and raised beach gravel which directly overlie the freshly-glaciated rock platform of Wright and Muff (1904). On the floor of the Irish Sea Basin southward directed tunnel valley systems (Whittington, 1977) and the internal structures of morainal banks along the coast confirm that deglaciation proceeded from south to north. In the south deglaciation began before 22 [36]Cl kyr BP (Bowen et al., 2002). Dated marine cores between the Isle of Man and the coast of County Down suggest that a marine seaway developed along the

central trough of the Irish Sea Basin by 23–24 cal ka BP (Kershaw, 1986). After this date the sea floor sediments in the northern basin are dominated by fine-grained, muddy sediments. The absence of coarse-grained sediments on the sea floor confirms that the ice sheet margins did not advance into the basin after early deglaciation.

Early deglaciation is also recorded from raised marine muds contained within erosional channels along the coast of south County Down at Kilkeel (Clarke et al., 2004). Channel profiles are graded to well below sea level, recording a marked fall in relative sea level associated with strong isostatic uplift. The timing of isostatic uplift is clearly related to widespread deglacierisation of all sectors of the continental shelf and the disappearance of at least two thirds of the ice sheet. Because the marine muds formed at a time of uplift and are dated to ~19 cal kyr BP they record a rapid rise in eustatic sea level. Radio-carbon dates from marine muds at Belderg Pier, County Mayo and Corvish, north County Donegal also show that these areas were deglaciated by ~21 cal ka BP. Finally, surface exposure ages using [10]Be from the Bloody Foreland moraine in northwestern Donegal suggests that ice had withdrawn from the continental shelf before 19.5 cal kyr BP (Clark et al., 2007b). Similar dates from raised marine beds indicate that different sectors of the ice sheet withdrew from the continental shelf at much the same time. Age-constrained events from the Irish Sea Basin show that the entire basin was deglaciated in a few thousand years because of intense marine downdraw (Eyles and McCabe, 1989; McCabe et al., 2005). Whether or not deglaciation rates were as rapid along other sectors of the ice sheet remains unknown.

Early deglaciation of continental shelf areas was quickly followed and accompanied by widespread terrestrial deglaciation. In southern counties extensive outwash trains along the major valleys record rapid ice sheet wastage which is also suggested by the absence of well-defined morainic halts. Much of the South Ireland End moraine stretching from Caher and Bansha at the eastern end of the Glen of Aherlow northeastwards towards the Wicklow Mountains consists of kettled outwash. General sedimentation patterns and ridge orientation reflect temporary halts in the general wastage northwestwards across the lowlands. The overall wastage towards westerly

centres of ice sheet dispersal is reinforced by the east to west esker system extending across the central Irish lowlands from The Derries west to Athlone (Sollas, 1896; Farrington, 1970). Although the esker ridges are time-transgressive their morphology records minor depositional ice-marginal halts as the ice margin retreated to the west. These patterns strongly reinforce the idea that centres of ice dispersal in the west were critical contributors to overall ice sheet growth and that they persisted throughout the full glacial cycle. The pattern of eskers formed during deglaciation especially to the north of the Slieve Bloom mountains strongly supports the eastward erratic carriage along the western flanks of the Slieve Bloom Range (Gallagher et al., 1998). There is little doubt from erratic and esker patterns that the ice margin retreated westwards across the central plain towards a major centre of ice sheet dispersal in Galway and along the western seaboard. In the northern and western sectors of the country it is difficult to reconstruct patterns of early deglaciation because later ice sheet readvances were widespread and may have removed evidence. However, subglacial bedform patterns show that the ice sheet contracted from coastal margins forming remanie lowland ice domes. In broad terms the distribution of continuous ribbed moraines across the north Irish lowlands is perhaps synonymous with the remaining ice domes. Clearly, widespread backwasting and downwasting during early deglaciation resulted in a loss of over two thirds of the ice sheet and led to terrestrial reorganisation of the remaining ice sheet masses.

6. **Readvances in the north and west.** Ice sheet flow-lines reconstructed from streamlined bedforms in northern and western Ireland clearly show that the remanie ice masses grew and readvanced between 18 and 16.6 cal kyr BP to form what is traditionally known as the Drumlin Readvance ice sheet (Synge, 1968). Terminal outwash at Killard Point representing the subglacial output from the readvance system can be traced almost continuously southwestwards to Athlone, a distance of 180 km (McCabe and Clark, 1998). Because the moraines are tightly age-constrained by AMS ^{14}C dates from interbedded or overridden marine mud beds, two main readvances are defined, reaching limits at Clogher Head and Killard Point/Dunany Point (Fig. 4.10). The main

areas of ice sheet dispersal in the north were the large lowland basins of Lough Neagh, Omagh and Lough Erne. Similar age-constrained ice sheet limits occur at Carndonagh and Ballycrampsey in north County Donegal indicating that the Donegal ice cap oscillated at the same time as the lowland centres of ice sheet dispersal. In western Ireland the limits of large ice lobes largely fed by lowland ice streams are well marked by moraines on the margins of major bays. On the northern slopes of the Ox Mountains, moraines at Lough Easky show that Donegal Bay was filled by ice flows originating from the Erne/Shannon lowlands and Donegal Mountains (Fig. 4.12a). Similarly the Furnace Lough moraine on the northern side of Clew Bay records the limit of ice flows which originated in Joyce's Country to the south (Fig. 4.12b). This ice dispersal centre was sufficiently active to produce ice flow lines 90 km in length from Joyce's country north to Killala Bay. Sixteen surface exposure ages from large erratics resting on both moraine systems show that final deposition of both moraines occurred ~15.6 cal kyr BP indicating that the associated ice readvance occurred earlier (Clark et al., 2007a,b). Spreads of ice contact, glaciomarine sediments exposed at Askillaun on the south side of Clew Bay, show the relative sea levels were high (>45 m OD) near the ice lobe margin. Similarly ice contact, fjord head deltas at the exits of the Kylemore valley and in Killary Harbour directly face the open Atlantic Ocean and are associated with shoreline heights up to 80m OD (Coxon and Browne, 1991; Thomas and Chiverrell, 2006). These heights imply deep isostatic depression driven by a thick ice sheet in western Ireland that persisted into the final readvance phase around 16 cal kyr BP. The degree and known patterns of isostatic depression confirmed from both eastern and western sectors of the ice sheet now show that thick ice centred on the extensive Irish lowlands was the main driver of isostatic loading rather than ice from restricted parts of the western Scottish highlands.

7. **Stagnation zone retreat.** Ice sheet decay from the limits of the Drumlin Readvance Ice Sheet limits (Killard Point Readvance) began at 15.6±0.3 cal ka BP in western Ireland and 15.5 cal ka BP on the northern margins of the Irish Sea Basin (Clark et al., 2007a). Ice retreat north from the Killard Point ice limit was accompanied by marine inundation of the subglacial

bedforms across the entire Ards Peninsula (Morrison and Stephens, 1965) and deposition of red marine muds in Strangford Lough at Rough Island (Fig. 4.13). The muds date to 15.6–15.0 cal ka BP (McCabe et al., 2005). When these dates are compared to dates from the earlier readvance limits at Killard Point they suggest that the ice margin persisted at its maximum extent for about one thousand years.

Stagnation zone retreat can be viewed as a massive, almost *in situ* collapse of the ice sheet accompanied by melting over the entire mass without any discernable pattern of frontal retreat. In the eastern sector of the ice sheet there are no frontal moraines or evidence for ice-marginal halts/re-equilibrations between ice sheet limits at Killard Point and the dispersal centres around Lough Neagh. This absence strongly points to widespread and rapid ice dissipation. Similarly in western and northwestern sectors there are no major moraines between readvance ice limits at Arduns–Carndonagh–Lough Easky–Furnace– Lough and centres of ice dispersal in the mountains and lowland basins. Spreads and hummocky zones of stratified drift in central Tyrone and adjacent valleys simply reinforce the idea the ice sheet disappeared rapidly and concentrated outwash within topographically confined lows as ice centres collapsed. Stratigraphically the event is recorded by subglacial esker ridges superimposed across subglacial bedforms and the extensive ice-marginal lake sediments in County Tyrone (Dardis, 1985b, c; Knight and McCabe, 1997).

Clark et al., (2007a) have pointed out that this phase of deglaciation occurred during the Oldest Dryas Stadial event about one thousand years prior to the onset of the Bølling interstadial. The deglaciation trigger may therefore be related to an earlier abrupt warming that is recorded in the GISP [18]O ice core record and from a marine core recovered 200 km east of the maximum extent of the BIIS margin off western Scotland (Knutz et al., 2007). This earlier warming is about one-quarter of the amplitude of the Bølling warming, which represented ~10°C of atmospheric warming over Greenland (Severinghaus and Brook, 1999). If scaled there may have been a 2–3° of atmospheric warming at 15.5 cal ka BP which may induce a 330–500 m rise of the equilibrium line altitude of the ice sheet causing the entire lowland-based ice sheet to be in the ablation area (Clark et al., 2007a).

Fig. 4.12a Hummocky morainic topography around the margins of Benbulbin Mountain, County Sligo. This morainic feature is part of a regionally extensive ice limit which has been traced along the northern exits of Glenade, Glencar and along the northern slopes of the Ox mountains to Lough Easky and Bunnyconnellon. The moraine occurs around 300 m OD and was deposited at the southern margin of a large ice lobe located in Dundalk Bay during the Killard Point Stadial (Clark et al., 2007a).

Fig. 4.12b Morainic ridges blocking the southern margin of Furnace Lough, Clew Bay. The moraine was deposited by ice flow moving across Clew Bay from the southeast. Bedforms modified by this ice flow are seen in the background to the south (Clark et al., 2007a).

Fig. 4.13 Rough Island, Strangford Lough County Down. Basal till occurring above beach level contains erratics of Ailsa Craig microgranite, Tertiary basalt, Cretaceous chalk and other igneous rocks of northern provenance deposited by ice flow to the south. This direction is consistent with drumlin orientations in the Ards Peninsula. A discontinuous stone line occurs between the till and the overlying regional drape of laminated red marine mud above the base of the spade. The marine mud records the initial late-glacial marine transgression of the County Down lowlands immediately after stagnation zone retreat from the Killard Point moraine, 30 km to the south. The transgression occurred around 15 cal ka BP (McCabe et al., 2007b). *By kind permission of GSNI*

The sensitivity of the lowland ice sheet to relatively small temperature changes partially explains why the ice sheet melted away rapidly. Perhaps some of the ice mass had already been downdrawn while relative sea levels remained high during deglaciation. However, as the Irish ice retreated southwards from the north coast towards the Lough Neagh lowlands, lakes developed along the ice margins in central Antrim. Retreat was rapid because subaqueous outwash from ice margins are interbedded with and overlain by thick, rhythmically-bedded sequences of lake clays exposed in the Bann lowlands at Vow. At this time Scottish ice readvanced into lake sediments forming the Armoy moraine (Shaw and Carter, 1980) (Armoy/Ballymoney/Lower Quilley) and an ice dam on the northern lake margin. The geological evidence shows that both ice sheets were slightly out of phase, one advancing while the other was retreating. Different ice sheet sensitivities probably explain this pattern because the accumulation-area ratio for lowland ice (Irish ice mass) will be lower than that for the ice mass based in the western highlands of Scotland. It can be argued that the lowland ice was much more responsive to a change in temperature because a small rise in the equilibrium line will affect most of the ice sheet and therefore effect general ablation. In contrast a small rise in the equilibrium line will not have as great an impact on a highland ice mass because most of the accumulation area is above the change in the equilibrium line. This potential energy difference between the two ice sheet sources suggests that as the Irish ice waned, the accompanying reduction in ice pressure and area along the north and east Antrim coast facilitated expansion of Scottish ice southwestwards to the Armoy moraine during final deglaciation.

8. **The Late Glacial and the Nahanagan event.** Originally Jessen and Farrington (1938) identified a period of relatively warm climate from organic, shallow-water mud containing fossils of the giant Irish deer (*Megaloceros giganteus*) followed by deposition of inorganic sediments by solifluction near Ballybetagh, County Dublin. Later Jessen (1949) developed a tripartite stratigraphic scheme with a relatively warm Allerød interstadial represented by dark organic detritus between two lighter beds representing soliflucted or cold climate deposits preserved in inclosed depressions. However, Watts (1985) demonstrated

that there were other distinct climatic events during the late-glacial, and that formal chronostratigraphic units are not applicable to the late-glacial because the drivers of climate and the geological responses are time-transgressive and involve lags. Watts (1977) reviewed the late glacial in terms of a loose narrative of events within a timeframework and correlated between sites using radiocarbon dating. Pollen diagrams from Woodgrange (Singh, 1970), Roddans Port (Morrison and Stephens, 1965), Ballybetagh and Dunshaughlin (Watts, 1977) and Coolteen and Belle Lake (Craig, 1973, 1978) in eastern counties have many common characteristics and similar sequences of events. These include scattered herbs at 13 ka, a *Rumex* phase from 13 to 12.4 ka, a Juniper phase with a peak of sea buckthorn, a grass phase with herbaceous communities with local birch copses ending at 10,900 years BP with soil instability and solifluction. The final phase is often termed the Younger *Dryas* Period, even though Dryas leaves are less abundant than those of the dwarf willow (*Salix herbacea*). The tundra flora is consistent with a diversity of Arctic plants with low percentage cover. The significant changes in vegetation during the late-glacial may also reflect millennial to sub-millennial climate changes and events that are recognised from other areas of Europe.

The term Nahanagan Stadial comes from moraines formed on the floor of cirque lake Lough Nahanagan between 420 and 520 m OD. (Colhoun and Synge, 1980). One moraine contained irregular masses of ice-shoved lake clays with pollen of *Graminae* and *Cyperaceae* with *juniperus*, *Rumex* and *Empetrum*. Conventional ^{14}C dates from these clays (11,600±260 and 11,500±550; Colhoun and Synge, 1980) provide a maximal age for ice advance across the deglaciated lake floor basin but the exact timing of the Nahanagan Stadial event has been questioned (Cwynar and Watts, 1989). Gray and Coxon (1991) consider that the cold spike was perhaps 3–400 years spanning the period between 11,000 and 10,500 years BP. Other evidence for Nahanagan Stadial activity include periglacial features, protalus ramparts, fossil rock glaciers and soliflucted debris (Wilson, 1990; Gray and Coxon, 1991; Anderson et al., 1998). Many corries around Ireland contain small moraines composed of freshly quarried blocky debris of possible Nahanagan age but none have been dated. Final glaciation within corrie forms in Ireland was small and certainly ice did not advance outside earlier large corrie moraines sited on corrie lips. Restricted ice development during the Nahanagan Stadial in Ireland contrasts markedly with that in western Highlands of Scotland and may record a sharp climatic gradient across the British Isles during the Younger *Dryas*.

The range and nature of the evidence

Surficial distribution and age constraints on the major glacigenic systems identified from field observations show that at least 95 percent of the landforms and known glacigenic deposits date to the last major glacial cycle in Ireland. Evidence for earlier events is extremely fragmentary and a result of the efficiency of ice sheet erosion and transport onto the adjacent continental shelf zones. Even within the last major glacial cycle there is a marked loss in preservation potential back in time towards the start of the cycle, where only six sites and some erratic trains in the entire island record ice sheet growth early in the major cycle. The available stratigraphies suggest that large ice sheets developed early and late in the cycle and that cool climate events between these expansions were on millennial timescales. Records are poor from the middle parts of the cycle and may include waxing and waning of significant ice sheets. However, the ice sheets which developed at the beginning and at the end of the cycle have a series of common features. Judging from isostatic records and distribution of deposits, both ice sheets covered most of the island with limits on the continental shelf. The earlier ice was more extensive and probably resulted in deeper isostatic depression in the Irish Sea area. Both ice sheets involved considerable areas of ice sheet dispersal in the north and west. Both ice sheets deeply depressed the land surfaces below contemporary eustatic sea levels resulting in raised glaciomarine and marine stratigraphic components. It is likely therefore that both were destabilised by marine downdraw during deglaciation especially along major coastal embayments and in the Irish Sea basin. The roles of ice streams draining the ice sheet and moving towards isostatically depressed coastal bays cannot be underestimated. Ice sheet anatomy from the end of the last major cycle is reasonably well age-constrained showing that ice sheet events were organised on millennial and centennial timescales (McCabe et al., 2007a). The ice sheet first formed in the west and eventually developed into a lowland ice sheet much larger in volume, extent and influence than generally supposed. The standard notion that ice from the Scottish highlands was the dominant agent controlling isostatic patterns in northern Britain and ice flows in the Irish Sea area cannot be sustained from age-constrained deglacial marine and raised beach records. The main reasons behind the standard notion are largely historical, and the mistaken impression that postglacial rebound records in Scotland are directly related to earlier crustal depression during the ice sheet maximum. It is suggested that the well-preserved, raised postglacial beach systems in Scotland are primarily a function of delayed isostatic uplift because ice masses remained later in Scotland and there was extensive ice sheet growth during the Younger *Dryas* event. In addition, uplift curves from northern Ireland suggest that crustal responses to ice sheet loading are linked to the millennial timescale oscillations and configuration of the ice sheet. Mobility and changes in ice sheet configuration were characteristic of ice behaviour during the last period of ice sheet growth together with widespread bedform overprinting. Finally, the recognition that the ice was oscillating and readvancing on millennial timescales means that terms such as the LGM are largely redundant in terms of the patterns of ice sheet activity (McCabe et al., 2007a). Specifically in western sectors of the ice sheet, the ice limits were more extensive prior to the standard dated LGM.

5

Glacial Bedforms

Introduction

Glacial bedforms comprise flow-transverse, flow-parallel and non-aligned morphological forms recording subglacial imprints of ice sheets (Fig. 5.1). Traditionally drumlins and other flow-parallel forms were thought to dominate the Irish drumlin belt, though satellite imagery now indicates that extensive areas of transverse or ribbed moraine with drumlinised summits are common (Fig. 5.2). Given the widespread distribution of subglacial bedforms right across north central Ireland (Fig. 1.4) and their influence on early road patterns and human occupation, it is surprising that our knowledge of their detailed morphology and origins is so poorly-developed. In some areas such as the Upper Lough Erne Basin, the intricate lake margins and the disposition of adjacent wetlands are totally determined by bedform patterns (Fig 5.3a,b). The size, form and gross orientation of many individual and chains of lakes across the north central lowlands is distinctly linked to local ridge orientation and spacing. Thus bedform distribution across at least half of the island is not only a measure of their intrinsic landscape importance but provides critical evidence for the scientific process of ice sheet reconstructions and the complex interplays between subglacial deposition, erosion, subglacial sediment transport and ice sheet limits (Fig. 5.4). Certainly bedforms not only impart distinct glacial grains and lineations to the landscape but also give many north central counties of the island landscapes which are often compared to a basket of eggs with parallel ridges controlling the number and density of streams, bogs, lakes, roads, fences and boundaries which run along the strike of the ridges (Fig. 5.5) (Close, 1864–67).

Drumlin formation and drumlinisation

Because subglacial bedforms are ubiquitous especially across north central Ireland it is interesting to look at how their role as glacial indicators has changed with time. Progress in the discovery of the history of the last glacial period began with Close's (1864–67) observations on parallel ridges of drift or unstratified boulder clay and the idea that they were produced by some widely and uniformly working agent producing ice flow lines. Later Wright (1912, 1920) used drumlin orientations to reconstruct local ice flow patterns. Perhaps one of the key developments resulting from field mapping was the proposal by Synge (1969, 1970) that a drumlin ice sheet existed and that the drumlin fields were bounded by discrete moraines (Fig. 1.4). Others like Vernon (1966) as part of the quantitative revolution of the time thought morphometric analysis was the key to interpretation. Traditional thought focused on drumlin orientations and ice flow but this did not provide historical perspectives into ice sheet history and could not add to the growing concepts of mobile ice sheets and dynamic subglacial systems. Therefore workers began to use facies analysis as a tool to record the internal geometry of drumlins (Dardis, 1982, 1985a). Eight main facies associations and depositional settings were identified together with variable drumlin morphologies and were thought to be part of a sequence of events that contributed to drumlin formation (McCabe, 1992).

An important distinction was made between drumlin formation (all processes which contributed to creation of the final drumlin form during ice sheet history) and drumlinisation (final subglacial streamlining processes) (McCabe and Dardis, 1989b). As far back as 1982 Dardis recognised that subglacial bedforms included a wide range of diverse forms and identified the presence of glacially-modified transverse ribbed or Rogen moraine in east County Tyrone. Using satellite imagery the wide

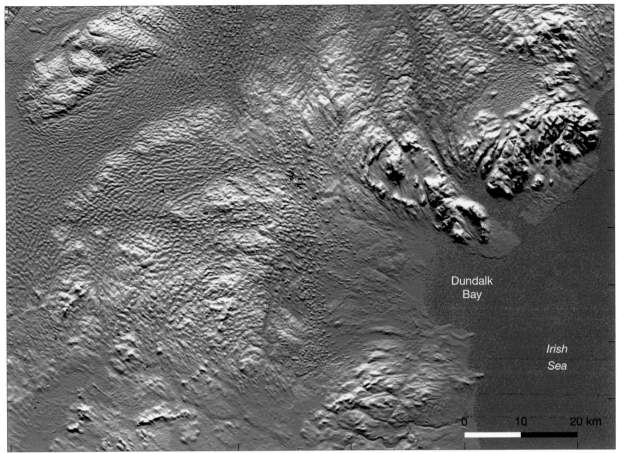

Fig. 5.1 Relief image showing the NE–SW aligned transverse ribbed moraine ridges in north central Ireland. Note that these ridges provide a grain to the landscape which is transverse to ice flow from the northwest. This is modified by a flow-parallel pattern where ice streams from Cavan, Armagh, Poyntz Pass and Banbridge overprint the older ridges (see Fig. 1.1). The most prominent ice flow lines converge on Dundalk Bay. Image derived using 90 m SRTM DEM. *By kind permission of GSNI.*

Fig. 5.2 Flow-parallel landforms (drumlins) along the crest of an earlier and larger flow-transverse ridge, Strangford Lough, County Down. Drumlins were eroded and streamlined by fast ice flow southwards (right to left) during the Killard Point Stadial (~16.5 cal ka BP). Debris generated during drumlinisation (final streamlining) was transported to tidewater ice margins at Killard Point by subglacial sediment transfer.

Fig. 5.3a Ribbed moraine in Upper Lough Erne, looking northwards to Trannish from The Lady Craigavon bridge.

Fig. 5.3b Drumlin swarm near Ballynahinch, County Down.

extent and basic patterns and some variants of ribbed moraine across the north central lowlands was recognised (McCabe et al., 1999). Complexity of bedform types and distribution was recognised, especially ice streams cross-cutting pristine ribbed moraines and the fact that dated ice-marginal moraines bordered bedform fields represent subglacial debris transfer towards tidewater ice sheet margins (McCabe and Clark, 1998). Associated shifts in centres of ice sheet dispersal had for long been recognised from north central Ireland (McCabe, 1969; Synge, 1969, 1970). Digital elevation models have extended this work and support the earlier hypothesis that major configuration changes occurred during the last glacial cycle and that several episodes of bedform

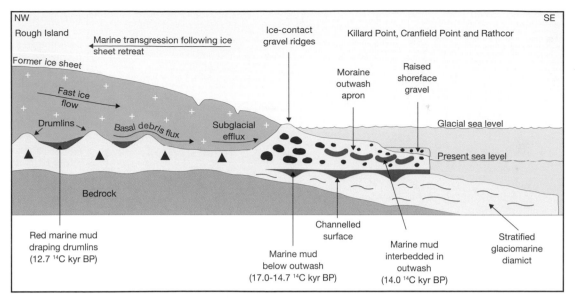

Fig. 5.4 Cartoon of the glacigenic system operating during the Killard Point Stadial in the northern Irish Sea Basin showing the relationships between subglacial flow-parallel bedforms (drumlins), dated marine mud, stratigraphy, terminal outwash and relative sea level (after McCabe et al., 1998). *By kind permission of GSNI*

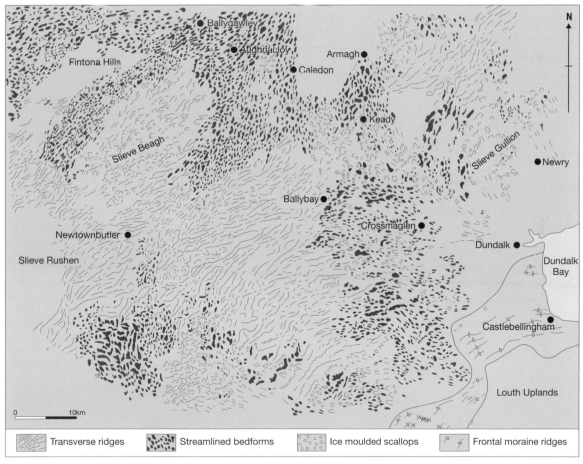

Fig. 5.5 Distribution of subglacial bedforms in north central Ireland derived from Landsat imagery (McCabe et al., 1999). At Keady a later ice stream from Armagh cut the transverse ridges forming a curved line of flow-parallel, streamlined ridges or drumlins ending near Dundalk Bay. Erosion and the subglacial transfer of sediment during drumlinisation provided the sediment to form about 20 frontal morainic ridges around the margin of the ice lobe grounded in Dundalk Bay. *By kind permission of GSNI*

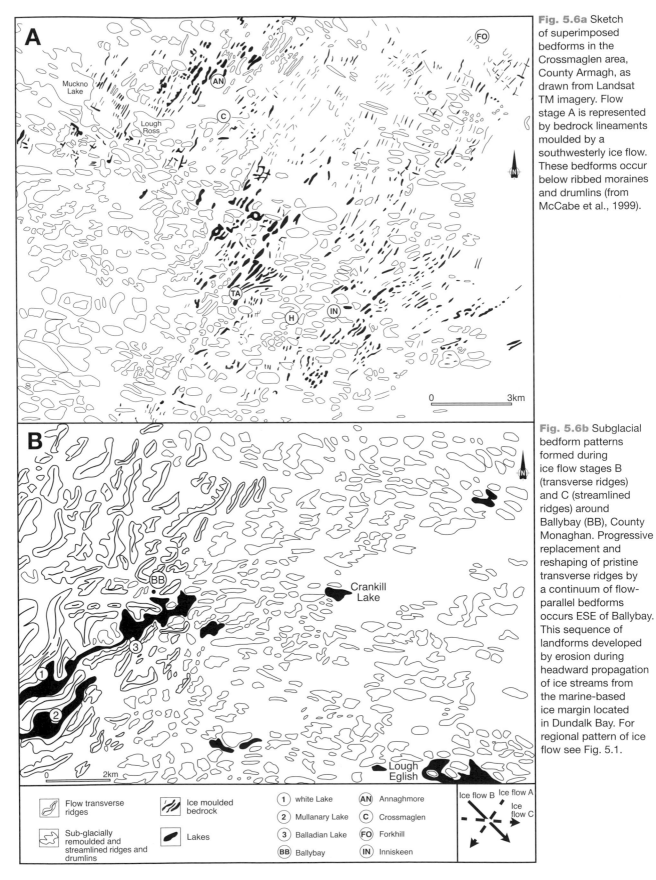

Fig. 5.6a Sketch of superimposed bedforms in the Crossmaglen area, County Armagh, as drawn from Landsat TM imagery. Flow stage A is represented by bedrock lineaments moulded by a southwesterly ice flow. These bedforms occur below ribbed moraines and drumlins (from McCabe et al., 1999).

Fig. 5.6b Subglacial bedform patterns formed during ice flow stages B (transverse ridges) and C (streamlined ridges) around Ballybay (BB), County Monaghan. Progressive replacement and reshaping of pristine transverse ridges by a continuum of flow-parallel bedforms occurs ESE of Ballybay. This sequence of landforms developed by erosion during headward propagation of ice streams from the marine-based ice margin located in Dundalk Bay. For regional pattern of ice flow see Fig. 5.1.

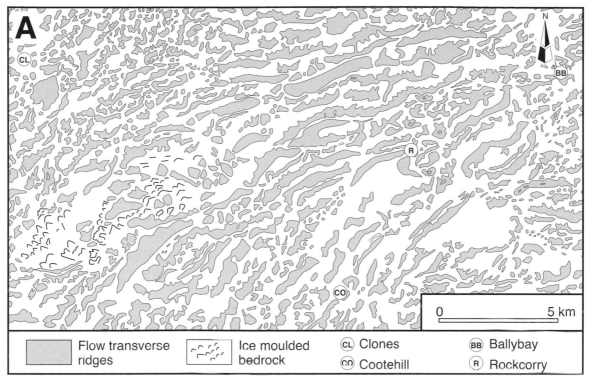

Fig. 5.7a Pattern of flow-transverse ridges around Ballybay. Ridge orientation is unrelated to bedrock strike.

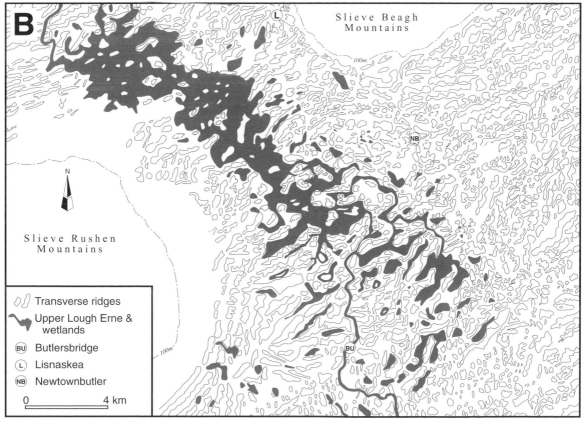

Fig. 5.7b Pattern of flow-transverse ridges in the Upper Lough Erne Basin. Lake and wetland boundaries are determined by the disposition of the transverse ridges, termed ribbed moraines. Ridges also lie contour-parallel around the Slieve Rushen and Slieve Beagh uplands.

formation occurred (Figs. 5.6a,b) (Clark and Meehan, 2001). More recently detailed characteristics of ribbed moraine (Dunlop and Clark, 2006) and non-aligned bedforms (Knight, 2006) have extended our knowledge of bedform characteristics not only across the island but in specific valleys towards the end of the last glaciation.

Pristine ribbed moraine

Pristine ribbed moraine occurs in areas where there has been little modification of the original landform by glacial overprinting or drumlinisation (Figs. 1.4, 5.7). It is the overall gross ridge morphology recognisable over much of the north central counties which strongly suggests that these ridges show planforms related to formative subglacial conditions. The main zone of pristine ribbed moraine ridges covers about 7000 km² across the north central Irish lowlands from the Ards Peninsula in the northeast to Strokestown in the southwest, a distance of about 185 km (McCabe et al., 1998; Knight et al., 1999; Dunlop, 2004) (Fig. 1.4).This zone is continuous and characterised by NE–SW trending ridges which give the entire landscape a surficial topographic grain which is independent of underlying geological structure. Smaller areas of ribbed moraine ridges occur in east Tyrone bordering the Lough Neagh basin and in the valleys between the compartmentalised highland blocks of central Sligo, northern Leitrim and western Fermanagh (Fig. 5.8). Morphological variants are common on the margins of the main zone of ribbed moraine because of later subglacial erosion, overprinting and headward extension of later ice streams (Figs. 1.1, 5.9).

Ribbed moraine landscapes, and in particular the pattern of lakes and wetlands of large parts of counties Leitrim, Roscommon, Cavan, Monaghan and Fermanagh are dominated by the orientation of these closely-spaced ridges. The ridges are a common subglacial bedform and are recognised by transverse linear forms, arcuate shape down-ice, anastomosing ridges, horns pointing down-ice, undulating tops, numerous transverse gaps and saddles, asymmetric transverse structure, though symmetry may also occur, linear crestlines, swales between ridges, sinuous to irregular plan outlines and a nested or crude jig-saw pattern across large lowland fields. These landforms are sometimes called Rogen moraine (Lunqvist, 1969; Bouchard, 1989; Aylesworth and Shilts, 1989; Hättestrand and Kleman, 1999).

The NE–SW ridge orientation across the main zone of ribbed moraine is emphasised by linearity of crestlines and the trend of the basal outlines of the ridges which can develop more intricate plan geometries (Fig. 5.6). This also provides an overall nested pattern of ridges and linear swales. Morphometric analysis of the ribbed moraine in north central Ireland has shown that these moraines occur at different scales (Dunlop, 2004). Classical scale ridges have typical dimensions of 300–1500 m long, several hundred metres wide and 10–30 m high. However, the examples from north central Ireland are much larger than those described in the literature and have been termed mega-scale ribbed moraine (Dunlop, 2004).

The ribbed moraine in Ireland exhibits many of the characteristic features commonly associated with this bedform elsewhere (Knight and McCabe, 1997a; Clark and Meehan, 2001). This is true of the various scales of ribbed moraine found across the region. In the case of the mega-scale ribbed moraine ridges, many have a straight to arcuate or planform morphology, are consistent in size with neighbouring ridges, are regularly spaced across the terrain and display many of the classic morphological features. However, close inspection of the mega-scale ribbed moraine ridges also reveals variations in morphology. For example, it is possible to identify drumlinised ridges, curved ridges that are concave up-ice, straight rectangular ridges and barchan-shaped ridges. In many regions the crests of mega-scale ridges are drumlinised and breaches that occur along the entire ridge are cut to the base of the ridge (Figs. 5.9a, c). These ridges therefore resemble transversally aligned mounds of sediment rather than intact ridges. Smaller ribbed moraine ridges located immediately south of Slieve na Calliagh situated 15 km northeast of Kells, County Meath are characterised by sequences of narrow, sinuous arcuate ridges which curve downstream. Variations in form within this moraine field include elongate straight ridges, rectangular forms and a few which are gently curved with their concave side facing upstream. The minor ribbed moraine ridges generally tend to be much shorter in length and as a result are usually straighter in form and often occur next to, or between classical-scale ridge fields. In other cases they form small isolated fields and may be superimposed on top of other glacial bedforms.

Longitudinal profiles of the ridges show that most of the mega-scale ridges have undulating crestlines. Smaller ridges possess linear and rounded crestlines whereas larger ridges show variable crestlines with

Fig. 5.8 Satellite image of the bedforms in the valleys between the compartmentalised upland blocks northwest of Lough Allen, County Leitrim. Note that in many areas ribbed moraine has been modified into a wide range of flow-parallel forms often with marked distal sediment tails. The original orientation of the ribbed moraine is marked by offset lines of streamlined ridges at right angles to the valley axis.

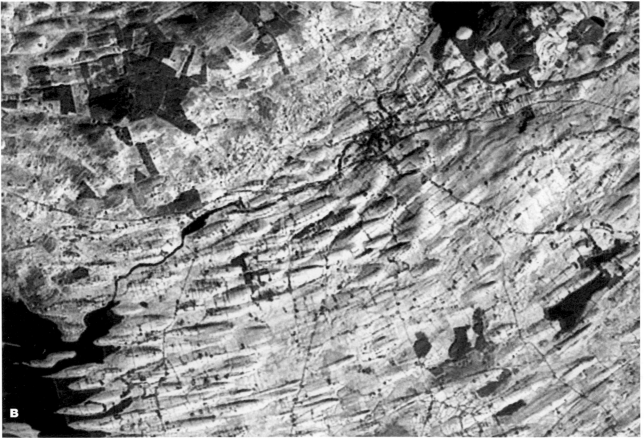

Fig. 5.9 Images of subglacial landform assemblages from County Leitrim. A) *(opposite top)* East to west transition from partially modified ribbed moraine with down-ice mouldings and tails into shield-shaped forms within a broad valley, near Drumahair. The shields are characterised by steep proximal slopes, elongated tails or down-ice projections and striated upper surfaces. The final ice sheet system involved erosion of transverse bedforms, sediment transport and moulding to the southwest and surficial grooving of resulting bedforms. B) *(opposite below)* Spindle-shaped and elongated streamlined drumlins recording the final ice flow which arcs from the Lough Allen area westward and northwestward through Lough Gara towards Tobercurry and the ice limit along the southern flanks of the Ox Mountains. Note that drumlin long axes determine the outline of Lough Gara towards the western side of the image. The village of Boyle is located at the top of the image. C) *(right)* Ribbed moraine orientated NE–SW on the low ground south of Slieve Anierin around Lough Scur. To the southeast (down-ice) the lateral continuity of the transverse moraines is disrupted and replaced by disarticulated ribbed moraine consisting of en echelon lines of individual drumlins. These features have steep slopes to the northwest, are fairly blunt and sometimes show sediment tails to the southeast. The original NE–SW pattern of the earlier transverse ridges is still recognisable even though deep, erosional troughs have modified or cut the original landforms.

shallow undulations (5–10 m) and occasionally deeper saddles (<20 m). Ridge heights are variable depending on the scale of the ridge but the larger forms are up to 60 m high and summits are not accordant. Analysis of transects totalling 100 km in length taken across elevation data show the ridges generally have an asymmetric cross-sectional profile though both ice-proximal and ice-distal slopes may be similar (Dunlop, 2004).

Ridge trends are not related to any structural trend in the underlying bedrock. Numerous small sections suggest that the ribbed moraines are composed of coarse and poorly-sorted massive diamict. In all cases sediments contain a high proportion (70%) of freshly quarried, angular clasts of local origin.

Bedrock highs within the zone of transverse ridges are striated and ice-moulded by NW to SE ice flows and record a regionally consistent ice flow pattern across the zone of transverse ridges. For example, this close field association northeast of Newbliss and west of Castleblaney strongly suggests that both (ridges and erosional marks) are genetically related to the same southeasterly ice flow (Fig. 5.5). The scale of the lineation grain together with ice moulding and striation patterns points to a subglacial origin.

Although the precise origins of Rogen-type moraine are not known (e.g. Fisher and Shaw, 1992; Kleman and Hättestrand, 1999; Dunlop, 2004) their presence is significant because it impacts on assessments of basal ice sheet

Fig. 5.10 Satellite image showing cross-cutting ribbed moraine around Cavan town. Note that different ridge orientations are associated with separate ice flows or ice pressure. These changes match striae patterns on the summits of adjacent uplands and are related to ice sheet reorganisation in areas of ice dispersal (McCabe, 1969) during the build-up of ice before the Killard Point Stadial. *By kind permission of GSNI*

dynamics and the relative chronology of bedform development during the last glacial cycle. The distribution of ribbed moraine across the north central Irish lowlands suggests formation in a zone located between ice dispersal centres in north-central Ireland, and ice-marginal positions south of this region. The overall distribution of ribbed moraine and indeed its morphological similarity over extensive zones shows that it could not have formed during LGM times because the ice mass configuration and ice flow lines do not match the mapped distributions of ribbed features. The range of morphological styles of the ribbed ridges, and different orientations in different areas, suggest they were formed in several phases associated with changes in subglacial thermal and hydraulic regimes early in the last deglacial cycle, possibly after the LGM (Knight et al., 1999). At this time previously deposited subglacial debris was reorganised morphologically into transverse ridges. It is noted that ribbed moraines form an orthogonal pattern of intersecting, cross-cutting

ridges around Cavan town in the Upper Lough Erne basin (Figs. 5.8, 5.10). This relationship provides evidence for shifting ice flow-patterns and migration of local ice sheet divides. Similar shifts in the ice sheet were proposed from the striae and drumlin patterns in the Lough Erne Basin (McCabe, 1969). These transverse ridges are generally found in a pristine condition in lowland areas adjacent to centres of ice dispersal and are cut or modified especially on the periphery of the bedform field by later ice flow during the Killard Point Stadial (Fig. 5.11). This relationship suggests they formed beneath a reduced ice sheet extent following early ice sheet withdrawal from the continental shelf and Irish Sea Basin before 20 cal ka BP. The available morphological and dating evidence suggests that the mapped planiforms of ribbed moraines formed beneath mobile ice masses which shifted around core areas of ice dispersal in north-central Ireland after early and extensive deglaciation from more extensive or possible LGM limits.

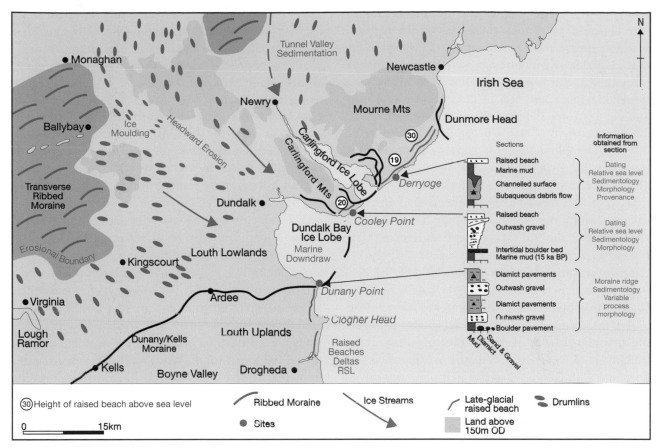

Fig. 5.11 Deglacial records from the northern Irish Sea Basin showing the patterns of ice flow during the Killard Point Stadial and the stratigraphic position of beds of dated marine mud. Note that the trend of the terminal moraines and outwash is at right angles to the drumlins recording ice flow. The subglacial sediment flux to tidewater ice sheet margins records erosion especially along the outer margins of the zone of transverse ribbed moraine. *By kind permission of GSNI*

Modified ribbed moraine

Post depositional morphological changes occur in at least five areas where distinct changes have been made on the original patterns within the zone of pristine ribbed moraine. Modification is marked along its entire southern margin and along western sectors. These changes record later ice movements from core areas towards coastal embayments occupied by tidewater ice sheet margins (McCabe and Clark, 1998). In core areas only the crests of the ribbed moraine ridges are striated.

First there is field evidence for modification of ribbed moraine by at least three major ice streams flowing southeastwards from centres of ice sheet dispersal in north central Ireland (Figs. 1.1, 5.11). A northwest–southeast glacial grain composed of streamlined ridges and drumlins cuts out entirely the northeast–southwest trending ribbed moraine along a 10 km-wide

tract around Keady (Fig. 5.1). Morphological changes resulted in disarticulated remnants of ribbed moraine some of which are streamlined and others moulded into down-ice hooks on ridge margins. This ice sheet flow line, termed the Armagh ice stream, is 70 km in length and extends from the southern margin of Lough Neagh towards limiting moraines around Dundalk Bay and swings through a 70° arc from south to east-south-east (Fig. 5.5). Smaller ice streams, tributary to the main Armagh ice stream, occur at Ballybay and Shercock. The SW–NE trend of the transverse ribbed moraines at Ballybay are mirrored by the 15 km long linear corridor of lakes between Cootehill and Ballybay (Fig. 5.6b). Non-streamlined ridges at Ballybay are replaced progressively (over 8 km down-ice distance) by flow-parallel bedforms and lakes (Eglish, Muckno) trending NW–SE (Fig. 5.6b) (McCabe et al., 1999). Bedforms along this flow-line include transverse ridges with crestline crosscuts, ridges showing surficial streamlining and sediment carryover to the east-southeast, comma-shaped forms with hooked tails, incomplete drumlins with truncated lee-sides similar to Blattnick moraines (Markgren and Lassila, 1980), asymmetric to symmetric barchanoid forms, and streamlined drumlins of classical, cigar and torpedo shapes. Both sets of directional overprinting are best described in terms of a form-process feedback involving erosion of transverse ridges (form obstruction) with leeside sediment transport and streamlining (process). The belief that this ice flow-stage involved headward erosion by fast ice flow into the pre-existing transverse ridges is based on the fact that bedform overprinting, drumlinisation and sediment streamlining is widespread on the eastern and southern margins of the transverse ridge field, and there is a general absence of intensive streamlining within the main transverse field. Bedform reorganisation on this scale may be related to changes in subglacial thermal regimes as warmer (wet-based) ice head-cut into drier-based ice that preserved the pristine form of many of the transverse ridges (Knight et al., 1999). Convergence of these ice flow lines into Dundalk Bay suggests that tidewater calving around the bay periphery, which initiated marine downdraw, propagation of fast ice flow back into the ice mass and sediment transfer to tidewater, was the main destabilising mechanism which eventually ended this ice flow stage (Fig. 5.11). This ice advance occurred during the Killard Point Stadial and is associated with prolonged ice growth across the north Irish lowlands and readvance to terminal limits around 14 ^{14}C kyr BP. The field evidence clearly records erosion, overprinting and sediment transfer processes along the ice streams (Fig. 5.4). The patterns of erosion and sediment transfer implies that regional-scale subglacial deformation and transfer as mobile sediment masses did not initiate sediment agglomeration during drumlinisation below these ice flows.

Second, modified ribbed moraine patterns are preserved on northwestern margins of the main area of ribbed moraine in the valleys between the compartmentalised Carboniferous uplands of County Sligo. Along the Diffagher valley between Belhavel Lough and Lough Allen, bedforms along a flow line 35 km long extend west into Ballysadare Bay. When this ice flow moved along the southern margin of Donegal Bay it formed a lateral moraine at about 400 m above sea level along the northern flanks of the Ox Mountains (Clark, 2007a). The original orientation (N–S) of the ribbed moraines was at right angles to the valley axis but later modification beneath ice flow westwards resulted in shield-like bedforms (Figs. 5.8, 5.9a). The relative positions of adjacent shields along a north–south axis resemble displaced lines of ribbed moraines. Displacement possibly of a kilometre or so is greatest along the valley axis. Shields average 1.5 km across by 1 km in length and are characterised by erosion marks or furrows which are parallel in adjacent landforms and along the displaced lines of former ribbed moraine. In most cases the eastern or upstream part of the landforms are much steeper than the downslope side which can be pointed or barchanoid in form. Along the valley axis the shields may persist or be reduced to isolated classical type drumlins. Southwest of Belhavel Lough the original ribbed moraines are much more attenuated and drumlinised, and related to a powerful ice stream from the northeast. This ice flow partially blocked the westward exit of the ice flow across the northern end of Lough Allen and slowed the westward ice flow along the Diffagher valley, thus preserving more of the original ribbed moraine morphology. From each valley margin former transverse segments extend downvalley towards the valley axis forming ridges curving west in the direction of ice flow (Fig. 5.9a). Ridges have been disarticulated into several segments, each of which has moved further downvalley with distance from valley margins. In addition the degree of streamlining and large-scale striation cuts (~0.5 km) increases towards valley axis. These ridges can be linked to other drumlinised ridges right across the valley showing a pronounced, westward facing convexity related to faster or more mobile ice flow along the valley axis.

Third, a lowland area of about 150 km² between Carrick on Shannon and Ballinamore to the southeast of the Slieve Anierin/Cuilcagh Mountains is dominated by tortuous drainage patterns, poor drainage and hundreds of small lakes and wetlands interspersed with substantial bedforms up to 25 m in height. On survey maps they seem to be disorganised but satellite images show two distinct trends. The earlier consists of SW–NE orientated ridges with slightly curved basal outlines, rounded crests and low sinuosity which are easily traced as a group for 20 km. In the northern part of the zone to the lee of Slieve Anierin between Leitrim and Ballinamore the ridges are disarticulated and eroded into isolated hills separated by deep (~20 m) flat-bottomed breaches with a southerly or southwesterly trend (Fig. 5.9c). Hills are characterised by steep, rounded slopes to the north and northwest and longer and much more gentle slopes to the southeast. The highest areas of the hills are rounded and curved summit areas situated immediately above the steep, northwestern facing slopes. Planiforms are rounded and distinct to the northwest but become diffuse on southern margins. Generally these bedforms are symmetric though about 20 percent are asymmetric with one pronounced tail trending southeast (Fig. 5.9c). Within 5–6 km to the southeast these hills which are modified ribbed moraines pass into well-streamlined classical, spindle and barchan forms immediately north of Mohill and Cloone (Fig. 5.9b). It is possible that the ribbed moraine which is partially modified records sticky subglacial flow possibly near an ice divide. The more streamlined lower part of the ice flow to the southeast could be a response to a more mobile ice associated with more meltwater at the ice–sediment interface.

Fourth, the lowland area (~60 m OD) around Strokestown, County Roscommon, is bounded on its eastern margin by the NNE–SSW trending Slieve Bawn ridge (~200 m OD). To the west of the ridge numerous lakes such as Kilglass Lough are aligned NE–SW parallel to the general trend of a nested set of subparallel ribbed moraines which are between 3 and 5 km in length, 0.2–0.3 km wide and 15–25 m in height. Planiforms are bulbous like a string of sausages. Within 2–3 km to the west the ridges have been modified into mega comma forms comprising a proximal, non-streamlined hillock which transforms southwestwards into a streamlined hook. In this area the lakes are now oriented parallel to the new streamlined forms. These bedforms cover an area of about 150 km² and after about 10 km pass into streamlined drumlins. It is probable that the marked development of commas or extended ridges only on the south side of each modified bedform is due to deflection of southward-moving ice by the Slieve Bawn ridge.

Fifth, the apparently chaotic distribution and patterns of lakes and wetlands across the Lough Erne lowlands is determined by ribbed moraine morphology, shape and orientation (Figs. 5.3a, 5.7b, 5.12). Sheets of thick subglacial tills, in places separated by interstadial deposits, dominate the basin and provide a glacial chronology for parts of the last cold stage (McCabe, 1987). In a similar way the overprinted subglacial bedform sets especially in the Upper Lough Erne basin between Cavan, Belturbet and Killeshandra can be used to evaluate the relative timing and palaeoenvironmental regimes of different ice flow events (Fig 5.14) (McCabe and Dunlop, 2006). In this area of around 100 km² ridge patterns and outlines are particularly clear because much of the landscape has been submerged by Lough Erne and the main subglacial lineaments are highlighted (Fig. 5.3a). A prominent southwesterly ice flow is recorded by NW–SE trending ridges west of Rivory and Belturbet (Fig. 5.14). These ridges extend at least as far south to Crossdoney where the present River Erne has breached a 6 km ridge at Tawlaght. NE–SW trending ridges around Redhills and Butlersbridge record a later ice flow to the southeast. These NE–SW trending ridges are continuous with the main area of ribbed moraine which extends eastwards into Monaghan, Armagh and Down.

Regionally, the upper parts of cross-cutting ribbed moraines have been drumlinised and striated by the last ice flow, forming tails of sediment on the southern margins of the ridges. This southward ice flow reached Lough Ramor and Kells. The regional patterns of the ridge long axes are consistent with early deglacial-stage ice flow directions. Different ridge trends and crosscutting can best be explained by a mobile ice sheet which shifted its centres of ice-sheet dispersal during the course of deglaciation. The final streamlining event occurred during the Killard Point Stadial. Although the morphological patterns can be attributed to subglacial changes during the course of the last deglacial cycle the sediments within the ridges were generated much earlier during the last cold stage. Superficial reorganisation of these sediments is therefore a signature for the last deglacial cycle which was characterised by migration of ice sheet divides and changes in basal thermal regimes between dry-based and wet-based ice. Major breaks in the morphological record shown by changes in bedform signatures may be related to climate signals and the operation of the coupled ice–ocean–atmosphere system in the North Atlantic (McCabe, 1996).

Drumlins

Because drumlins cannot be observed *in statu nascendi* it is argued that analysis of their internal structure presents different problems to the general processes which probably created the external geometry of bedforms over large sectors of the Irish lowlands (Fig. 5.13). Most drumlins are therefore form-discordant. A range of field evidence from north central Ireland shows that external drumlin morphology is mainly erosional in origin rather than formed by subglacial deformation. For example, the Derryvree drumlin from County Fermanagh is carved out of two distinct till sheets separated by *in situ* interstadial deposits (Figs. 4.8,

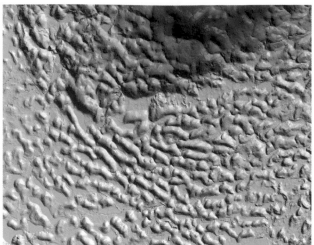

Fig. 5.12 *(left)* Pristine ribbed moraine in the Upper Lough Erne Basin between Lisnaskea and Newtownbutler, County Fermanagh. The high ground on the northeastern part of the image (top right) rises to 240 m at Teiges Mountain. The image is 14.5 km across.

Fig. 5.13 *(below)* Satellite image of Donegal Bay showing subglacial bedforms and directions of ice flow onto the inner continental shelf from centres of ice sheet dispersal in the Erne and Omagh basins and Donegal Mountains. Note the presence of cross-cut ridges near Donegal town and large shield-like forms cut in rock towards the mountains. On the north side of the bay around Doorin Point and St. John's Point classical and isolated drumlins dominate (Knight and McCabe, 1997).

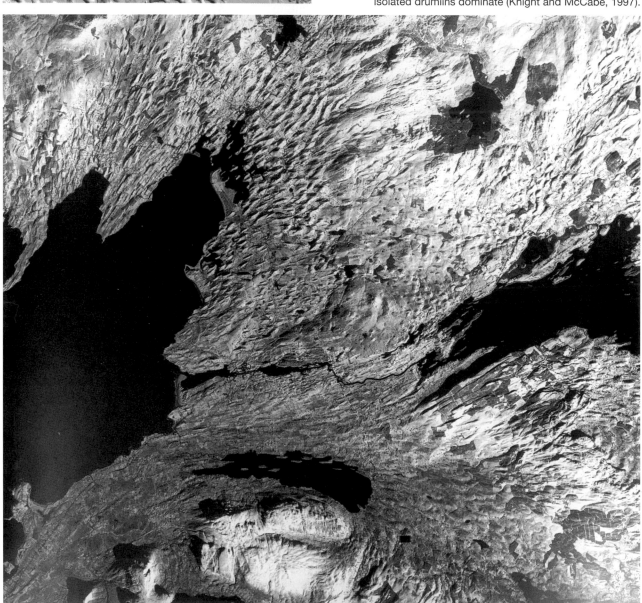

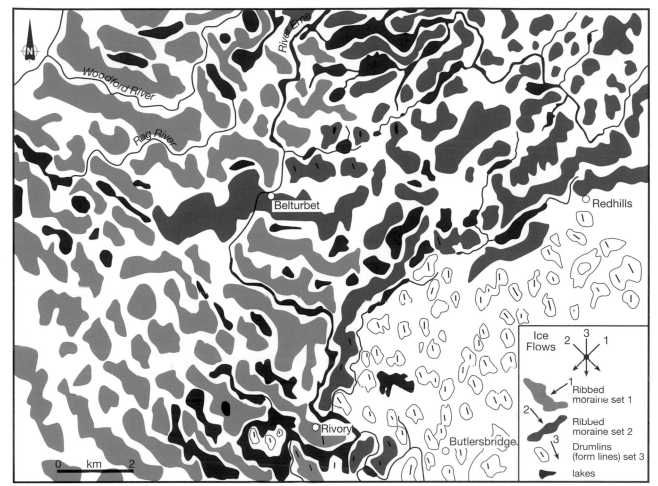

Fig. 5.14 Two major sets of ribbed moraines on either side of the river Erne, around Belturbet, County Cavan. Both sets are mutually exclusive. Although the most recent set of ribbed moraine in the southeast has been extensively drumlinised by ice flow southwards, the former trend (NE–SW) of the ribbed moraine is still visible. Drumlinisation occurred when a major ice stream extended headward (northward) from ice limits at Virginia and Kells. Preservation of the earlier-formed ribbed moraine suggests that the area immediately west of the river Erne was characterised by sticky ice spots where subglacial meltwater was not widespread.

4.9) (Colhoun et al., 1972), drumlins in north County Down can consist entirely of rock and drumlinisation of ribbed moraine shows clear evidence of subglacial erosion across the entire north of Ireland. More specifically, it is now more accurate to describe the traditional drumlin belt as a zone of drumlinised ribbed moraine rather than a classical drumlin field. The implication of this widespread field association is that the thick (~20–35 m), compact basal tills which have for long been regarded as the typical sediment of drumlin cores were deposited prior to drumlinisation. In most cases therefore they occur within earlier formed ribbed moraine and in some cases record multiple events prior to formation of ribbed moraine morphology (Fig. 5.15). In these cases large-scale sediment agglomeration and subglacial sediment formation has not contributed to the final

streamlining event known as drumlinisation. The term drumlin formation describes the total combination of erosional and sedimentation events which contribute to the internal geometry and external form of a drumlin (Dardis, 1985a).

Because drumlins in north central Ireland are largely erosional landforms they are therefore form-discordant. Their internal geometry may contain stratigraphy recording earlier glaciations/non-glacials, coarse-grained blocky tills inherited from ribbed moraine and newer facies formed during the actual drumlinisation processes. Therefore the general investigations on drumlin long axes and ice flow patterns yields little information on the range of subglacial settings which contributed to drumlin formation (e.g. Hill, 1973; Finch and Walsh, 1973). However, facies analysis has been used to identify

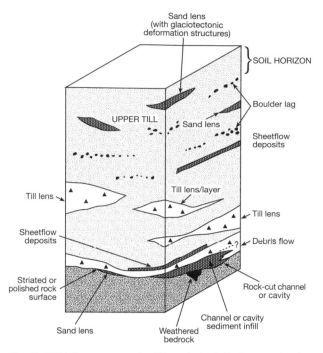

Fig. 5.15 Major components of the subglacial lodgement/meltout basal till facies association in drumlins, central Ulster (Dardis, 1985).

particular sedimentary settings from individual and closely-spaced groups of drumlins (Dardis, 1985a). These local models can then be evaluated within a depositional systems approach and related to larger-scale events, ice sheets and more general models of ice sheet activity. This field-based approach recording lithofacies geometry is an important investigative tool because drumlins are not simple landforms and contain deposits recording ice sheet history, prior to streamlining. All of these elements cannot be viewed directly from glaciological theory. However, Synge's (1969) concept of a Drumlin Ice Sheet centred in north central and northwestern Ireland provides a working model which describes drumlin patterns, final directions of ice flow and the moraines bordering the areas drumlinised. More recently, McCabe and Clark (1998) have shown that the last phase of drumlinisation occurred during a major southeasterly ice advance from the Irish lowlands. The outer limit of this readvance is marked by a terminal zone of ice contact landforms stretching for over 150 km from Killard Point in the northeast to Dundalk Bay, Ardee, Kells and Mullingar to the southwest. At Killard Point, County Down, marine mud interbedded with the terminal outwash dated to ~14 [14]Ckyr BP is considered to be synchronous with the final phase of drumlinisation because the debris which arrived at tidewater represents subglacial erosion and sediment transfer during ice sheet readvance.

Theory predicts that diversity in drumlin form and structure exists because basal processes may include features such as basal debris release sometimes in areas of pre-existing drift, pre-existing transverse landforms, temporal and spatial changes in meltwater production, meltwater pressure gradients, porewater gradients, shearing, compression, fast ice flow, low effective pressures, and a range of external forcing factors including marine downdraw. Some of these features which have contributed to drumlin formation have been identified and are grouped into seven main facies associations (FAs).

FA1 (Forms with a core of pre-existing, dated drift): The cores of drumlins near Maguiresbridge, County Fermanagh contain a lower till overlain by basins of organic silts (Colhoun et al., 1972). In the main the organic silts are undisturbed but at one site they are compressed by the final ice movement southwards (Fig. 4.8). The upper till postdates 30 ka BP and records the last ice sheet oscillation in the Erne Basin. Substantive sections in older drifts are found in large lowland basins adjacent to the major centres of ice dispersal in north central Ireland suggesting that they are selectively preserved there and eroded from the periphery of the drumlin ice sheet where ice flow was faster.

FA2 (Subglacial diamict facies): Innumerable road cuts through drumlins consist mainly of homogenised basal till with ancillary inter-till beds of sand/silt and stratified diamict (5.15) (Dardis, 1985a). Sandy beds can be interpreted as ephemeral sheet flow though there are gradations between stratified and massive diamicts. Dardis (1981) concluded that this facies variability was linked to changing conditions at the ice–sediment interface which was variable from site to site. He also showed that the thickness of this facies association decreases outward from the centres of major depositional basins (Lough Erne, Lough Neagh) towards the periphery of the drumlin-forming ice sheet. Generally it is difficult to age constrain either the multiple till beds which occur towards the base of drumlin stratigraphies or to identify the conditions necessary for agglomeration of thick basal till. However, the close spatial association of similar scale drumlins cut across older ribbed moraines strongly suggests that the till facies observed within drumlins dates mainly from earlier events. Furthermore, there is little internal evidence either for glaciotectonic deformation or pervasive deformation of the thick (20–30 m) sediments observed from fresh road cuts (McCabe, 1993). These observations strongly

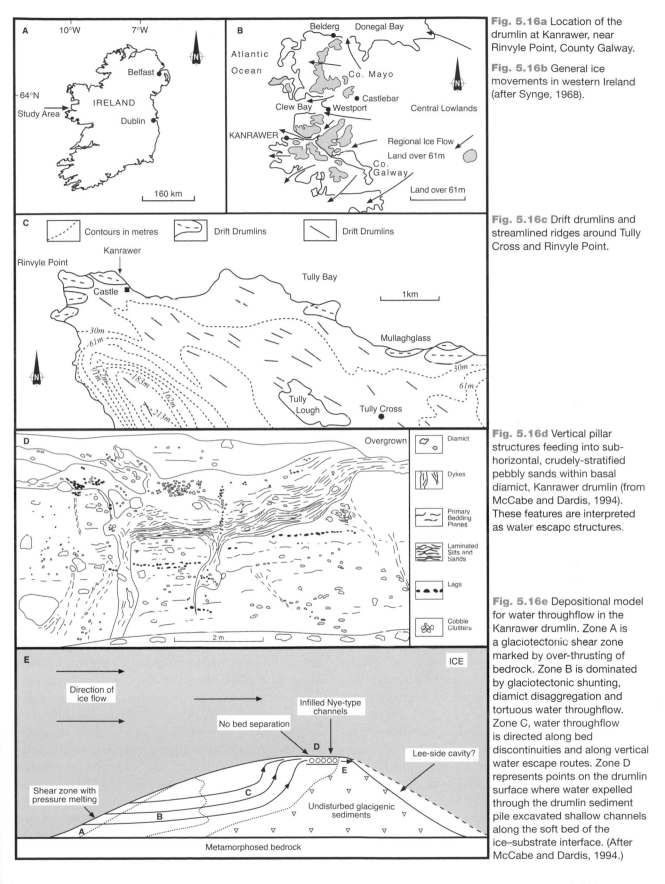

Fig. 5.16a Location of the drumlin at Kanrawer, near Rinvyle Point, County Galway.

Fig. 5.16b General ice movements in western Ireland (after Synge, 1968).

Fig. 5.16c Drift drumlins and streamlined ridges around Tully Cross and Rinvyle Point.

Fig. 5.16d Vertical pillar structures feeding into sub-horizontal, crudely-stratified pebbly sands within basal diamict, Kanrawer drumlin (from McCabe and Dardis, 1994). These features are interpreted as water escape structures.

Fig. 5.16e Depositional model for water throughflow in the Kanrawer drumlin. Zone A is a glaciotectonic shear zone marked by over-thrusting of bedrock. Zone B is dominated by glaciotectonic shunting, diamict disaggregation and tortuous water throughflow. Zone C, water throughflow is directed along bed discontinuities and along vertical water escape routes. Zone D represents points on the drumlin surface where water expelled through the drumlin sediment pile excavated shallow channels along the soft bed of the ice–substrate interface. (After McCabe and Dardis, 1994.)

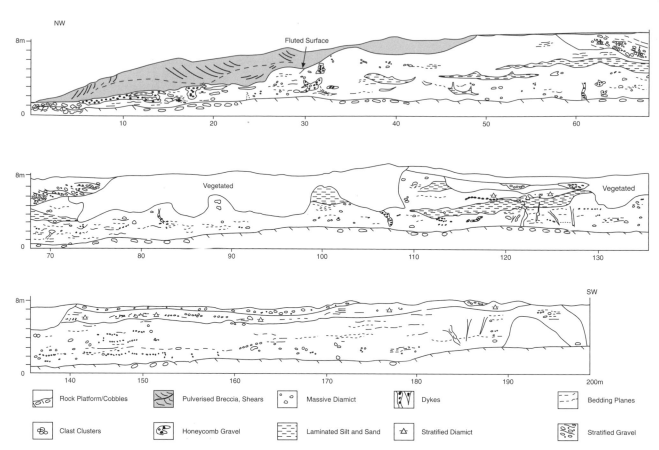

NW

8m

Fluted Surface

10   20   30   40   50   60

8m

Vegetated   Vegetated

70   80   90   100   110   120   130

SW

8m

140   150   160   170   180   190   200m

	Rock Platform/Cobbles		Pulverised Breccia, Shears		Massive Diamict		Dykes		Bedding Planes
	Clast Clusters		Honeycomb Gravel		Laminated Silt and Sand		Stratified Diamict		Stratified Gravel

Fig. 5.17 The Kanrawer drumlin exposure (after McCabe and Dardis, 1994).

support Dardis's model which emphasizes that drumlins were eroded from whatever sediment and rock combinations were available. Widespread deformation of subglacial sediment cannot explain either the internal facies geometry in drumlins or the direct erosional relationship between drumlins and underlying ribbed moraine.

FA3 (Water transfer facies): Four main facies occur in the Kanrawer drumlin which is one of a line of isolated drift drumlins situated on heavily ice-scoured bedrock of the Tully Peninsula, County Galway (Fig. 5.16) (McCabe and Dardis, 1994). The glacial grain of this coastal tract was sculptured by ice streams moving westwards from the lowlands onto the continental shelf (Fig. 5.16c). A coarse-grained, compact massive diamict containing glacially-facetted cobbles and boulders comprises about 85 percent of the section and rests on striated rock (Fig. 5.17). Texturally it is similar to basal tills though it contains a range of primary sedimentary features including stacked beds, laterally impersistent bed discontinuities, sandy stringers, gravelly horizons, rapid changes in textural variability and clast clusters which

are more characteristic of mass flows (Middleton and Hampton, 1976; Lowe, 1982; Schultz, 1984). However, fabric data suggests that at least some of the diamict beds are a result from direct debris release. Possibly these mixed characteristics reflect very localised subglacial sediment release followed by resedimentation by mass flow. Proximal exposures show post-depositional intrusive facies consisting of integrated honeycomb-like fills of open-work and matrix-supported cobble to pebble gravel (Fig. 5.18). Contact between gravel and diamict are sharp and irregular to rounded with evidence for erosion of the subjacent diamict. Westwards along section the gravelly pods and lenses decrease in size and are replaced by discrete wavy to planar bedded silty-sand lenses up to 3 m in length. These structures are closely related to wedge-shaped clastic pillar structures which cut across primary bedding in the diamict. In other cases the pillars either feed into the basal parts of subhorizontal silty-sand intrusions or rise up from their upper surfaces (Fig. 5.16d). Smaller scale, concave up dish structures are common throughout the distal parts of the section. Brecciated bedrock, up to 3 m thick,

Fig. 5.18 Massive diamict with honeycomb openwork gravel structures, formed as closed-conduit flow by high effective water pressures generated on the proximal slopes of the Kanrawer drumlin.

truncates the massive diamict only along the proximal or eastern slope of the drumlin. It is made up of pulverised sheets of angular rock fragments set in a coarse gravelly matrix with no glacially-facetted clasts. Prominent stacked thrust planes rise westwards providing a distinct structure to the breccia (Fig. 5.17). Finally, the top of the drumlin has been eroded by several small channels infilled with stacked, amalgamated gravelly beds.

A depositional model for the Kanrawer drumlin must not only explain the internal sedimentary relations but the fact that the drift drumlins occur as a linear association across a glacially-scoured rocky lowland. This field relationship suggests that initial sediment agglomeration may be related to a conveyor-belt type debris transport route parallel to ice flow, similar to those identified from modern glaciers (Gustavson and Boothroyd, 1982, 1987). The final drumlin form is clearly form-discordant (erosional) supporting the hypothesis that the linear group of drumlins have been eroded from deposits which were released into a subglacial conduit or possibly a re-entrant in the ice margin. Switches from

subglacial cavity deposition to erosion and streamlining are highly likely given the fact that ice sheets were extremely dynamic during the last cold period especially where they oscillated on the inner margin of the continental shelf.

The localised presence of brecciated bedrock along the proximal drumlin flank is a form–process relationship which post-dates diamict formation involving upthrusting of cataclastic debris possibly driven by high porewater pressures (Fig. 5.16e). The adjacent honeycomb network of gravel structures in the diamict is clearly related to water-throughflow driven by glaciotectonic shunting and diamict disaggregation concomitant with a high meltwater flux, creating interconnected passageways within the diamict (Fig. 5.18). Distally the horizontally-bedded lenses and sills are attributed to lower energy water flow, though vertical foliated structures may record fluidisation and forceful water escape (Fig. 5.16d) (Lowe and Lopiccolo, 1974). Shallow Nye-type channels cut on the distal drumlin surface may also record the exit of porewater rising through the sediment pile (Fig. 5.16e). The marked proximal to distal changes in the scale of water transfer facies seem to belong to the same general phase of water escape which seems to have operated during drumlinisation when form obstructions were common. The Kanrawer stratigraphy demonstrates that sequential combinations of subglacial processes (basal sediment release, mass flow, shearing and emplacement of breccia, tectonic shunting, high porewater pressures, development of water transfer facies and meltwater escape using Nye-type channels) contributed to drumlin formation rather than widespread subglacial sediment deformation.

FA4 (Overridden ice-marginal subaqueous facies): On the northern margin of Galway Bay a small group of drift drumlins rests on the subglacially-scoured granite surface (Fig. 5.19a). This group of drumlins are sited at the head of a major coastal embayment which conducted ice flows southwestwards onto the continental shelf. The bulk of the drumlins are composed of stratified sediments and are similar in scale and topographic position to other linear assemblages of four to nine drumlins (Barna, Spiddal, Kanrawer, Ballyconneely) which are morphologically different but parallel to adjacent erosional marks. Four major, largely conformable, facies are present in the main exposure at Derryloney (Fig. 5.19b, c). The base of the section consists of laminated and massive mud grading up into muddy diamict containing pods of pebbles and isolated cobbles. Occasional channels

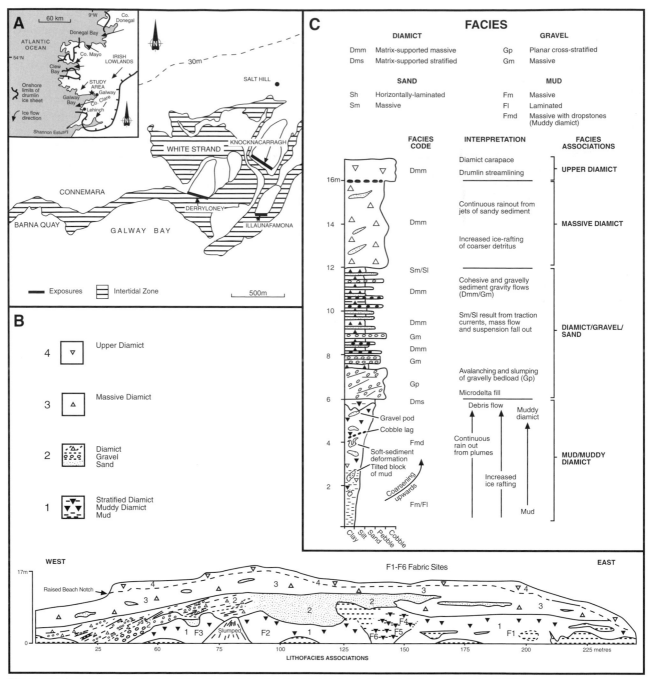

Fig. 5.19a Location of the Barna group of Drumlins and general ice flows (after McCabe and Dardis, 1989)

Fig. 5.19b Stratigraphy and geometry of lithofacies associations in the Derryloney drumlin, Galway Bay

Fig. 5.19c Generalised facies log from the central part of the Derryloney drumlin, Galway Bay (after McCabe and Dardis, 1989). Facies code is from Eyles et al., 1983. *By kind permission of SEPM (Society for Sedimentary Geology).*

cutting the mud are in-filled with stratified diamict. This facies is overlain by low-angle, planar cross-stratified gravel filling channels eroded into the muddy diamict. Abrupt textural changes are common with massive and laminated sand, which contain a range of soft sediment deformation structures. The gravel is overlain by stacked beds of massive diamict separated by thin sandy stringers. All of the stratified facies are overlain by a carapace of massive diamict across the entire section (Fig. 5.20). At most localities the carapace is strongly discordant with underlying facies and may contain mega-clusters of glacially-facetted boulders.

Fig. 5.20 Undeformed, stratified facies comprising the core of the Derryloney drumlin, Galway Bay.

The interbedded, flat-lying beds comprising the drumlin cores seem to have been part of a spread of sediment deposited by a range of processes around the head of the bay because similar facies are repeated within several drumlins in the group (Fig. 5.19c). The light-coloured mud is texturally similar to rock flour which was finally deposited by rain out from turbid plumes. The transition into muddy diamict may reflect increasing amounts of coarse debris from either ice-rafting or small channels. Gravelly pseudonodules in the mud suggest the latter process operated. Channelling of the muddy diamict surface and deposition of sand and gravel are probably related to discrete meltwater inputs across the site. The thin diamict beds are typical of resedimentation by mass flows. This interbedded association of gravel, sand and diamict is best explained by deposition on flat slopes fed by heterogeneous debris from a glacial efflux. The sharp contact with the overlying massive diamict reflects a change in the dominant depositional process rather than depositional environment. Numerous sandy stringers within the massive diamict records supression of bottom current activity and an increase in rain-out. Interbedded relationships between the diamict and main stratified facies indicate contemporaneity.

Interbedded stratified facies sequences forming the drumlin core are best explained by changes in the focus and strength of glacigenic input at the head of Galway Bay (McCabe and Dardis, 1989a). However, these sequences are not typical of either high-energy deposition in subglacial conduit or subaerial environments. Possible environmental settings include either a re-entrant or embayment if the ice margin was grounded locally or near the exit of a conduit when the ice margin was inactive and meltwater flow decreased. Given the fact that ice margins were extremely dynamic along terrestrial/marine interfaces during the last deglaciation, either of the scenarios outlined above are possible. Deposition at tidewater is the more likely because there is dated evidence that deep isostatic depression occurred along the western seaboard for thousands of years around the LGM and ice-marginal adjustments are then likely to occur in response to isostatic compensation, mass balance and changes in RSLs (McCabe et al., 2007a). Finally, the form-discordant till carapace records drumlinisation. Elsewhere in eastern Ireland stratified deposits form terminal outwash immediately outside drumlin ice sheet limits (Fig. 5.11).

FA5 (Tectonic and cataclastic facies): Deformed bedrock is sometimes streamlined into low amplitude, flow-parallel drumlins across lowland areas underlain by flat beds of Carboniferous limestone (Williams and McCabe, in prep). At Inishcrone on the eastern margin of Killala Bay competent beds of Carboniferous limestone are overlain by a 10 m thick tectonic stratigraphy developed from subjacent bedrock (Fig. 5.21). Facies from the base up consist of displaced and rotated beds of limestone, thrust bed-parallel detachment of limestone slabs,

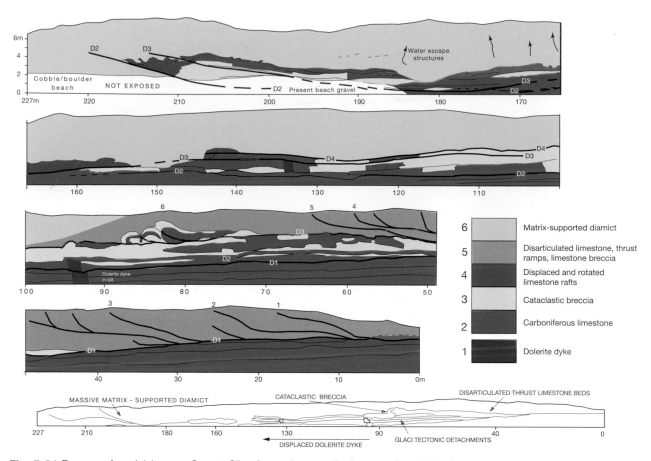

Fig. 5.21 Exposure from Inishcrone, County Sligo from a low amplitude streamlined ridge forming part of the drumlin swarm at the head of Killala Bay. The rock and sediments record subglacial glaciotectonic deformation of Carboniferous limestone which was facilitated by high porewater pressures. Together these processes are capable of generating the coarse fraction of basal till facies as rising thrust ramps are planed off by overriding ice (Williams and McCabe, unpubl).

thrust ramps cutting intact limestone beds, cataclastic debris and comminution of bedrock, and thrust ramps cutting both the cataclastics and fine-grained diamict which formed to the lee of the rock pile (Figs. 5.22, 5.23, 5.24). Deformation mechanisms are mainly slip and brecciation along beds or on branching shallow-angle thrust ramps. Thrust duplication involving comminution and brecciation by cataclasis is the prevalent process within the uppermost fault zones. The process of hydrofracture of intact limestone beds is associated with abnormally high fluid pressures generated during tectonic movement northwards which is also recorded by a displaced Tertiary dyke within the limestone (Fig. 5.25). The processes of hydrofracturing, cataclasis and thrusting occurred at much the same time as deposition of the fine-grained basal diamict because thrusts of limestone breccias occur with the basal diamict.

Deformation and sectional shortening of the limestone involved concurrent thrust faulting, cataclasis

and hydrofracture brecciation. These combined processes are capable of generating a great deal of *in situ* debris at the ice/rock interface by tectonic thickening. In addition the fact that the rising thrust planes are abruptly truncated along the upper surface of the drumlin shows that blocky debris has been continually skimmed from the surface of the growing tectonic pile into overriding ice. Because the internal structure is strongly form-discordant the final drumlin shape is erosional in origin. Observations indicate that the combinations of subglacial processes outlined above are commonly identified from exposures around Killala Bay where the bedrock has been extensively brecciated. It is likely that these processes rather than simple rock plucking are important in generating the coarse fractions found in many basal tills. In addition these processes, especially tectonic stacking and sectional shortening, operate over short distances and result in considerable thicknesses of freshly quarried debris. In

Fig. 5.22a Breccia composed of angular blocks of crushed limestone developed between intact rafts of limestone that have been shunted northwards, Inishcrone.

Fig. 5.23a Disarticulation, compression and folding of limestone beds, Inishcrone. Note that the folds occur on top of largely intact rafts of limestone which were shunted northwards with the same sense of movement as the fold vergence.

Fig. 5.22b Folded beds of Carboniferous limestone, Inishcrone. Note that the entire exposure records about 50 percent compression of the original flat-lying beds.

Fig. 5.23b Northward rising thrust ramps consisting of disarticulated beds of Carboniferous limestone from the south side of the Inishcrone exposure.

Fig. 5.22c Detail of crumpling, rupture and formation of breccia as competent beds or rafts of limestone fail after moving at slightly different speeds and directions, Inishcrone.

Fig. 5. 24 Shear structure along bed of disarticulated limestone within matrix-supported diamict towards the distal (northern) end of the Inishcrone exposure.

the Irish lowlands basal tills are derived locally and have not travelled very far, possibly a few kilometres. These mechanisms seem capable of providing the main source of coarse debris because simple plucking cannot occur beneath a blanket of debris unless the debris is continu-

ally removed. The sequence at Inishcrone shows that detachment of rock slabs can occur at depths greater than 6 m below a tectonic pile, and this section formed within 20 km of the probable ice margin. In more central areas of ice sheet dispersal, porewater pressures

Fig. 5.25a Tertiary dolerite dyke within Carboniferous limestone beds truncated by limestone rafts which were shunted northwards during glaciotectonics, Inishcrone.

Fig. 5.25b The truncated top of the Tertiary dyke transported 60 m northwards between largely intact rafts of limestone, Inishcrone.

would be higher and overburden pressures greater and capable of generating even thicker tectonic piles. It is also evident that there is not a space problem within the tectonic system because bedrock has simply been shortened and thickened with concomitant bedrock loss to overriding ice.

FA6 (Subglacial channel facies): The cores of drumlins located within the subglacial tunnel valley system between Lough Neagh and Newry contain extensive beds of undeformed stratified sediments (Dardis and McCabe, 1983) (Fig. 5.26). The Poyntzpass channel is a rock-cut, Nye-type network thought to be more extensive throughout central Ulster. Areas bordering the channel are dominated by large rock drumlins and ribbed moraine especially to the west of the channel axis. The orientation of the 'sand-cored' drumlins is slightly different to the north–south trend of the large tunnel valley. Several exposures in drumlins situated in different parts of the channel system contain different facies sequences in their sand cores. At Jerrettspass on the floor of the main channel the core of a torpedo-shaped drumlin consists of a basal prograding coarse-clastic deltaic sequence (Fig. 5.27). Bottomsets are mainly fine-grained, horizontally-bedded silts and clays containing some isolated cobbles. These grade upwards into toesets which are mainly type A and B ripple drift lamination and a range of other primary sedimentary structures (Fig. 5.28a). Overlying foresets are large-scale trough-shaped, avalanche front cross-stratified gravels dipping southwards between 15 and 30°. Gravelly topset beds are discontinuous and concentrated at the southern end of the section. A thick basal till facies with weakly-defined foliation is gradational from the topset beds and is cut by small channels infilled with laminated diamict. The surface of the entire drumlin is marked by a clast rich, massive till with a much coarser matrix than the lower tills. Drumlins situated on the flanks of the channel or in smaller channels have more complex internal successions. At Tandragee the drumlins consist of an irregular rock basement overlain by compact massive till which grades into stratified diamict and cross-stratified gravels especially in the lee of rocky knobs. The stratified core consists of undeformed massive and cross-bedded sands which are overlain by a thick carapace of basal till (Dardis and McCabe, 1983). Nearby the stratified core of the Glebe Hill drumlin consists mainly of alternating beds of diamict and rhythmically-bedded sand, silt and clay (Fig. 5.28b).

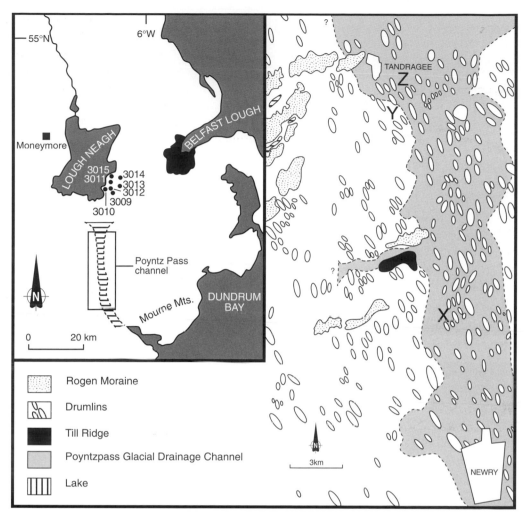

Fig. 5.26 Location of the Poyntzpass tunnel valley. Note that it opens northwards and acted as a drainage conduit form the major centre of ice sheet dispersal in the Lough Neagh basin. The morphology of the feature is characterised by sand-cored, streamlined drumlins along the valley floor and ribbed moraine along the flanks of the feature. The position of the Jerrettspass, Glebe Hill and Tandragee drumlins is indicated by X, Y and Z respectively.

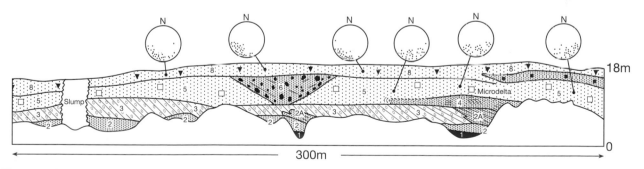

Fig. 5.27 Sedimentary facies of the Jerrettspass drumlin. 1. bottomset beds; 2. toeset beds; 3. foreset beds; 4. topset beds; 5. lower till facies; 6. sediment flow facies; 7. massive till facies; 8. upper till facies. (After Dardis and McCabe, 1983.)

The variability in the internal sedimentary successions has been attributed to a range of energy levels within a hierarchical subglacial channel system. For example, the sediments at Jerrettspass are sedimentologically similar to gravelly megadunes that migrated along the main channel axis whereas the other cores with alternating beds of diamict, sands and rhythmites are more representative of mass flows, cohesionless flows and turbidity

Fig. 5.28a Sand core of the Jerrettspass drumlin, County Down.

Fig. 5.28b Sand core of the Glebe Hill drumlin, consisting of rhythmically bedded silt, sand and diamict.

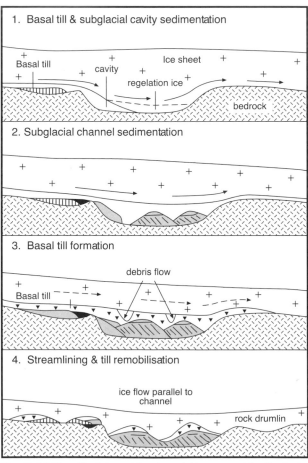

1. Basal till & subglacial cavity sedimentation

Basal till   cavity   Ice sheet
                      regelation ice
                                    bedrock

2. Subglacial channel sedimentation

3. Basal till formation

debris flow

Basal till

4. Streamlining & till remobilisation

ice flow parallel to channel

rock drumlin

Fig. 5.29 Simplified depositional model to account for the facies arrangements in the sand-cored drumlins within and adjacent to the Poyntzpass tunnel valley (from Dardis and McCabe, 1983).

currents within smaller channels or cavities (Fig. 5.29). Once the channel system opened it seems to have been self-maintained by meltwater flow. However, the formation of the stratified cores relates to a phase of reduced meltwater production well below the high flow regimes expected when the tunnel valley was initiated. If this interpretation is correct then the stratified cores represent lower energy flows when the ice sheet retreated from the Irish Sea. At this time the ice sheet had shrunk by about two thirds of its original size by marine downdraw and climate change. Lowered energy and geomorphic activity are linked to lowering of ice sheet profiles prior to later ice buildup and drumlinisation. This phase of ice sheet inactivity and relatively low meltwater production is recorded by the presence of similar stratified cores within drumlins adjacent to other rock-cut channel systems across much of the ice sheet in central Ulster (Dardis, 1982). Because the stratified cores show transitional relationships with overlying basal tills the latter probably record passive release of debris leading to blocking of subglacial channel networks and initiation of sheetflow (Fig. 5.30). These new basal conditions may well have

facilitated later drumlinisation and sediment moulding. The presence of undeformed stratified cores in drumlins therefore provides critical stratigraphic evidence for changes in ice sheet basal conditions and configuration immediately prior to final drumlinisation which occurred during Heinrich 1 times in eastern Ireland (McCabe and Clark, 1998).

FA7 (Lee-side stratified facies): About fifty drumlins with lee-side stratified facies have been found within a 400 m² zone south of Lough Neagh (Dardis et al. 1984). In all cases stratified deposits occur only on the southern (downglacier) flanks of either barchanoid or whaleback drumlins (Fig. 5.31). In most cases the stratified sequences infill embayments excavated in the lee sides of barchanoid forms and are superimposed directly on whaleback forms. The field evidence and lee position of the stratified deposits show that they are not glacially overridden proglacial outwash but show interbedded relationships with a thin carapace of basal till. Detailed logs from disused gravel pits illustrate the range of lee-side sedimentary processes which operated in lee-side cavities (Figs. 5.32, 5.33). The lec-

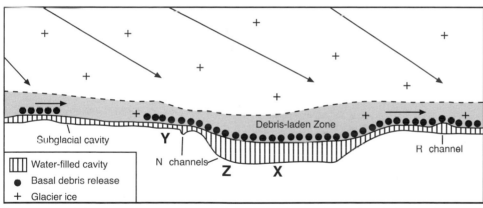

Fig. 5.30 Simplified reconstruction of major ice/substrate and ice/water interface environments in the vicinity of the Poyntzpass tunnel valley prior to drumlinisation. The bulk of subglacial cavity sedimentation occurred before deposition of subglacial basal till deposition. X, Y and Z mark the approximate positions of the Jerrettspass, Glebe Hill and Tandragee drumlins in relation to the main channel (from Dardis and McCabe, 1983).

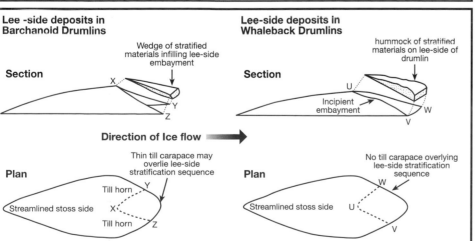

Fig. 5.31 Stratigraphic position of lee-side stratification sequences in barchanoid and whaleback drumlins (from Dardis et al., 1984).

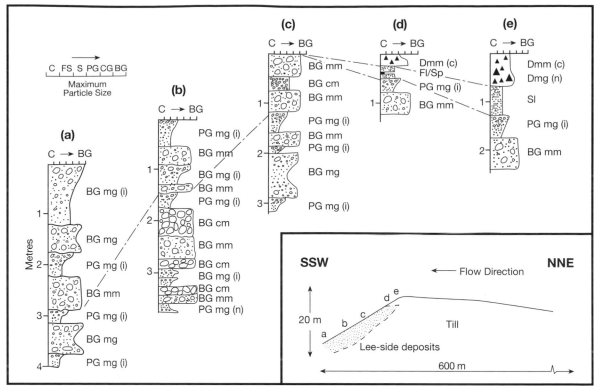

Fig 5.32a Lee-side proximal to distal facies relationships in the Mullantur drumlin, County Armagh. C=mud, FS=fine sand, S=medium and coarse sand, PG=pebble gravel, CG=cobble gravel, BG=boulder gravel (from Dardis et al., 1984).

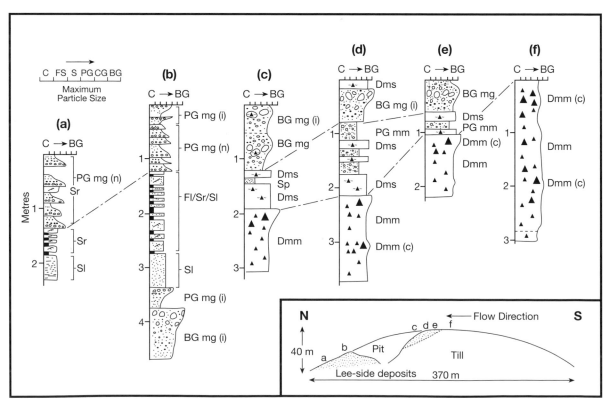

Fig. 5.32b Lee-side proximal to distal lithofacies relationships in the Derrylard drumlin, County Armagh (from Dardis et al., 1984).

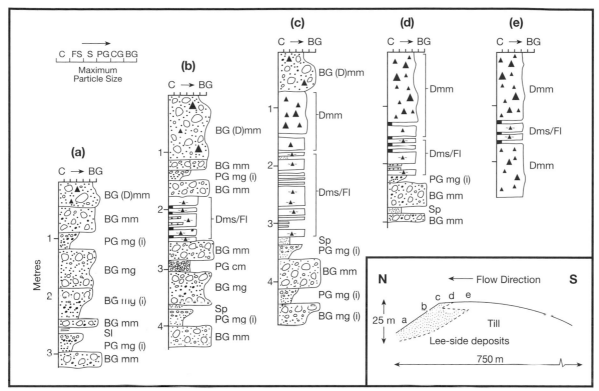

Fig. 5.33a Lee-side proximal to distal facies sequences in the Buncrana drumlin, County Donegal (from Dardis et al., 1984).

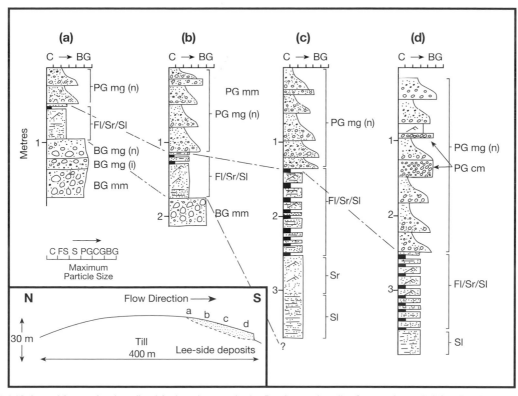

Fig. 5.33b Lee-side proximal to distal facies changes in the Derrinraw drumlin, County Armagh (after Dardis et al., 1984).

side deposits at Mullantur are interbedded with basal till and show proximal to distal changes, including the development of stratification downslope from disorganised gravel into normal and inverse-to-normal graded beds (Fig. 5.32a). At Derrylard the transition from till into lee-side deposits is marked by stratified diamicts interbedded with pebbly gravel (Fig. 5.32b). This prograded sequence coarsens upwards and shows a downslope facies transition from cohesive flows to rhythmically-bedded sand and mud. At Buncrana the exposure is characterised by a well developed proximal to distal transition from stratified diamict and mud into boulder gravels draped by basal till (Fig. 5.33a). The disorganised nature of the boulder facies suggest deposition by high density flows close to points of sediment input (Lowe, 1982). Lee-side deposits partly superimposed on whaleback forms tend to contain greater amounts of parallel laminated sand, pebbly gravel and avalanche front structures. Far-travelled erratics are common and are only occasionally found in the till below.

Morphostratigraphic data show that the lee-side sequences are *in situ* and formed on the protected lee-side face of till protuberances and record deposition in water-filled cavities (Fig. 5.34). Sequences at Derrylard and Buncrana suggest that sediment transport across the drumlin occurred as a high-density fluid mix composed of boulders suspended in pebbles and some sand. Textural changes and grading downslope correspond closely to proximal/distal changes (transitions from high to lower density flows) observed from coarse-grained, submarine conglomerates (Walker, 1975). Coarsening upwards sequences from sand to pebbly gravel common in whaleback forms may record either progradation or velocity zoning and increased water depths towards the distal parts of the cavity.

Stratigraphic evidence shows that cavity formation is a form/process relationship that was penecon-

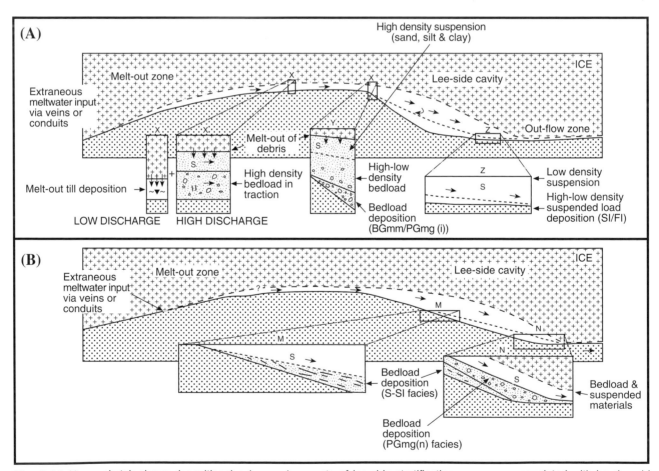

Fig. 5.34 Upper sketch shows depositional palaeoenvironments of lee-side stratification sequences associated with barchanoid drumlins and inferred hydrodynamic conditions. Lower sketch records depositional palaeoenvironments associated with whaleback drumlins (from Dardis et al., 1984).

temporaneous with drumlinisation. This sedimentary system consisting of erosion by hydraulic processes across the drumlin form and deposition in lee-side cavities provides evidence that the final processes that shaped drumlins were water/sediment mixtures rather than subglacial deformation by ice itself. It is difficult to estimate the influence of ponded water on ice sheet dynamics, but the widespread presence of water may be linked to periodic triggers in ice sheet movement, especially if outflow was impeded.

FA8 (Remoblised/superimposed facies): Dardis (1982, 1985a) demonstrated that some drumlins display a slightly hummocky topography which is linked to deposition after drumlinisation. The facies is normally coarse-grained gravel and stratified diamict and is particularly significant in central Ulster where resedimentation occurred along drumlin flanks which were submerged in proglacial lakes. In other areas characterised by regional ice stagnation drumlin morphologies are often modified by local meltwater deposits.

The Drumlin Ice Sheet

Synge's idea of a Drumlin Ice Sheet in north central and western Ireland bounded by extensive ice-marginal moraines and outwash focuses attention on the information that glacial bedforms can contribute to knowledge of glacial history. The final subglacial imprints on the landscapes of north central and western areas undoubtedly occurred after a major deglaciation after 22 kyr BP that depleted the ice sheet by about two thirds from its maximum extent on the continental shelf. The resulting ice masses and centres of ice dispersal were located mainly in north central and western lowlands. Complexity both in bedform patterns and internal facies geometries can be related mainly to subsequent ice sheet history covering about five thousand years of deglacial history. The timeframe (~20–15 calka BP) involved therefore implies that the reduced ice sheet was dynamic and geomorphologically active at a time when high magnitude and rapid climate signals were received from the North Atlantic ocean. The variability in bedform type and internal structure which once was considered incapable of rationalisation can now be interpreted as different subsystems of a dynamic Drumlin Ice Sheet. The latter term is still considered a useful descriptor because the main area of drumlinisation is approximately synonymous with the reduced ice sheet in the north of the island.

Advances in radiocarbon age constraints of deglacial history come from marine beds deposited around the margins of the reduced drumlin ice sheet (McCabe et al., 2005). These, together with bedform evidence from satellite imagery, provide the age basis for the sequential series of events and glaciological changes which occurred during the history of the drumlin ice sheet. The events used to model 'drumlin' ice sheet events are identified from both regional geomorphic evidence and from drumlin stratigraphies. The drivers for events may be either internal or externally forced and are often linked to rates of operation of geomorphic processes and whether or not the ice sheet was geomorphically active. The main elements of the model include:

1. Drift deposits older than 30 calka BP occur near centres of lowland ice dispersal and generally are a significant part of bedform stratigraphies. At Maguiresbridge these older tills are widespread and in excess of 15 m thick. Their apparent absence elsewhere probably means that former glacigenic detritus has been reworked by later ice. It is extremely difficult to assess when the very thick (<80 m) subglacial deposits agglomerated but may have begun as the ice sheet grew towards Late Glacial Maximum limits prior to 25 calka BP. Because blocky tills and vast amounts of coarse angular to edge-rounded debris are observed from countless small sections it is likely that compressional forces and high porewater pressures leading to enhanced vertical stacking of sediment contributed significantly to thick localised diamict sequences. The sequence from Inishcrone provides evidence for tectonic thickening of local rock debris prior to streamlining.

2. Cosmogenic dates (Bowen et al., 2002) and marine muds from County Down show that ice had withdrawn from its maximum limits into north central Ireland before 19 calka BP when isostatic uplift was underway (Clark et al., 2004). In general ice sheet margins were terrestrial and located kilometres from the coast. Variable striae on the Slieve Beagh uplands on the eastern margin of the Lough Erne basin suggest that ice flow directions changed during the last glacial cycle (McCabe, 1969). This evidence for change is supported by the orientations of prominent ribbed moraine across the lowlands of Upper Lough Erne which record a change in ice flow from the northeast to northwest (Fig. 5.14). Because most of the ribbed moraine field concentrated across the

entire northern lowlands (Ards Peninsula southwest to Roscommon) is pristine and mainly unmodified, it is argued that it developed beneath the initial or early stage drumlin ice sheet. This hypothesis is supported by marked changes in the orientations of ribbed moraine ridges found on the periphery of the Lough Neagh Basin. To the west of Dungannon N–S and NW–SE ridges dominate but the orientation changes to NE–SW in County Armagh and E–W ridges occur to the north of the basin. These orientations are at right angles to radial ice flow out of the basin and form a pattern recording the position of the former Lough Neagh ice dome.

Crosscutting and superimposed ribbed moraines record shifts in the local centres of ice dispersal marked by changes in ice flow of 20 to 30 degrees. However, across much of the lowlands the orientation of ribbed moraine is regionally consistent, depicting uniform ice flow. It is within the areas of ribbed moraine that subglacial diamict sequences are thickest. Because the restricted ice sheet had insufficient energy to generate up to 80 m of sediment regionally, it is suggested that it reorganised pre-existing sediment into flow-transverse ridges. This probably occurred when the ice sheet was growing after early deglaciation and when basal conditions were dry and minimal meltwater was present.

3. At this time when ice-marginal configuration and meltwater production was substantially reduced, major conduits draining the ice masses were probably dominated by low energy flows rather than pipefull conditions. It is likely that stratified sediments began to accumulate in tunnel valleys and conduits near the ice sheet margins. The stratified cores in drumlins in central Ulster, Poyntz Pass, Spiddal, Barna and Ballyconeety may owe their origin to the relatively low energy glacial conditions at this time mainly along the periphery of the drumlin ice sheet.

4. Between 18 and 16 cal ka BP two ice sheet readvances drumlinised the entire belt of ribbed moraine across the north central lowlands (McCabe et al., 2007b). Where ice sheet margins readvanced major moraines mark ice sheet limits. These include the Killard Point moraine which can be traced southwestwards from County Down across the lowlands of Louth and Meath. Where ice advanced into the large marine embayments of western Ireland linear, com-

plexly-thrust moraines were formed on the north side of Clew Bay and on the south side of Donegal Bay. Smaller ice-pushed moraines of this general age occur in north Donegal at Ballycrampsey and Moville. Characteristically these ice-marginal ice sheet readvances overran and incorporated marine muds especially in Dundalk Bay where the deposits are tightly age-constrained. At Killard Point marine muds are interbedded with the terminal outwash and record deep isostatic depression when eustatic sea level was low. Clearly, the glacial readvance system consisted of subglacial sediment transfer to tidewater ice margins during the erosional processes of drumlinisation. The greatly increased geomorphic activity and sediment output during the readvance is recorded by the largest exposure of glacigenic debris currently known in Ireland which was deposited at Askillaun where the ice readvanced northwestwards into Clew Bay. The subglacial imprints at the head of Clew Bay record drumlinisation across earlier streamlined ridges forming barchanoid drumlins with their horns facing northwest, the final direction of ice flow (Fig. 4.11).

5. Drumlinisation of ribbed moraine with varying degrees of modification occurred across most of the area covered by the drumlin ice sheet (Fig. 5.35). Tracing ice flow lines across the island suggests that drumlinisation was more or less synchronous. Former stratigraphies have been preserved during the readvance and in some central sectors of the ice sheet which were sticky and slow moving, former ribbed moraine patterns are still evident. For example, in the Upper Lough Erne and Upper Shannon basins cross-cutting ribbed moraine ridges are preserved but show tails of sediment dispersal southwards. Stratified sediments in tunnel valleys and subglacial conduits were overridden and carapaces of tills created. Towards ice sheet centres most internal drumlin geometries were inherited from former ribbed moraine sediments which are generally coarse-grained and massive. The final modifications during late stages of drumlinisation include sediment steamlining with down-ice tailing, lee-side cavity sedimentation, water through-flow facies and other features due to form obstruction once drumlins had formed.

6. The most significant change during the evolution of the drumlin ice sheet occurred when the swathe of

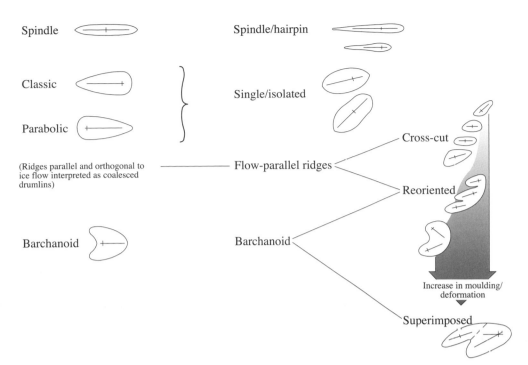

Fig. 5.35 Drumlin morphological types around Donegal town (from Knight and McCabe, 1997 and Hanvey, 1988).

ribbed moraine across the north of the island was drumlinised. The cross-cutting relationship records a change in basal ice conditions at the onset of ice readvance during the Killard Point Stadial. This change involved a shift from a cold, dry-based ice sheet during ribbed moraine formation to warmer and wetter-based conditions during drumlinisation. The latter is also characterised by headward erosion by ice streams that ended at tidewater into the fields of ribbed moraine farther inland (Figs. 5.1, 5.5). Current models from the northern Irish Sea basin show that marine downdraw was an important process which induced headward propagation of ice streams which extended into the heart of the drumlin ice sheet (Fig. 1.1) (Eyles and McCabe, 1989a). A well-defined suite of bedform changes occur down-ice along the paths of ice streams which cut out entire sections of ribbed moraine (Fig. 5.11) (Knight et al., 1999; McCabe et al., 1999). The cutouts and range of streamlined landforms are particularly

significant along the Armagh ice stream which may be traced for 70 km from the southern Lough Neagh dispersal centre in a curved path southeastwards to the limiting moraines on the margin of Dundalk Bay.

7. Because the geological record discriminates between an initial accumulation-fed advance and a later advance accelerated by marine downdraw it is sometimes difficult to relate landscape changes directly to climate system reorganisation. Nevertheless millennial timescale ice mass variability appears to have been characteristic of amphi-North Atlantic ice sheets during the last glaciation (Bowen 1994; Fronval et al., 1995). The ice sheet oscillations which also occurred around Heinrich 1 times in northern Ireland record drumlinisation and fast ice flow and may be correlated with some of the millennial-timescale climate changes in the circum-North Atlantic (Knight and McCabe, 1997b).

6

The Irish Sea Glacier

Introduction

The Irish Sea Basin (4000 km^2) extends from the northern approaches of the Celtic Sea in the south, to the North Channel separating northern Ireland from southwestern Scotland (Fig. 6.1). A central trough connects St. George's channel in the south with the North Channel in the north with depths of >250 m along the Beauford Dyke. Shallow (<50 m) platforms bordering coasts contain Quaternary sediments less than 50 m thick whereas thicknesses of >300 m occur in the central basin (Whittington, 1977). At a maximum the basin is 200 km wide immediately south of the Isle of Man and closes northward into the 20 km wide North Channel graben. The importance of the Irish Sea Basin lies in the fact that it occupies a central position between the two main islands in the British Isles. Therefore, during major glaciations it was generally occupied by a large ice stream fed by converging ice flows from centres of ice sheet dispersal sourced from neighbouring land areas (Fig. 6.2). This relationship means that the basin acted as a major ice conduit. It drained the composite British ice sheet, it acted as a route for sediment transfer onto the continental shelf, it influenced ice sheet dynamics on adjacent land areas though marine downdraw and ice wastage, it acted as a sediment sink during deglaciations and it sourced ice flows that moved onto adjacent land areas. For these reasons the significance of the 'Irish Sea Drifts' will include reference to other sites around critical parts of the basin.

Although correlation of beds or events from different parts of the basin was always a popular methodology (Mitchell et al., 1973) the approach was flawed because of the absence of age constraints on deposits and events. A related problem has been the interpretations of heterogeneous glacigenic successions preserved around basinal margins. Traditionally the umbrella terms of Irish Sea Drifts and Irish Sea Tills were used to describe mud, muddy diamict and fine-grained diamict beds which were often interbedded with stratified deposits. In general these successions are confined to a narrow coastal margin forming low cliffs on the upper shoreface which immediately suggests that they are genetically related to glacigenic deposition from ice located in the basin itself rather than terrestrial ice. The perceptive work of Lewis (1894) concluded that the complex facies in the basin reflect deposition partly during ice advance, partly during ice retreat and partly when the sea was washing against an ice wall with areas covered by floating rafts of ice. Later workers recognised that some so-called shelly deposits were carried inland. In exposures where fine-grained Irish Sea deposits are interbedded with coarse-grained, locally derived gravel or diamict a general model was invoked based on the varying strengths of Irish Sea and local ice masses. Implicit assumptions in this model are the existence of, and sediment deposition from a terrestrial ice sheet, a glacioeustatically lowered sea level and a belief that the timing and style of ice retreat was wholly controlled by climate. There is little doubt that glacially thrust sediments occur hundreds of metres above sea level and that sediment spreads have been thickened by ice push, leading to the duplex stratigraphy seen in coastal exposures at the St. Bees, Bride and Ben Head moraines. However, most deposits around the basin are relatively undeformed and characterised by interbedded relationships, facies variability and marked facies transitions often within channelled geometries (Fig. 6.3).

Research methodologies

Historically, the Quaternary of the Irish Sea Basin was based largely on the description of type sections (Mitchell et al., 1973) which contain lithological units which are then correlated away from the reference area as mappable units. This scheme over-emphasises the vertical succession of lithological types and is essentially

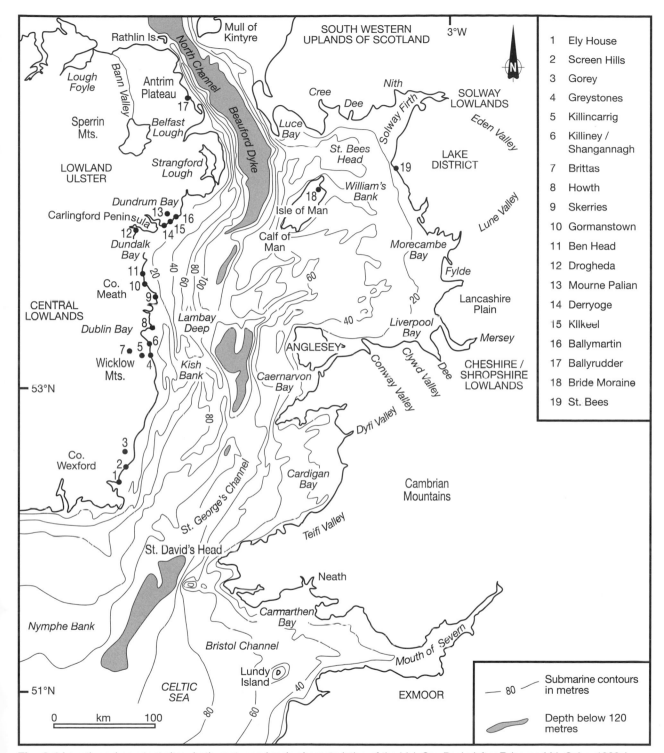

Fig. 6.1 Locations, important sites, bathymetry and main characteristics of the Irish Sea Basin (after Eyles and McCabe, 1989a).

a layer cake approach to stratigraphy. Correlations of individual units away from stratotypes using a variety of lithological criteria has resulted in a proliferation of local stratigraphic names for units, the origins of which are poorly understood. Because glacigenic environments can be sedimentologically complex, bed-for-bed correlations are especially ineffective and locally-defined units may have little meaning within a basinal context.

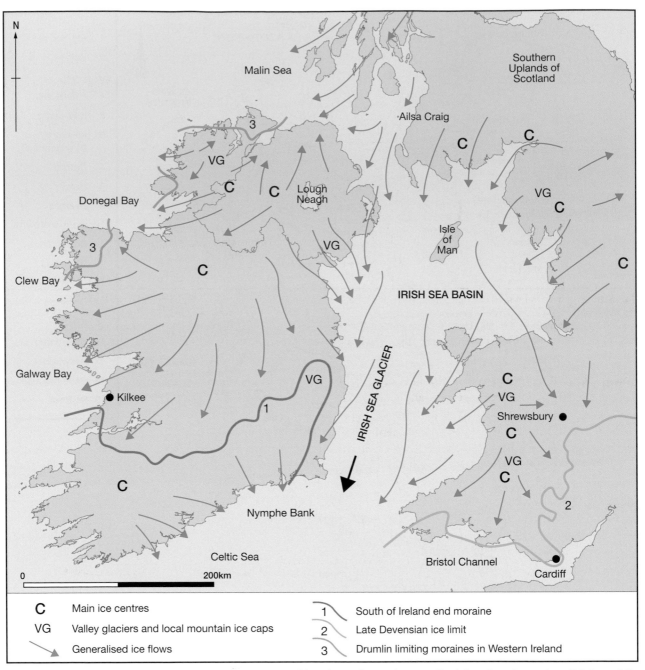

Fig. 6.2 Various Late-Devensian ice limits, some centres of ice sheet dispersal, generalised flow lines in western Britain and the postulated path of the Irish Sea Glacier (after Eyles and McCabe, 1989). *By kind permission of GSNI.*

Development of seismic techniques and subsurface exploration reveals the sedimentary patterns of the broad-scale, three-dimensional stratigraphy within large sedimentary basins. Results demonstrate the importance of lateral accretion during deposition and stresses spatial changes in facies types and thickness resulting from sediment bodies building out across the basin. Factors including changes in relative sea level,

water depths, sediment supply, subsidence and position of the ice front or major glacial effluxes give rise to sedimentary packages (sequences) bounded by unconformities (sequence boundaries) recording major breaks in sedimentation. Eyles and McCabe (1989a) adopted this depositional systems approach based on detailed facies studies, followed by identification of depositional sequences and their three-dimensional geometry and

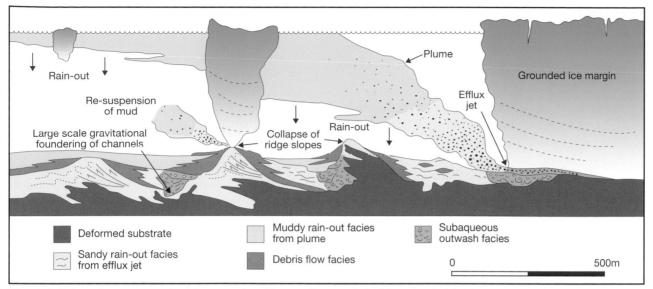

Fig. 6.3 Schematic representation of a retreating tidewater ice margin building morainal bank complexes. Sandy rain-out facies deposited from efflux jets form thick drapes in the ice-proximal part of the model, between the ice margin and inner push moraines. Muddy facies deposited from plumes blanket and flatten out underlying topography. Debris flows result from depositional oversteepening in areas of very rapid rain-out near the efflux jet, or as a result of collapse of growing structural highs (after Eyles and McCabe, 1989a).
By kind permission of GSNI.

arrangement within the larger depositional system operating in the Irish Sea Basin. Sea level changes, tectonics, climate forcing, position of ice sheet margins, activity of ice sheet margins, rates of ice retreat, sediment supply and the activity of ice sheet flow-lines from ice source areas all contributed to facies types/sequences formed in glacigenic settings (Fig. 6.3). Walker (1984) provides introductions to facies analysis methods based on careful documentation of facies types, contact relationships, depositional geometry, age and palaeontological data combined with analogue information from both modern and ancient environments. Basically, dynamic stratigraphy is a lumping process, integrating facies and sequences into genetically-related packages which contrasts with other investigations based on splitting of genetically-related sequences and disregards their sequence context.

The glacigenic depositional systems recognised from the Irish Sea Basin together define a distinct event stratigraphy that records changes in relative sea level caused by ice sheet loading and crustal downwarping (Eyles and McCabe, 1989a). Inherent in this model is the presence of subglacial deposits in some areas though their presence is restricted on most basinal margins. Recent radiocarbon and cosmogenic dates confirm the original comment by Eyles and McCabe (1989a) that the sediments represent the record of an extremely short-lived deglaciation event perhaps a few thousand

years in duration shortly after 22 cal ka BP (Bowen et al., 2002; McCabe et al., 2005). During this time the ice margin retreated rapidly from south to north along the basin with successive ice fronts depositing wedges of sediment that offlap northwards. This deglaciation pattern was coeval with marine incursion into the basin and formed discrete depositional packages which have a high preservation potential because the ice margin wasted northwards (Fig. 1.2).

Depositional systems within the event stratigraphy

The glacial geology of the basin is mainly an event stratigraphy recording the entry of marine waters into a glacioisostatically depressed basin from ~22 to 15 cal ka BP. However, there are much older glaciomarine deposits preserved between regional till beds near basin margins at Drogheda which record deep isostatic depression during an earlier deglaciation possibly during OIS 3. The early deglaciation and withdrawal of ice from the Celtic Sea is marked by an extensive system of back-filled subglacial channels scoured across the Nymphe Bank (Blundell et al., 1971). Ice retreat northwards was driven by a succession of calving bays where the ice loss generated ice streams which reached deep into terrestrial portions of the ice sheet. Ice sheet flow lines pass directly offshore along the south coast with no evidence for any deflection or subsequent overriding. Striae along the eastern coast support a general ice flow southwards. A classification of the eight main depositional systems found in the basin includes:

1. The Irish Sea ice stream transported shell fragments and erratics of northern provenance inland up to 20 km and up to ~700 m on the eastern flanks of the Wicklow Mountains. Basal tills sometimes containing marine shell fragments occur along the coast at Gormanstown, in north County Dublin at Howth and inland as far as Brittas (Hoare, 1975) and record onshore ice movements from the Irish Sea. In general tills associated with onshore ice movements contain a suite of far-travelled erratics including Ailsa Craig microgranite, Cretaceous flint and chalk, Tertiary igneous erratics and marine muds.

2. Retreat from the Celtic Sea area into St. George's Channel is marked by the largest Quaternary ice-contact delta complex in the British Isles known as the Screen Hills (Thomas and Kerr, 1987; Eyles and McCabe, 1989a).

3. A scoured subglacial topography is infilled with up to 100 m of glaciomarine sediments (Whittington, 1977). Seismic records show that the Irish platform contains sinuous, locally-overdeepened steep-sided valleys that drain coastal margins of the Irish Sea towards the central trough (Fig. 6.4).

4. Morainal banks include arcuate belts of hummocky, undulating topography up to a few kilometres in width or spreads of sediment marking the former positions of ice margins which reached tidewater. Each complex may represent a temporary halt or re-equilibration of the Irish Sea Glacier margin as it withdrew north. Typically they are associated with shallow water deposition and to the lee of bedrock highs which acted as pinning points. Along the coast of counties Dublin and Meath several moraines record ice-marginal halts, tectonics and thickening of proglacial outwash spreads. In some cases the

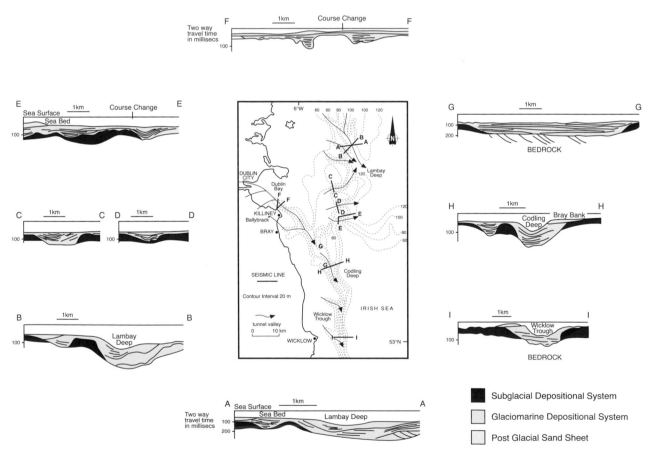

Fig. 6.4 Seismic profiles from the central portions of the Irish Sea Basin showing arrangement of subglacial and glaciomarine depositional systems.

term 'morainal banks' is broadened to include relatively thick (~15 m) spreads of sediment formed in basins such as the Mourne Plain, vacated by slowly-retreating or static ice margins. The basal sediments of the Mourne Plain record the early deglaciation of the northern Irish Sea Basin as ice margins retreated inland towards centres of ice dispersal.

5. Large subaerial channels cut into early deglacial sediments of the Mourne Plain record rapid postglacial uplift because most of the ice sheet had been downdrawn. Beds of marine mud infilling these channels at Kilkeel record a global sea level pulse around 19 cal ka BP when the basin was open to marine influences through St. George's Channel.

6. Extensive beds of marine mud occur on the margins of Dundalk Bay recording shallow marine conditions from 19 to <16 cal ka BP and deep isostatic depression late into the deglacial cycle in the northern basin (McCabe et al., 2005). A unique example of an intertidal boulder platform occurs at Cooley Point on the southern margin of the Carlingford Peninsula.

7. Ridge moraines partially composed of overridden marine muds and diamicts occur along the coasts of County Down and Louth and record two major ice sheet readvances into the northern basin between 18 and 16 cal ka BP (McCabe et al., 2007b). These moraines occur in front of the main drumlin swarms across the north Irish lowlands and represent the output of subglacial sediment transfer during drumlinisation.

8. Marine limits, raised beach indicators and emergent facies sequences are common elements associated with readvance ice sheet marginal deposits and often display synchroneity (Stephens and McCabe, 1977). These features represent the final record of basin emergence after at least 15,000 years of isostatic depression and oscillating ice sheet margins covering much of the coastline.

Marine shelly faunas

The fact that most of the ice-contact sediments exposed along basin margins are associated with sediment delivery to tidewater margins from subglacial meltwater effluxes means that shelly faunas are most likely derived from pre-existing deposits overrun by ice. Although local facies sequences are different at sites including the Skreen Hills, Skerries and Derryoge the marine molluscs are all fragmented and only the most robust shells survive. In most cases the largest shell fragments (~1–2 cm) occur in pebbly gravel facies and the smallest (~0.2) are present in fine-grained diamict or muddy diamict. Shell fragments are so abundant at Derryoge and Skerries that it is almost certain that the subglacial meltwater erosion occurred across shell banks on the sea floor. Most shelly gravel currently exposed is related to point sources and direct erosion of older shell banks whereas subjacent fine-grained diamict or mud do not contain in situ shells. This is a further indicator that the macrofaunas are derived. The presence of complete, derived and abraded marine shells in both fine- and coarse-grained gravel successions testifies to transport and deposition by meltwater. Because this pattern is repeated basinwide it compliments the idea of an event stratigraphy recording the entry of marine waters and rapid retreat of the ice as a tidewater margin.

The Wexford Manurial Gravels have long been known to contain a suite of molluscan shells which include Crag species (McMillan, 1964). About 133 different species have been found in sections between the river Slaney and the coastal cliffs of east Wexford. The most extensive exposures occur in the ice-contact delta of the Screen Hills which was sourced from several subglacial effluxes delivering shells and sediment southwards and southeastwards into the sea as ice decayed into St. George's Channel. McMillan concluded that the shells are mixed and have both northern and southern affinities. Crag shells such as *Neptunea contraria* form a small but significant element in the derived population. Other disused gravel workings representing meltwater discharges from the Irish Sea glacier at Killincarrig, County Wicklow, contain similar derived assemblages.

Shell samples from gravels and diamict sediment submitted for radiocarbon dating have been collected from sites at Skerries, Derryoge and Ballyrudder. These sections are all part of the event stratigraphy of the basin and stratigraphically are part of the last deglacial depositional cycle. However, all the radiocarbon dates obtained were greater than 40 kyr BP and infinite (McCabe, 1995b).These dates confirm the idea that the shells are derived and much older than the depositional sequences in which they occur. Similarly amino-acid analysis on *Arctica islandica* shells from Derryoge showed that the assemblage was of mixed ages, with the youngest faunal elements being Middle Midlandian in age (Bowen, unpubl.). Therefore the basinal distribution of shelly sediment, its local sedimentology, its association with subglacial meltwater erosion and delivery to tidewater and various age estimates using different techniques shows that the shells now found in deglacial sediments have been derived from older facies on the floor of the Irish Sea. Similar conclusions have been suggested for shelly deposits on the Mull of Kintyre in the northern part of the basin (Sutherland, 1984).

Marine microfaunas

Extensive studies from about thirty sites around the basin show that two major types of microfaunal assemblages occur in fine-grained facies (mud, muddy diamict) which often occur in morainal banks, delta complexes and ice-pushed moraines. One assemblage contains mixed microfaunas and record erosion by subglacial meltwaters and ice from the floor of the Irish Sea. The preservation of the temperate species is generally poor, while that of the Arctic species is good and often very good. The almost exclusively foraminiferal fauna is dominated by *Elphidium clavatum, Haynesina orbiculare, Cibicides lobatulus* and *Cibicides fletcheri* (Fig. 1.6). Although this biofacies is typical of shallow-water, Arctic environments where glaciers are in retreat, the assemblage also contains poorly-preserved temperate species including *Ammonia batava, Ammonia falsobeccarii, Elphidium crispum* and *Elphidium* species. In general the temperate species are more common in southeastern Ireland than elsewhere in the basin. The size and condition of tests show clear evidence of transport and recycling but in most cases foraminiferal assemblages show little change in vertical successions (Austin and McCarroll, 1992). Uniform distributions of foraminifera in vertical section cannot be easily explained by deposition as basal till or forms of melt-out till because the typical assemblage contains two discrete biofacies which undoudtedly have different sources. It is more likely that warm elements eroded by meltwater from older sediments on the sea bed are recycled by suspension and redeposition with *in situ* cold elements formed during ice retreat. Evidence for mixing and current-sorting are common within modern microfaunas in the southern Irish Sea today where larger specimens are worked into sand and smaller ones are resuspended and redeposited into muds in mud-traps, such as the Lambay Deep (Haynes, 1964). The recycling of older material, together with penecontemporaneous reworking of the cold water species, is consistent with

meltwater pumping towards ice margins. This faunal data generally supports the event stratigraphy which emphasises the role played by retreating tidewater glaciers and high relative sea level during deglaciation of the basin. At a detailed level it is argued that the processes of rain-out, suspension sedimentation and density currents which are responsible for the largely conformable sedimentary sequences are also in sympathy with species distribution and species ecology. Rapid facies changes are inferred to mark changes in local depositional processes which are consistent with reworking along tidewater margins and explains the absence of any clear vertical zonation in biofacies assemblages. Comparisons with microfaunas that accompanied the rapid deglaciation of the Svalbard–Barents Sea margin suggests that the dominance of *Elphidium clavatum* assemblages in the Irish Sea Drifts is a signature for a major meltwater event(s). There is evidence for substantial fluxes of meltwater into the basin during deglaciation including regional expression of subglacial erosional marks, ice-directed channels and tunnel valley systems, subaqueous spreads and major incisions on the sea floor (Fig. 6.4) (Wingfield, 1989). The presence of large meltwater fluxes during deglaciation is not in doubt but the controlling mechanisms may involve either climate change or rising sea level that progressively broke the seals around subglacial meltwater reservoirs, resulting in catastrophic drainage to successive ice-marginal positions.

The second association occurs in muds that form part of a stratigraphy that records ice-free and ice sheet readvance events towards the end of the deglacial cycle. These muds tend to be red in eastern County Down and grey in Dundalk Bay and in south County Down. The mud beds occur below, within and above glacigenic deposits (Fig. 5.4). The microfaunas are dominated (85–95%) by the foraminifera *Elphidium clavatum* and the ostracod *Roundstonia globulifera* (5–10%) that shows intact instars. These are opportunistic biocoenoses occurring in contemporary Arctic-subarctic areas recently vacated by tidewater glaciers (Hald et al., 1994). Accessory forms include miliolids, polymorphinids, Lagena, *Quinqueloculina seminulum* and distinctly cold to very cold water forms including *Cytheropteron dimlingtonense* and *C. montrosiense*. Shell tests are characterised by intact glossy preservation together with a range of sizes and an absence of temperate forms. In the case of the foraminifera the final aperture is often intact. These features together with fragile foraminifera species (*Pseudopolymorphina novangliae, Legena clavata, Oolina/Fissurina* spp.) and articulate valves of *Cytheropteron* species with a range of juveniles show that these microfaunas represent *in situ* biocoenoses. These *in situ* microfaunas, which are compositionally distinct from other mixed assemblages that are commonly redeposited at ice sheet margins by reworking, therefore provide an opportunity to develop a ^{14}C chronology for sea level fluctuations and ice sheet readvances in the northern part of the basin. In most cases the radiocarbon dates are from monospecific samples of *E. clavatum*. An important aspect of this dating approach is that together with the lithostratigraphy it provides a coherent timeframe for palaeoenvironmental changes occurring before, during and after ice sheet readvances in the north Irish Sea area (Fig. 5.4). Generally these marine muds are marine rather than glaciomarine and contain only occasional pebbles with little evidence for ice-rafting.

Lateral margins of the Irish Sea Glacier

Farrington (1949) recorded till associated with ice advance from the Irish Sea Basin 24 km inland on Saggart Hill and erratics of Irish Sea provenance occur up to 550 m on the northeastern Wicklow mountains. These limits were generally regarded as dating from the penultimate glaciation though cosmogenic dates from the Wicklows suggest that most surficial landforms postdate 18–19 kyr cal BP when deglaciation began (Ballantyne et al., 2006). Distributions of granitic erratics from the mountains and erratics of Irish Sea provenance clearly show that during the last glacial maximum the Wicklow ice dome was encircled by and confluent with thick ice moving from the Irish Midlands and Irish Sea basin. The combined ice flows moving along the eastern flanks of the mountain were ~700 m thick and formed the lateral margin of the large composite Irish Sea Glacier. Directional indicators south of Dublin show consistent and powerful ice flow southwards subparallel to the present coast into the Celtic Sea which may also reflect the presence of ice within the mountains (Fig. 6.1). The position of the zone of confluence between mountain and Irish Sea ice has been used to reconstruct the later ice sheet decay patterns on the eastern flanks of the mountains. These reconstructions depict continuous ice sheet margins based on inferred links between isolated gravel and deltaic deposits and overflow channels (Charlesworth, 1928b, 1937; Warren, 1993). Furthermore, it is implicit in these models that the ice margin downwasted and backwasted in a controlled, regular manner. Static models of this type do not explain the field relationships between the sedimentary fills in basins on low ground to the east of the mountains and the series of spectacular meltwater channels incised into bedrock highs between Bray and Gorey (Fig. 1.4). A series of deltas composed largely of gravelly foreset beds around 300 m OD between Glencullin and Glencree were deposited by meltwaters draining south and east from the mountains into lakes formed between the Irish Sea glacier and rising ground on the flanks of the mountain. The presence of coarse-grained, angular granite detritus and fresh joint blocks transported downslope shows that the mountain ice had withdrawn some distance while the Irish Sea ice occupied the low ground to the east of the mountains, late in the deglacial cycle.

At its maximum the marginal zone of the Irish Sea Glacier covered an irregular bedrock topography with a local relief of ~250 m with many topographic elements transverse to the regional north to south ice flow. The large meltwater channels which occur between Bray and Gorey are part of a regional meltwater system that cannot be explained by classical proglacial models but by ice sheet control on meltwater constrained beneath the ice sheet margin (Fig. 6.5). About twelve channels are characterised by similar morphologies, topographic settings and meltwater flow patterns forming a unique landform system for around 70 km (McCabe and O'Cofaigh, 1994). Channels are isolated cuts across bedrock, they are steep-sided resembling railway cuts, they have up and down long profiles, they have sudden open intakes and exits hanging above adjacent ground, they tend to maintain a fairly constant cross-profile and are often incised across bedrock ridges transverse to ice flow (Fig. 6.6). Several occupy and modify older large topographic lows. The most striking feature is that all channels record north to south ice-directed drainage and individual channels are separated by rock basins partially infilled with sediment. The main characteristics and north to south pattern of the linear suite of channels shows that they are subglacial in origin and are part of a regional ice-directed meltwater system (Hoare, 1976). The variable height of channels along the system is also consistent with a southward sloping ice surface and an up and down hydraulic gradient. The overall position of the channel system may be related to subglacial macroscale boundary conditions where the margin of the glacier encountered rising ground and rock elements transverse to ice flow. Fast-moving ice further out in the basin encountered little resistance to movement and was separated from slower-moving ice

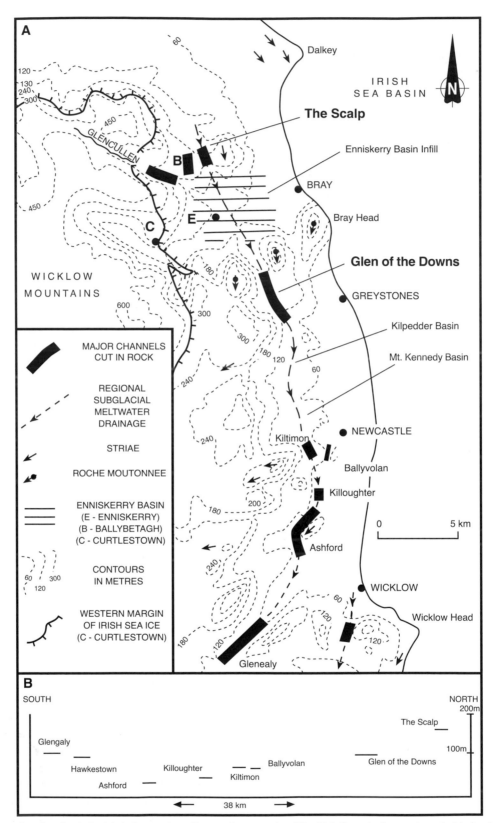

Fig. 6.5 Pattern of regional subglacial drainage along the western margin of the Irish Sea glacier, east-central Ireland. The north–south aligned subglacial channels are separated by topographic basins. Lower diagram shows the longitudinal profile of the system (after McCabe and O'Cofaigh, 1994).

Fig. 6.6 The Scalp subglacial meltwater channel cut across the east–west Ballybetagh/Shankill ridge, at the northern rim of the Enniskerry basin, County Wicklow. The ice-directed channel transported limestone and granite boulders southwards across the subglacial delta at Killegar. See geology map in Fig. 6.7.

along mountain flanks. Resulting north–south subglacial shears probably occurred in this zone and focused initial channel development.

The operation of a subglacial meltwater system is also supported by the sedimentary infills which occur in the rock basins between the main rockcut channels (Fig. 6.7). The Enniskerry basin infill for example lies astride the northern part of the meltwater system and traditionally has been used for reconstructing discrete ice sheet events (Farrington, 1944). Because the height of the foresets (155 m OD) defining a lake level is much higher than that of the intake of the exit channel (100 m OD) on the southern basin margin (The Glen of The Downs) the lake cannot be interpreted in terms of proglacial logic. The sediments infilling the basin consist of subaqueous fan gravel and large-scale foresets overlain by gravel and sandy bedforms formed by density flows. These largely conformable sequences are often overlain by coarse-grained basal till, tectonised by southward moving ice and overthrust or deformed with a southerly sense of shear. The upper and deformed part of the infill is consistent with the concept that the deposits formed in a subglacial lake that reached the ice roof. This interpretation is strengthened by the transport of granite

eroded from the northern rim of the basin and transported from the Scalp subglacially in an esker channel across the lake sediments at Killegar (Fig. 6.8, 6.9). Significantly 60–80 percent of clasts are derived from Carboniferous outcrops at least 5–10 km to the north possibly associated with constant pumping into the basin along subglacial conduits. It is sometimes difficult to reconcile high energy channel erosion coeval with lower energy sedimentation. One explanation may be that basin sediments formed during low energy phases when the overall system was in decline. A more likely scenario suggested from southerly-directed palaeoflows in the basin is that through-flow is maintained along an energy gradient which to some extent is supported by the relative absence of fines in the basin (Fig. 6.8).

An interesting problem concerns subglacial lake development in this particular glaciological setting. Theory predicts that the location of a regional meltwater system should be related to the lowest point of a water equivalent line where it shallowed towards the ice margin along the flanks of the Wicklow mountains (Rothlisberger and Lang, 1987). It is therefore envisaged that initial lake genesis occurred to the lee of the Ballybetagh rock ridge which was transverse to ice

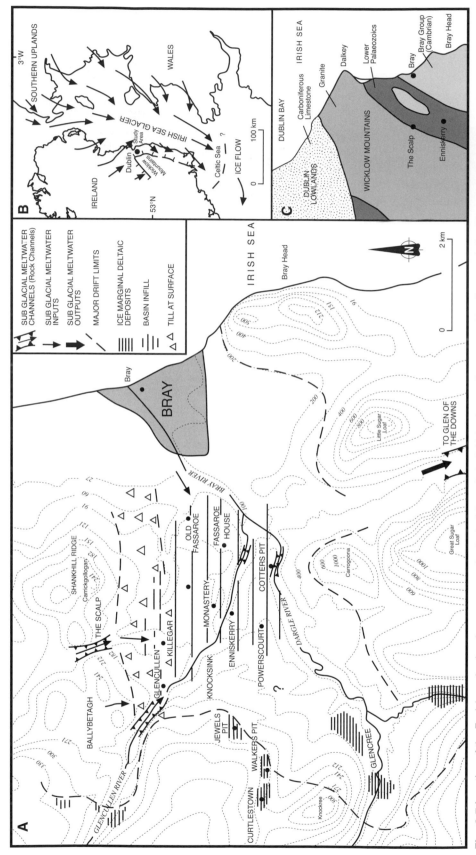

Fig. 6.7 Location of the Enniskerry basin, place names and extent of the subglacial lake. Insets show general ice flows and simplified geology of the basin (after Charlesworth, 1963).

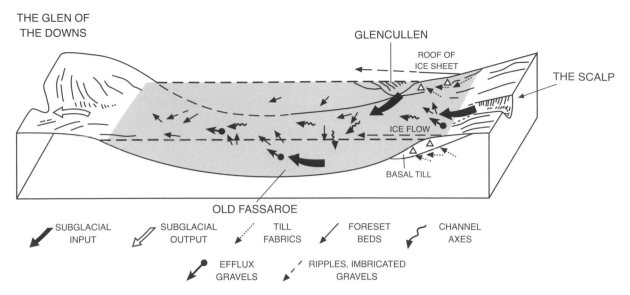

THE GLEN OF
THE DOWNS

GLENCULLEN

ROOF OF
ICE SHEET

THE SCALP

ICE FLOW

BASAL TILL

OLD FASSAROE

SUBGLACIAL
INPUT

SUBGLACIAL
OUTPUT

TILL
FABRICS

FORESET
BEDS

CHANNEL
AXES

EFFLUX
GRAVELS

RIPPLES, IMBRICATED
GRAVELS

Fig. 6.8 The main subglacial inputs and exits of the Enniskerry lake basin and the measured palaeocurrent directions within the basin which have either southerly or basinward-directed components (after McCabe and O'Cofaigh, 1994).

Fig. 6.9 Subglacial efflux sequence of granite boulder gravel eroded and transported by meltwater from the Scalp meltwater channel. The gravel is a tightly packed, largely openwork deposit with crude lines of boulders, boulder cluster bedforms, edge-rounded boulders and occasional pebbly drapes.

flow. Abrupt pressure contrasts on either side of the ridge effected lee side uncoupling on the assumption that the ice flux was sufficiently fast to prevent lee side closure. This is consistent with regional field evidence of marine downdraw effecting fast ice flow. The low pressure cavity would be expected to draw in water from subglacial sources which would enhance ice roof melt. Regional hydrostatic gradients may also enhance subglacial melting by vein and cavity formation into the main basin, the lowest point in the system. A corollary to the model is that enhanced ice deformation on the stoss (north) side of the Ballybetagh/Shankill ridge

would lead to vastly increased meltwater pressures, thus ensuring pressure conduit flow during incision of the Scalp meltwater channel (Fig. 6.6). This would be important because the Scalp is located along a structural weakness which was exploited by the high meltwater pressures and preferential conduit flow. The field evidence therefore provides the links between subglacial drainage into and out of a topographically or efflux defined basin and in all cases the erosional channels are typical of Nye-type channels.

This reconstruction of regional subglacial meltwater activity emphasises that the main channels are inlet and outlet Nye-type features reflecting ice-directed north to south meltwater pressure gradients beneath the margin of the Irish Sea Glacier. The pattern of field evidence does not support older concepts of overflow channels, proglacial channels and multiple glacial events. The field evidence requires vast, even catastrophic, fluxes of meltwater showing that the fluvioglacial landforms must date from the early deglaciation of this area after 22 cal ka BP. In a wider context this deglacial event including formation of prominent features such as the Glen of the Downs channel (cut in >120 m of rock and 1.5 km long), are high magnitude events and these are consistent with rapid ice wastage and the event stratigraphy in the rest of the Irish Sea Basin.

Deglaciation of the southern basin

The southern basin is defined from the offshore limits in the Celtic Sea in the south to the Boyne estuary in the north, a distance of 240 km. Exposure surface dates place the initial stages of deglaciation shortly after 22 cal ka BP (Bowen et al., 2002). Eyles and McCabe (1989a) argue from sedimentological evidence that tidewater glacier margins effectively removed ice from the basin in a few thousand years. Sediments recording deglaciation from this time slot are identified from eight coastal areas which are described from south to north.

1. The Nymphe Bank

The approximate southern limit of the Irish Sea Glacier at 51° 34′ N in the northern part of the Celtic Sea occurs where a thick (~10 m) till sheet ends and is replaced by isolated patches of drift to the south (Fig. 6.1) (Garrard, 1977). At this time the ice which moved south across the present coastline of Waterford and Wexford was coeval with the larger ice stream moving southeastwards down St. George's Channel (Fig. 10.6). On the Nymphe Bank there are unconformities or deep channels infilled with Pleistocene sediment which is generally massive but with laterally impersistent reflections at its base and more stratified towards the top (Delantley and Whittington, 1977). The channel morphology includes enclosed basins 10–20 km long and around 45 m deep which consistently trend NE–SW. They cross-cut underlying structural trends and are similar to subglacial channels. Channel orientation, characteristics and km scales all suggest they formed as the ice sheet began to decay when large volumes of meltwater were released. The scale of these features is consistent with catastrophic ice

sheet disintegration accompanied by marginal retreat and backfilling of the subglacial channelled topography by massive and stratified glaciomarine spreads. It is also evident that this type of ice decay may have been triggered by isostatic depression and rising sea levels at a time when the ice sheet flow lines were longest and sensitive to external influences, especially marine downdraw.

2. The Screen Hills delta

The largest (150 km²) tract of proglacial marine sediments exposed in the Irish Sea Basin occurs between Kilmuckridge and Curraghcloe, east county Wexford (Fig. 6.10, 6.11a–e). The Screen Hills is an area of rolling relief up to 70 m high which is often characterised by steep slopes, discontinuous ridges and enclosed depressions. Extensive coastal exposures show a general coarsening-upwards sequence of mud, sand, gravel and diamict influenced by ice-thrusting and gravitational loading (Thomas and Summers, 1984). A few kilometres south of the ice-contact delta complex the flat ground at Ely House is underlain by laminated to stratified sandy diamict peppered with clasts up to cobble size which deform the underlying stratification (Thomas and Summers, 1982). In many cases draped lamination or onlapped beds continues across the top of the larger clasts and records the action of traction or density currents immediately after the clasts were emplaced. These characteristics record a non-glacial environment dominated by density currents, sediment rain-out from a water column and dropstones from floating ice masses detached from shore ice or bergs. This body of water clearly extended south from the Screen Hills moraine towards the Celtic Sea and field relations between the Ely House deposit and the Screen Hills moraine to the north suggest contemporaneity. Because the ice pressure and ice flow came from the north it is difficult to see how an extensive freshwater lake could form in an area facing south into the Celtic Sea Basin. A more likely explanation is that the deposits at Ely House are glaciomarine and this is entirely consistent with current estimates of ice thickness farther north and attendant isostatic depression (Fig. 6.12).

Thomas and Summers (1983) identified four major stratigraphic elements from coastal exposures (Fig. 6.10). Lowermost marine muds are massive, commonly weakly laminated, contain lonestones and a mixed assemblage of marine microfaunas characterised by different depth requirements, degree of preservation,

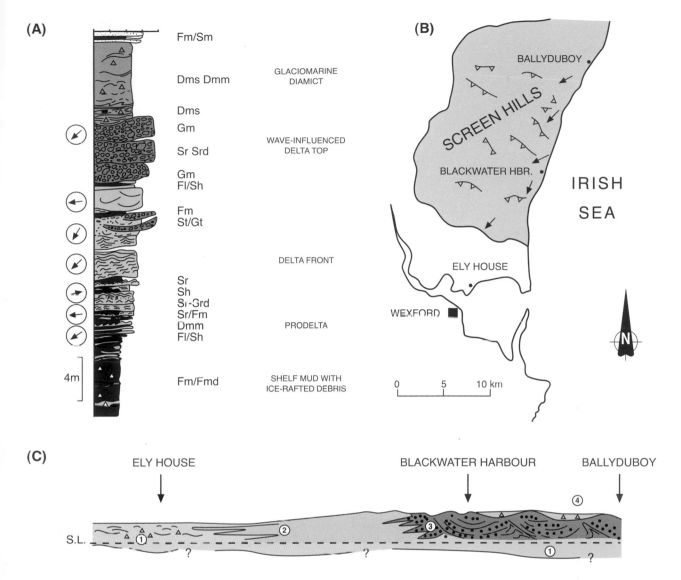

(A)

Fm/Sm

Dms Dmm — GLACIOMARINE DIAMICT

Dms
Gm

Sr Srd — WAVE-INFLUENCED DELTA TOP

Gm
Fl/Sh

Fm
St/Gt

DELTA FRONT

Sr
Sh
Sr-Grd
Sr/Fm — PRODELTA
Dmm
Fl/Sh

4m

Fm/Fmd — SHELF MUD WITH ICE-RAFTED DEBRIS

(B)

BALLYDUBOY

SCREEN HILLS

BLACKWATER HBR.

IRISH SEA

ELY HOUSE

WEXFORD

0   5   10 km

N

(C)

ELY HOUSE   BLACKWATER HARBOUR   BALLYDUBOY

④

S.L. ①   ②   ③   ①

?   ?   ?

Fig. 6.10 Generalised lithofacies log from Screen Hills, southeast Ireland showing a coarsening-upward shallow marine sequence deposited on the margins of a southward prograding, glacier-fed braid delta, capped by proglacial diamict facies (after Eyles and McCabe, 1989a). Section is deformed by large-scale gravitational loading structures and ice thrusting. Location map showing ice thrust ridges is redrawn from Thomas and Summers, 1984. Schematic stratigraphy shown between Ballyduboy and Ely House.

evidence of transport and both temperate, cosmopolitan, cold-water and extinct species (Thomas and Kerr, 1987). They noted that the cold faunas were less abraded than the other specimens which may suggest that part of the microfauna is *in situ*. It is significant that Thomas and Kerr (1987) observed that the muds grade into and are conformable with the overlying coarsening upwards sequence (>12 m) of laminated silts and clays, laminated and rippled fine sands and thin beds of diamict exposed at Knocknasillogue (Fig. 6.11a). This element is indicative of prodelta sediment with abundant ice-rafted debris because it contains climbing and superimposed

ripples deposited from density currents or quasicontinuous turbidity currents, bidirectional cross-beds, lonestones, wave-ripples and trace fossils. These facies grade into wedges of pebbly-sand and gravel that tend to occur at four or five points along the exposure. The gravelly foresets beds dip southeastwards and record the position of discrete inputs of meltwater from subglacial effluxes which prograded over the prodelta area (Fig. 6.10).

The presence of localised thrusting within the sediment package is a function of local ice-marginal pushing rather than extensive polyphase deformation (Thomas

Fig. 6.11a Weakly-laminated diamictic mud formed by rain-out, ice-rafting and subsidiary bottom current activity, Screen Hills, County Wexford.

Fig. 6.11b Stratified sandy diamict formed from debris flows and ice-rafting, Screen Hills.

Fig. 6.11c Laminated diamict containing lonestones showing evidence for soft sediment deformation and sagging, Screen Hills.

Fig. 6.11d Pebbly gravel loaded into rhythmically bedded silt and sand resulting in water escape structures.

Fig. 6.11e Sheared diamict from the ice-contact part of the Screen Hills delta, County Wexford.

and Summers, 1984). Locally the gravelly foresets grade into tabular bodies of trough-bedded sand and massive gravel which appear to be the product of the migration of megaripples under unidirectional flows above wave base (Bourgeois and Leithold, 1984). These facies interfinger with braided-river gravel recording transport southwards. The thickest braided-river gravel sequences often form broad and asymmetrical sag basins recording gravitational loading of rapidly-deposited gravel prograding over marine muds. Diapiric deformation of the mud is closely associated with the loading structures. The succession is completed by a coarse-grained, sandy diamict supporting a highly variable number of clasts. The diamict is interbedded with gravel facies, is non-erosive at its base and shows a well developed drape geometry that thickens in topographic lows. It is massive but is made up of amalgamated beds, contains weakly-ordered clast fabrics and randomly distributed steeply-dipping clasts. These characteristics can be attributed to the rain-out of sandy suspended sediment from efflux jets, close to a tidewater margin and related reworking by marine currents and downslope resedimentation. Because the diamict occurs as discontinuous lenses and sheets across the top of the delta complex it probably records the partial submergence of the braid delta

Fig. 6.12 Laminated diamict formed mainly by density currents Ely House, County Wexford. Note the presence of lonestones dropped from floating ice indicating the presence of open marine conditions several kilometres immediately south of the Screen Hills delta.

and renewed ice-proximal sedimentation across a fan surface of variable relief. The drape-like stratification of the diamict and thickening in topographic lows is a function of sediment gravity flows rather than ice advance and subglacial till deposition on the delta surface. Overall the local patterns of ice thrusting and sediment deformation are typical of an ice-contact delta where sediment accumulation buttresses the ice margin leading to a decrease in ice wastage, followed by minor advances onto ice-proximal slopes. These events increase local relief and promote sediment gravity flow away from topographic highs, the facies pattern noted across the surface of the delta.

3. Morainal bank at Greystones

Most work on the records of dissipation of Irish Sea ice from the eastern flanks of the Wicklow Mountains use models of ice dammed lakes and overflow channels for reconstruction (eg. Charlesworth, 1937; Warren, 1993). Charlesworth (1937) reconstructed ten continuous lines depicting ice downwasting by joining channel positions, gravelly mounds and deltas between Arklow and Bray. This methodology is flawed because it is based entirely on morphology and presupposes that ice wastage was regular and continuous. The absence of moraines and the presence of restricted glacigenic deposits including diamicts and muddy sediment have been noted along the coast of east Wicklow north of the Screen Hills suggesting that rapid ice retreat occurred. Exposures of diamicts between Cahore Point and Clogga have been interpreted as older drift largely because they contain granite erratics derived from the Leinster Mountains to the northwest or that they contain muddy sediment (Synge, 1964).

The large sedimentary apron (2 km²) on the south or lee side of the large rock ridge (240 m OD) at Bray Head, County Wicklow is significant because it is the only major depocentre recording a halt in ice retreat between The Screen Hills and Bray (Fig. 6.13). Its position immediately to the lee of the rock ridge suggests that it is a re-equilibration feature formed when the ice margin halted or pinned on the rock ridge. Sedimentary variability and the critical position of the spread emphasise this close genetic link between glacigenic deposition and the position of the ridge. Four major lithofacies associations (LFAs) (mud, sand, diamict/gravel, gravel) are identified on the basis of sedimentary contacts, two-dimensional sediment geometry, range of facies, facies relationships, vertical position of facies and dominant sedimentary characteristics (Fig. 6.14). Analysis of these facies indicates sediment accretion to the lee of the ridge from a subglacial point efflux in a glaciomarine

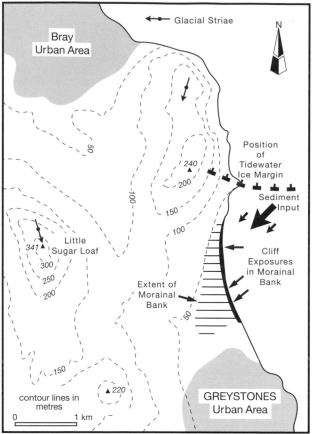

Fig. 6.13 Topographic setting of the moraical bank exposures north of Greystones, County Wicklow. The moraical bank is sited to the lee of the rock ridge of Bray Head and the inferred location of meltwater and sediment input. Most of the area above 100 m is drift free and ice scoured.

setting (Fig. 6.3) (McCabe and O'Cofaigh, 1995). Accessory erratics (Ailsa Crag microgranite, Cretaceous chalk, flint, tertiary basalt) in all facies are consistent with a general northern provenance as well as the dominant limestones and quartzites.

LFA1 occurs at the base of the 1.6 km long exposure consisting of interbedded massive mud, laminated mud and muddy diamict (Fig. 6.15a,b). Subtle changes in grain size, sandy partings and pebble lags along scour surfaces are pervasive and testify to current winnowing at a variety of scales during plume sedimentation. Horizontal surfaces cut across the mud bank are overlain by chaotic, coarse-grained gravel lenses containing edge-rounded mud clasts. The stratigraphic position of the gravel entombed in mud indicates rapid deposition from hyperconcentrated sediment flows as the mud bank grew. Concurrent scouring of the mud occurred by jet flow extension across the mud bank producing mud clasts and gravel deposition by fallout from the plug flow

of the jet. LFA2 consists of parallel laminated sand (70%) and laminated mud infilling depressions on the eroded mud surface. Lateral continuity of facies suggests deposition from meltwater-driven, high-density sandy flows. The presence of pebbles and thin lenses of diamict may record local gravity flow off topographically higher parts of the mud bank. LFA3 is a tabular drape of diamict (80%) and gravel resting on LFAs 1 and 2. The stacked diamict beds are texturally variable showing rapid lateral facies transition with massive pebbly gravel beds. Repeated gravelly lags show that deposition occurred within shallow channels. Sedimentary features in the diamicts including massive/crudely-stratified stacked beds, outsized clasts, projecting clasts, and weakly-developed fabrics are typical of debris flows. Interbedded massive gravel represents weakly-channelised flows with some local turbulence and size sorting. LFA4 occurs within channels at the top of the sequence as low angle, cross-stratified, mainly pebble gravel. The gravel is crudely stratified though locally cobble clusters show long-axis imbrication transverse to flow. Collectively the gravels represent mass flow in channels with some higher energy flows showing evidence for traction current activity. The entire sequence is subaqueous in origin and did not evolve with shallowing into a Gilbert-type delta probably because it formed rapidly. The absence of any grounding line ice-push structures and boulder gravel also shows that the ice grounding line remained pinned on the rock ridge to the north with the variable meltwater jet operating to the lee of the ridge.

Marked sediment aggradation along tidewater ice fronts occurs best at halts during ice-marginal recession when basal networks can establish, enlarge and focus sediment point sources. The sediments at Greystones are therefore dependent on ice-marginal re-equilibration along the Bray Head ridge. Analogy with rates of moraical bank growth from Alaska suggests that debris supply was restricted possibly to tens of years or less (Powell, 1990). The model of rapid growth of the morainic shoal, the absence of glaciotectonic structures and a delta surface supports the idea of a halt in the ice margin without climatic control. Sedimentary variability within the sequence relates to changes in the dominant processes which are related to changes in the strength and activity of the efflux jet. LFAs 1 and 3 comprise most (>90%) of the sequence including deposition of a mud bank followed by stacked beds of diamict/gravel. Muds and muddy diamicts of the mud bank represent initial deposition from turbid plumes and ice rafting whereas

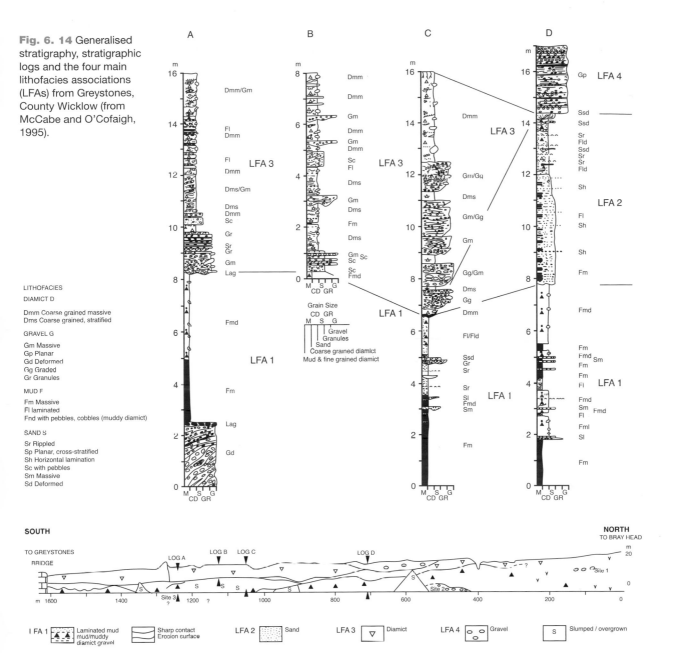

Fig. 6. 14 Generalised stratigraphy, stratigraphic logs and the four main lithofacies associations (LFAs) from Greystones, County Wicklow (from McCabe and O'Cofaigh, 1995).

LITHOFACIES

DIAMICT D

Dmm Coarse grained massive
Dms Coarse grained, stratified

GRAVEL G

Gm Massive
Gp Planar
Gd Deformed
Gg Graded
Gr Granules

MUD F

Fm Massive
Fl laminated
Fnd with pebbles, cobbles (muddy diamict)

SAND S

Sr Rippled
Sp Planar, cross-stratified
Sh Horizontal lamination
Sc with pebbles
Sm Massive
Sd Deformed

the interbedded diamict/gravel facies are a response to unstable, quasicontinuous sediment pulses associated with the plug flow of a jet. It is the continuity and rapid facies changes within the diamict/gravel facies that points to rapid change in the nature of sediment supply, possibly related to tunnel blockage and collapse, accelerated calving, mass flow of heterogeneous debris downslope and an unstable jet. Therefore the stratigraphic couplet of fine-grained sediments overlain by coarse-grained diamicts can, in certain topographic settings, provide a signature for ice-marginal re-equilibration followed by rapid decay.

4. Glaciomarine Facies within subglacial tunnel valleys, Killiney Bay

For around 150 years facies variability from the extensive coastal exposures between rock headlands at Killiney and Bray has led to a plethora of methods to subdivide and therefore explain the glacigenic succession (Mitchell et al., 1973; Hoare, 1977). Interpretations range from the traditional tripartite division (lower and upper till separated by stratified sediment) of the Geological Survey (Synge, 1963) to more complex schemes involving up to eight tills recording ice sheet oscillations (Hoare, 1977). However, these interpretations do not

Fig. 6.15a Conformable facies sequence of mud, diamict and gravel from the morainal bank at Greystones, County Wicklow. Note that the surface of the muddy diamict at the base of the section is channelled and overlain directly by stratified sediments.

Fig. 6.15b Detail of laminated and massive mud beds from the base of the morainal bank at Greystones.

explain the channelled sediment geometry, interbedded facies and facies continua, facies variability, the position of the sediments within a topographic low and the wider relationship of the sequence to an adjacent tunnel valley system immediately offshore (Figs. 6.4, 6.16). These characteristics have been resolved into a model which depicts deposition along a western tributary of the tunnel valley system which extended inland to Ballybrack (Fig. 6.17). Using sedimentological criteria the sequence is resolved into four lithofacies associations recording meltwater pumping of heterogeneous sediment along the tunnel valley towards tidewater (Fig. 6.18) (Eyles and McCabe, 1989b).

LFA1 consists of gravel-filled, cross-cutting channels banked against the bedrock sidewall of the Ballybrack channel below Killiney station. Channels are infilled by conformably stacked beds of massive and crudely-bedded boulder gravel and massive sand. B-axis imbrications record palaeoflows from the west and the stacked U-shaped channel margins recording cut and fill are very similar to rapid vertical aggradation found in closed esker conduits (Gustavson and Boothroyd, 1987). The deformation structures observed at the base of the section are similar to structures produced by gravitational foundering in response to high rates of sedimentation and repeated cutting and filling. LFA2 consists of weakly-laminated and massive muds containing isolated clasts, steeply-dipping clasts and clast clusters forming the base of the exposed section. Similar fine-grained facies are traditionally identified as tills even though they are stratified, contain steeply-dipping clasts and outsized clasts. These attributes are consistent with fallout from sediment plumes and delivery of larger clasts from ice-rafting (Mackiewicz et al., 1984). LFA3 infills an undulating erosion surface cut across LFAs 1 and 2. It is discontinuous with laterally-impersistent beds of sand/cobble admixtures and cross-bedded gravel. A range of massive, laminated and rippled sands with some muddy horizons containing lonestones infill shallow channels or drape gravelly highs. The intimate association of lonestones, sandy drapes, muddy beds, cross-beds and more massive or disorganised gravelly beds indicates a subaqueous origin with some ice-rafting. LFA4 consists of interbedded weakly-graded gravel, matrix-rich gravel and massive diamict facies contained within broad overlapping channels. These attributes, together with a variety of grading patterns in coarse gravel, are explained by reference to a sediment gravity flow origin for all these facies (Walker, 1975; Postma, 1984, 1985). Interbedded matrix-rich gravel, graded gravel and diamict can be attributed to downslope slumping with the transformation of debris flows into fully turbulent flows as a consequence of dilution and acceleration downslope. (Postma, 1985). Large (~1 m across) clasts within diamict beds were either freighted during flow or were derived from floating icebergs. Perhaps the most striking features of the sequence are the gravel-filled dykes (3–4 m high) which penetrate the diamict beds (Fig. 6.19). Some dykes radiating from the base of gravel beds indicated simultaneous loading and injection in response to rapid emplacement of gravel on a saturated diamict substrate. Adjacent dewatering structures also testify to rapid accumulation of poorly-sorted, saturated and unstable sediments on steep depositional slopes.

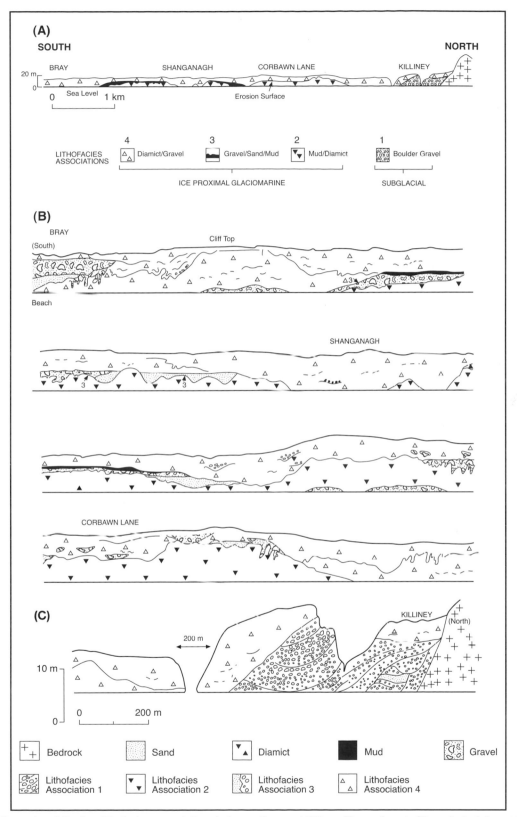

Fig. 6.16 Geometry of the four lithofacies associations between Bray and Killiney. The sediments fill a subglacially-cut tunnel valley (the Ballybrack valley). The precise age relationship between lithofacies association 1 and the other facies is not known with certainty (after Eyles and McCabe, 1989b).

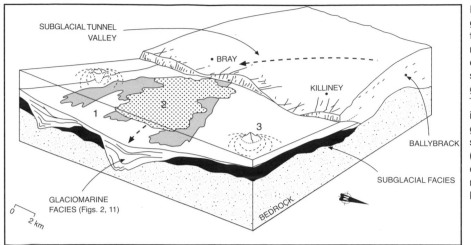

Fig. 6.17 Depositional model for glaciomarine sediments infilling the Ballybrack tunnel valley. Rapid retreat of a tidewater ice margin by calving followed a rise in relative sea level generated by regional glacio-isostatic downwarping. Tunnel valleys exposed by rapid ice retreat were infilled by sediment gravity flow facies (1), mud from suspended sediment plumes (2) and ice-rafted debris (3). The entire sequence was subject to modification by water escape and localised ice thrusting.

DIAMICT
Dmm   Matrix-supported massive
Dms   Matrix-supported stratified

SAND
Sh   Horizontally-laminated
Sr   Ripple cross-laminated

GRAVEL
Gm   Massive
Gp   Planar cross-stratified
Gg   Graded

MUD
Fm   Massive
Fmd   Massive with dropstones

Fig. 6.18 Graphic logs from sites A, B and C shown in Figure 6.16. Arrows represent palaeocurrent directions and numbers identify lithofacies associations shown in Figure 6.16.

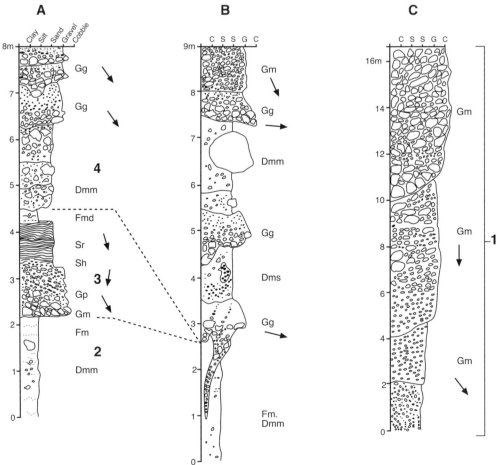

Fig. 6.19 Large gravel-filled dykes at the base of crudely-bedded gravels emplaced partially by loading. Some of these features acted as escape routes for water trapped in sediments lower in the succession. Shanganagh, County Dublin.

The valley systems in this area together with channels in Dublin Bay (-120 m OD) converge with a major NNE–SSW tunnel valley system (~100 km long) offshore from Lambeg Deep to the Wicklow trough (Fig. 6.4) (Whittington, 1977). Overdeepening of these valleys into bedrock records erosion by powerful meltwater streams under high hydrostatic pressure. The sedimentary characteristics of the four lithofacies are consistent with rapid emplacement of heterogeneous sediments by meltwater driven episodes. These were of a pulsatory nature because of the number of erosion surfaces and channelled geometries in the exposure. Drapes within the deposits are indicative of short breaks in deposition whereas large loading structures record rapid deposition. These changes are consistent with the drastic changes which were taking place as the Irish Sea Glacier withdrew, resulting in episodes of sedimentation and erosion. The local depositional environment may have been akin to a glacier-fed fjord-head delta subject to continuous downslope slumping. The description of a Pliocene fan delta that prograded by mass flow processes is highly appropriate and suggests that a similar

depositional environment occurred at the exit of the tunnel valley (Postma, 1984, 1985). It is argued that the glaciomarine section preserved in the Ballybrack valley is representative of the larger depositional system and offshore seismic records described by Whittington (1977). The general channel orientation and flat-lying reflectors of seismic sections show no evidence for steep depositional surfaces that may record moraines or halts in the ice retreat, and indicate that channels were backfilled from the northwest from ice margins close to the present coast. Sediment delivery on this ice sheet scale must have involved resedimentation by meltwaters which is consistent with the common presence of northern erratics in all four facies associations.

5. Moraine ridges in north County Dublin

Fluvioglacial sediments deposited during the withdrawal of the last ice sheet across the lowlands of north County Dublin are discontinuous and range from isolated ridges to subdued gravelly spreads (Hoare, 1975). Hoare (1975) proposed four stillstands by joining isolated deposits suggesting that the ice margin trended approximately west to east. This conclusion is supported by ice flow indicators which show that the last ice sheet from the north Irish lowlands deposited limestone tills across this coastal sector (Stephens and Synge, 1958; Hoare, 1975). Below this till sheet as far north as the Boyne valley multiple till facies containing marine shell fragments and erratics of northern Irish Sea provenance are commonly preserved inland to Slane. The age of this onshore ice advance from the Irish Sea basin is unknown.

6. Emergent sequence at Skerries

Along the margins of the Irish Sea Basin glacigenic sequences preserved or deposited in topographic lows tend to show a wide range of facies types, often intricately interbedded. At Skerries a thick (>12 m) glacigenic sequence containing three major facies associations is restricted to a narrow north–south topographic low between Hampton Hall and Barnageera House. LFA1 consists of massive clast supported diamict beds that grade into laterally extensive beds of cobble and pebble gravel. Sedimentary structures including a wide range of grain sizes, stacked/amalgamated beds, massive structure, isolated large clasts, inverse grading patterns, clast projections, coarse-tail grading and clast clustering are all consistent with a debris-flow origin (McCabe et al., 1990). The amino acid ratios for *Arctica islandica* shell fragments contained in the gravels show that the assemblage is mixed, ranging in age from Late Devensian to Oxygen Isotope Stage 5e or earlier. However, the youngest ratios and the presence of Stage 3 elements suggest that shells living in the Irish Sea around 22 cal ka BP were incorporated into the shelly gravel (McCabe et al., 1990). LFA2 drapes the gravel and consists of massive and laminated mud with discontinuous wisps and laminae of sand. Like the underlying gravel, the mud contains a mixed marine microfauna showing evidence of reworking by meltwater transport. The boreo-arctic element tends to be less abraded than the thicker walled temperate microfauna. The stratigraphic position and tabular geometry of the mud is similar to ice-proximal mud deposited from sediment plumes after the ice margin vacated the site. A sharp erosion surface across the mud is marked by a boulder lag and amalgamated beds of poorly-bedded cobble and pebble gravel (LFA 3) contained within shallow scour-like features. LFA3 therefore records winnowing, a coarsening upwards trend and mass flow of gravelly sediment. The most likely source for these gravels is the dissected fan between Barnageera House and Hacketstown immediately to the west of the present coastline. The presence of discrete openwork gravel beds may record wave sorting, and resedimentation and poorly-sorted gravel within scours were probably formed by mass flow and mixing along channel scours following storm events.

This facies sequence (debris flow/mud drape/shallow-marine gravel) is similar to that described from glacio-isostatically depressed areas that have experienced strong rebound after ice retreat (Andrews, 1978). The lowermost diamict and gravel assemblage is typical of disorganised, interbedded coarse-grained sediments recording ice-marginal retreat and reworking of glacial and local angular debris. As much as 100 m of isostatic uplift could occur in a few thousand years or less and elevate deepwater muds into high-energy, wave-dominated settings where they are truncated by erosion and overlain by coarse-grained shoreface gravels which may have a fan delta origin. Overall this sediment spread records withdrawal of the tidewater ice margin northwards in this part of the basin, which is consistent with relative sea levels higher than present near the exit of the Delvin channel a few kilometres to the north on the border of Dublin and Meath. The stratigraphy and absence of evidence for distinct readvance moraines in this area is also consistent with continued tidewater ice wastage and formation of localised sediment spreads in topographic lows.

7. The Ben Head moraine

This moraine can be traced continuously from the coast of eastern county Meath westwards and then southwestwards along the south side of the upper Boyne river to Summerhill and beyond (McCabe, 1973). At Summerhill this NW/SW ice limit is associated with deposition in ponded water and formation of the Trim eskers (Synge, 1950). On the coastal lowlands the east–west trending feature comprises a moraine ridge bounded by steep ice-contact slopes to the north and an outwash spread sloping south to Gormanstown. The top of the moraine consists of coarse-grained, poorly-sorted gravels thrust along a nested set of shear planes which rise steeply southwards indicating ice pressure from the north (Fig. 6.20). Large-scale diapiric muddy intrusions through the outwash and intrusive dykes within the tills suggest that the entire sequence of underlying tills and stratified deposits reflects the presence of deeper seated glacio-tectonic structures. The lower parts of the outwash consist of parallel laminated silts and clays indicating

Fig. 6.20 Coastal exposure in the Ben Head moraine, County Meath. The lower sediments have been truncated probably by the sea and are overlain by gravel and sand. These sediments grade distally into rhythmites but have been deformed into a suite of southward-rising shear planes at Ben Head which constitute the crestline of the moraine. This moraine continues inland to the Hill of Tara and Galtrim.

deposition in ponded water. No *in situ* marine micro-faunas were identified though the current coastal exposures were probably kilometres removed from the more easterly parts of the original depositional system out in the basin.

The lateral continuity of this moraine west into the Irish Midlands suggests it represents an important halt in the general ice retreat northwards. Its eastern position in a coastal lowland free of rock control, its location backed against rising ground around Tara Hill to the west and the influence of glaciotectonic thickening on the coast suggests it represents a readvance of the ice sheet margin during deglaciation of the western Irish Sea Basin. Most sedimentological and faunal indicators currently available suggest that the preserved part of the moraine is terrestrial rather than glaciomarine. If this is the case then it is argued that an increase in the rate of isostatic uplift resulted in a fall in relative sea level which led to a decrease in the loss of mass at the ice sheet margin. In addition there is very little evidence for significant climate forcing at this time (~20 cal ca BP) from North Atlantic records.

8. The Donacarney moraine

A large NW/SE trending ridge composed of gravelly debris occurs on the south side of the lower Boyne estuary. It seems to be associated with the final ice readvance from the Irish Sea Basin into the lower Boyne valley. The position of the moraine across the estuary and its orientation is different from the earlier west/east ice limits at Ben Head suggests a restricted ice advance

from the basin itself rather than from an Irish source. This interpretation is further strengthened by the fact that the areas of coastal county Louth and lower Boyne valley to the north are underlain by till facies of Irish Sea provenance. It is also notable that multiple till facies occur within the rock gorge of the lower Boyne immediately north of the limiting moraine at Donacarney. At least four different till facies are banked against the southern wall of the Boyne at Drogheda. These tabular beds are separated either by undeformed stratified deposits or by sharp planar erosional surfaces. Because they are derived from different dominant bedrock lithologies the facies may record several minor oscillations of the ice from directions between east and north.

After the ice wasted from the Donacarney limit, meltwater could escape from the Boyne catchment directly into the Irish Sea basin. The multiple terraces along the Boyne developed at this time and formed moraine/outwash couplets or delta terraces into a high relative sea level between 7 and 19 m OD. Prior to this stage meltwater was blocked from escaping east down the Boyne valley and was diverted south into the upper Nanny valley forming the spectacular rock-cut gorge at Slanduff. The small raised delta at the exit of the Nanny channel and outwash at Mosney records part of this event. It is not known whether or not the ice advance to Donacarney was glaciomarine or terrestrial. However, there is little doubt that the ice readvanced southwestwards over the sea bed and incorporated vast amounts of sea flood muds and shell banks. Theory predicts, however, that at this stage of deglaciation when about two thirds of the ice sheet had already disappeared, isostatic uplift was rapid and a sea level lowstand was already developing immediately prior to the well-documented global eustatic sea level rise 19,000 years ago (Clark et al., 2004).

Deglaciation model for the southern basin

Subglacial bedform patterns across the north Irish lowlands show that remanie ice domes remained after the ice lost over two thirds of its mass and extent and vacated the coasts of the Irish Sea Basin (Fig. 4.10). This widespread event involving ice-marginal retreat of 300 km from its southern limits in the Celtic Sea into the Ulster hinterland is now known as early deglaciation of the BIIS. Exposure dates from Waterford show deglaciation occurred after 22 cal ka BP though exposure dates from the Wicklow Mountains to the north suggest a later start. A global meltwater pulse recorded from muds within channels cut during an earlier lowstand show conclusively that the coast of southeast Ulster was deglaciated before 19 cal ka BP. This timeframe of less than 3000 years supports the original hypothesis of Eyles and McCabe (1989) that ice vacated the basin in a few thousand years. The main driving mechanism effecting ice wastage was fast ice flow towards tidewater ice sheet margins which emphasises the role played by isostatic disequilibrium. There is no doubt that a general climatic warming associated with an increase in meridional overturning in the North Atlantic Ocean also contributed. The sedimentary and geomorphic evidence also supports rapid ice wastage. The presence of ice-contact spreads and the general absence of linear push moraines strongly points to catastrophic ice retreat. Most thick ice-contact glacigenic sequences in the western basin occur to the lee of rock ridges which acted as pinning points that stabilised the ice margin and increased sedimentation from point sources. Exceptions to this landform pattern occur in the lower Boyne valley and Meath areas where small push moraines are traceable across country. These elements have formed quite late in the deglacial cycle and probably reflect increasing amounts and rates of isostatic uplift which caused a substantial drop of relative sea level in the northern basin and decreased ice loss. By this time much of the Irish Sea Basin must have emerged forming shallow muddy marine sea floors possibly with scattered islands. However, the BIIS had withdrawn inland before 20,000 years BP and glacigenic sedimentation was replaced by paraglacial sedimentation and sediment reorganisation in the southern basin.

Early deglaciation of the northern basin

Introduction

Ice-directional indicator patterns show that the Irish Sea Basin was a conduit which drained adjacent land areas and did not sustain any independent centres of dispersal. As such, the ice stream was generally thinner than on adjacent land masses and therefore very susceptible to any change either in ice supply or wastage. This sensitivity and rapid patterns of ice wastage which characterised the southern basin still operated when ice margins contracted into the northern basin, because the same climate and dynamic factors driving ice sheet recession and early glaciation still operated. However, the presence of thick glacigenic successions show that extensive depocentres developed on basin margins because ice wastage slowed in response to shorter flow lines and topographic constraints. In the Mourne Plain and on the coast of south County Down these depocentres have a high preservation potential because they have not been eroded by the later ice sheet readvances from the north Irish lowlands which reached the flanks of the Mourne mountains (Figs. 5.11, 6.21). The glacigenic sequences of the Mourne Plain belong to the early deglaciation phase because they are cut by channels formed during a sea level lowstand response to rapid isostatic uplift. Mud infilling the channels is ~19,000 years old and is a result of global eustatic sea level rise (Clark et al., 2004).

The Mourne Plain occurs on the low ground between the southern margin of the Mourne Mountains and the extensive seacliffs cut in glacigenic deposits around Ballymartin and Kilkeel (Fig. 6.21). During early deglaciation there was no sediment in this basin because ice flows from the Irish midlands, through Carlingford Lough and cols in the Mourne Mountains, and from the northern Irish Sea Basin were coeval and generally moved southwards along the axis of the basin (Stephens et al., 1975). The basin is unique because the topographic high ground of the Mourne mountains must have influenced the configuration of the ice sheet margins as the ice wasted (McCabe, 1986). Initially, it was the separation of the ice sheet into three masses, which retreated west towards Carlingford Lough, north into the mountains and northeastwards on the margins of the Irish Sea basin, that formed successive margins to the depocentre

Fig. 6.21 The Mourne Plain near Kilkeel, County Down, looking northwestwards. Most of the area in the foreground is underlain by glacigenic deposits, mainly glaciomarine diamicts, marine muds, ice-marginal gravel and emergent beachface gravel and sand. The Mourne Plain acted as a major depositional sink as ice wasted along the flanks of the Mourne Mountains which are seen in the background.

(Fig. 6.22). Formerly the depocentre was sea floor with a contemporaneous marine cliff now buried inland. It is argued that the glacigenic sequences which accumulated from successive ice-marginal inputs are coarse-grained glaciomarine deposits for a number of reasons. First, they are pre-dated and truncated by late-glacial beach facies indicating emergence. Second, they are cut by channels related to emergence of a sea floor during early post-glacial isostatic emergence. Because these channels are infilled by marine mud formed during a global meltwater pulse when global eustatic level was low, the area was still deeply isostatically depressed and open to flooding from the Irish Sea. Third, the pattern of ice sheet margins during decay left the depocentre open to direct influences from the Irish Sea. Fourth, the detailed sedimentology and facies sequences at Ballymartin, Kilkeel and Derrryoge are typical of sedimentation from retreating tidewater glaciers.

Ballymartin coast

This coastal exposure is dominated by diamict beds interbedded with stratified facies from centimetre to metre scales (Fig. 6.23). Historically, this sequence was used to reconstruct discrete ice sheet glaciations because facies containing erratics of different provenance were noted in section (Mitchell et al., 1973; Stephens et al., 1975). This layer-cake stratigraphical approach also assumed all fine-grained, laminated mud/diamict to be terrestrial tills and evaluated the deposits in terms of umbrella descriptors such as Irish Sea Till. Clearly, a basic description of the sediments and their geometries would aid reconstruction of depositional environments, rather than preconceived concepts, static glaciation models and bed-for-bed correlations.

The most complete sequence occurs at the northeastern end of the exposure where four main facies, consisting of three diamicts overlain by gravel, are exposed (Fig. 6.23). At beach level up to 1 m of silt-rich, massive diamict contains dispersed clasts, marine shell fragments and far-travelled erratics of northern Irish Sea provenance including Cretaceous chalk, flint and Ailsa Craig microgranite. It consists of thin to lenticular beds (5–10 cm thick) separated by stringers of sand or silt. An overlying diamict extends along the exposure continuously as a tabular bed (3–4 m thick) for about 20 m and then thins and grades laterally into thin beds of granite

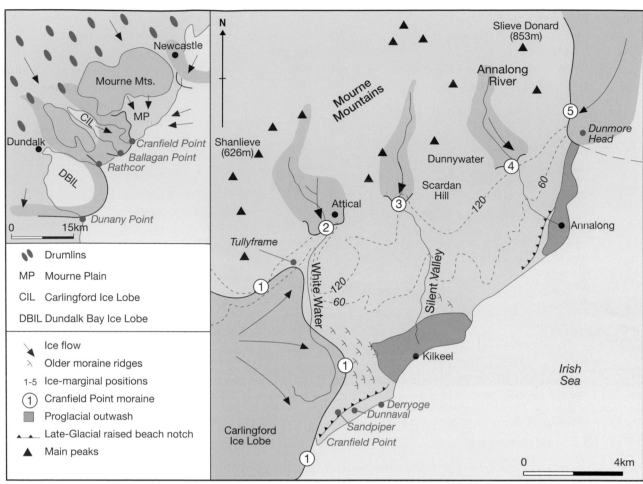

Fig. 6.22 Location of the Mourne and Carlingford Mountains and their influence on the ice-marginal configuration and position of ice lobes during the Killard Point Stadial. Note that the topographic barrier of the mountain masses created an ice-free area on the Mourne Plain on the south (lee) side of the mountains during the early deglaciation when ice withdrew from the Irish Sea Basin. *By kind permission of GSNI.*

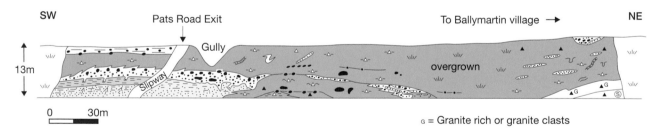

Fig. 6.23 Interbedded glacigenic facies at Ballymartin, south County Down (see Fig. 6.27 for lithofacies key). *By kind permission of GSNI.*

granules which become interbedded with the overlying fine-grained diamict. The main exposure of this diamict is massive, consisting of amalgamated beds of granite cobbles and boulders set in a matrix of granite granules. Some coarse beds show coarse-tail grading patterns locally. Clast shape is variable from glacially facetted to subrounded and edge-rounded. Up to 8 m of fine-grained diamict directly overlies the granite facies along a sharp contact. It is characterised by bed stacking, centimetre-scale laminations, laminated scour infills, interbedded sand and silt, dispersed pebbles, variable grain size, discontinuous sandy to gravelly lenses, clast lines, and fold noses on a metre scale (Fig. 6.24a, b). Clasts are mainly derived from Silurian bedrock though the large clasts (~1 m) which depress underlying laminae are invariably granite. These large lonestones are draped by silty or sandy laminae (Fig. 6.24c). Small vertical pipes up to 3 m high similar to water escape structures occasionally truncate primary beds. The upper part of the

Fig. 6.24a Fine-grained, laminated diamict beds with isolated clasts at Ballymartin, County Down. Note the slight colour differences which are related to grain size.

Fig. 6.24b Close-up of laminated diamict from Ballymartin deposited mainly from density currents.

Fig. 6.24c Large lonestone (dropstone) within laminated diamict showing depression of bedding below the clast.

stratified diamict is cut by a 4 m wide channel infilled by crudely-bedded gravel. Bed junctions are often amalgamated or marked by sudden changes in grain size. Individual beds show a variety of grading patterns including normal, inverse and inverse to normal grading. Other beds are disorganised and mainly clast supported.

The most significant element in this section apart from the differences in erratic content is that all beds thin westwards and become intricately interbedded, which denotes contemporaneity. Similarly, variable clast provenance within the sediment pile suggests that the input from different ice margins varied spatially in response to topographic control on ice sheet configuration. These inputs are associated with textural diversity, especially within the diamicts where granitic detritus tends to be coarse and bouldery and input from Irish Sea sources is fine-grained and shelly. There are no major sequence breaks, glaciotectonics, intraformational organic horizons, weathering features or desiccation phenomena which could be used to subdivide the sequence. Interbedding at both centimetre and metre scales within the sequence together with close genetic relationships between all facies show that although depositional processes varied in intensity, the overall depositional environment did not. The flat-lying and channelled geometries of the diamicts, mud, sand and gravel are very similar to those found at grounding line morainal banks near tidewater effluxes (Hunter et al., 1996). Because the facies are truncated locally by late-glacial raised beach deposits and landforms it is argued that they represent an emergent glaciomarine spread deposited on an isostatically depressed surface on the margin of the Irish Sea Basin. It is noteworthy that the glaciomarine spread cannot be linked to any topographic form or moraine in the subjacent land area, and records infilling on basinal margins when ice masses decayed during early deglaciation.

Facies variability within the sequence is explained by a range of subaqueous processes which are also observed in contemporary and ancient glaciomarine settings (Fig. 6.25). For example, granite clasts occur in three common situations. In the eastern part of the section the coarse-grained diamict is characterised by amalgamated beds, coarse-tail grading, weakly-developed beds, clast clustering and discontinuous cobble lines, all indicating deposition by mass flow. Because the nearest granite outcrops are 3–4 km distant from the site it is probable that high energy flows delivered the erratics to the site from a nearby efflux system.

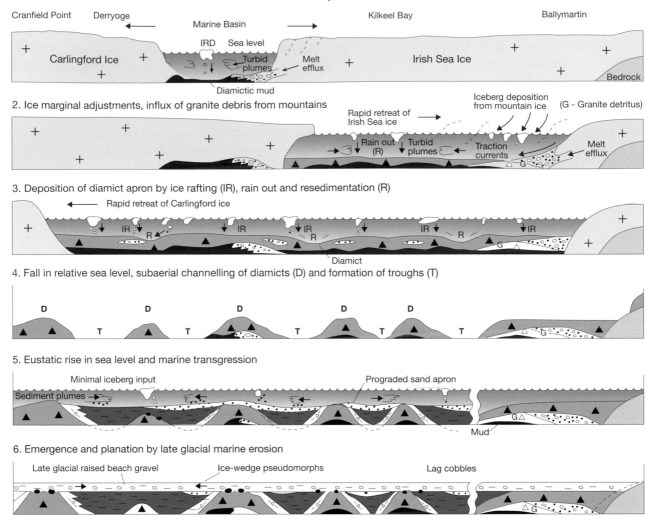

1. Separation of composite ice sheet and formation of marine embayment

Cranfield Point   Derryoge   Marine Basin   Kilkeel Bay   Ballymartin

IRD   Sea level

Carlingford Ice   Turbid plumes   Melt efflux   Irish Sea Ice

Bedrock

Diamictic mud

2. Ice marginal adjustments, influx of granite debris from mountains

Iceberg deposition from mountain ice   (G - Granite detritus)

Rapid retreat of Irish Sea ice

Rain out (R)   Turbid plumes   Traction currents   Melt efflux

G

3. Deposition of diamict apron by ice rafting (IR), rain out and resedimentation (R)

Rapid retreat of Carlingford ice

IR   R   IR   IR   R   IR   R   IR

G

Diamict

4. Fall in relative sea level, subaerial channelling of diamicts (D) and formation of troughs (T)

D   D   D   D   D

T   T   T   T   T   G

5. Eustatic rise in sea level and marine transgression

Minimal iceberg input   Prograded sand apron

Sediment plumes   G   Mud

6. Emergence and planation by late glacial marine erosion

Late glacial raised beach gravel   Ice-wedge pseudomorphs   Lag cobbles

G

Fig. 6.25 Postulated sequence of events affecting the Mourne Plain, south County Down during early deglaciation. Ice-marginal retreat was influenced by the topographic barrier of the Mourne Mountains and glaciomarine sediments accumulated on the Mourne Plain. These diamicts were channelled when isostatic uplift occurred and marine muds were deposited in the resulting troughs following a eustatic rise in sea level. Final emergence resulted in truncation of the succession and formation of raised shoreface beach gravel. Note that the Mourne Plain area lay beyond the later ice sheet readvance limits at Cranfield Point. The associated raised beach does not occur within these readvance limits. *By kind permission of GSNI.*

In contrast, within the fine-grained diamict granite material is present as stringers of quartz-rich granules or granite sand suggesting erosion, winnowing and resedimentation. In most cases the largest clasts (~1 m) which depress the beds in the fine-grained diamict are granite and these cannot be freighted in thin flows or currents which formed the bulk of this diamict. The lonestones are therefore attributed to ice rafting which contrasts with the source and origin of small pebbles in the diamict which are predominantly derived from local Silurian rocks and emplaced by mass flow. Although the fine-grained diamict is clearly interbedded with granite

and other facies at the western end of the section, its own internal primary sedimentary structures are also indicative of subaqueous deposition. These include stacked beds, small fold noses, sand lenses and streaks, discontinuous clast lines, sand infilled channel scours, variable lamination and drapes. The sediment pile therefore evolved with only minor sequence breaks and simply records changes in the intensity of the processes operating. Depositional processes include suspension from sediment plumes (silt/clay laminae), mass flows (thick resedimented beds), density currents (sandy stringers), channel cut and fill (fine-grained/laminated

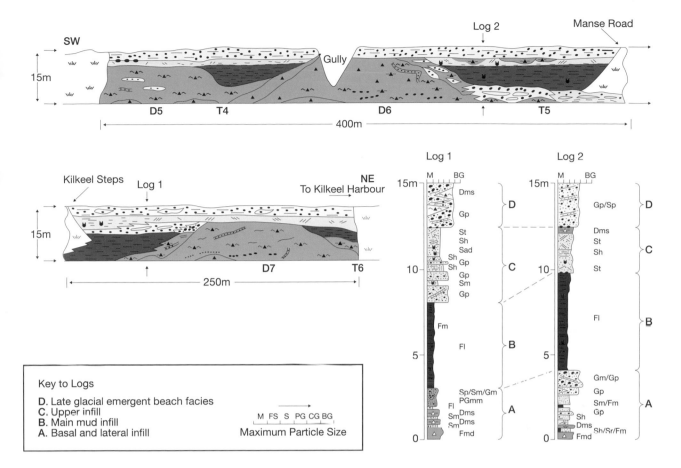

Fig. 6.26 Kilkeel section and logs (see Fig. 6.27 for lithofacies key). *By kind permission of GSNI.*

sand) and ice-rafting (lonestones). This grouping of processes is typical of tidewater settings away from direct meltwater venting and jet activity.

Kilkeel and Derryoge

The sediments of this coastal segment are unique because they provide stratigraphy and age-constrained evidence for at least four regional phases of major environmental changes during early deglaciation. These include glaciomarine conditions during ice recession (diamicts), rapid uplift (subaerial channelling), global meltwater pulse (marine transgression/mud) and final emergence (late-glacial raised beach gravel). Along the 3 km long exposure seven mesa-shaped highs (D1–7) composed of diamict are separated by troughs (T1–6) infilled with marine mud (Figs. 6.26, 6.27). These elements are truncated by a continuous, horizontal erosion surface and a tabular body of gravel forming an extensive spread of raised, late-glacial beach (Fig. 6.28). The postulated sequence of events contributing to the formation of the Mourne Plain include:

Phase 1 (Diamict deposition)

The mesa-shaped highs consist of stacked, massive, texturally variable sandy beds which are defined either by the presence of sharp contacts or by bundles of undeformed silt/sand laminae or by lenses of granules. Stratification is emphasised by discontinuous lines of clasts and massive gravelly lenses and ribbons. At Derryoge a 3 m thick sequence of poorly-sorted cobble and pebbly gravel beds grades up into variable stratified diamict. It contains abundant marine shell fragments including specimens of *Turritella and Arctica islandica* and a wide range of erratics of northern Irish Sea provenance. Clasts from the diamict are mainly (~85%) derived from the local Silurian country rock but up to 8 percent consist of Carboniferous limestone derived from outcrops at the mouth of Carlingford Lough. Thin washes and stringers of granitic detritus record a third source of sediment input from the Mourne Mountains, four kilometres to the north. Occasional vertical pipes lined with gravelly debris extend from the base to the top of the section. As at Ballymartin the largest clasts are generally granites.

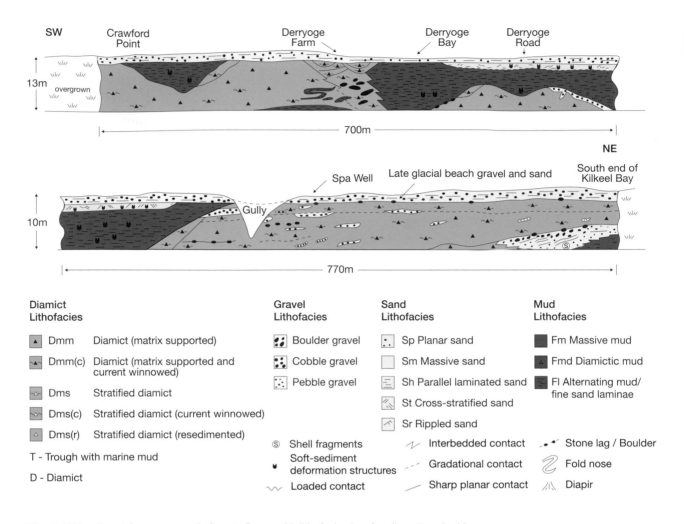

Fig. 6.27 Section at Derryoge, south County Down with lithofacies key for all sections in this area. *By kind permission of GSNI.*

The general sedimentary characteristics of the diamict are also similar to those at Ballymartin and are best explained by mass flow processes. The degree of stratification and presence of more gravelly flows is related to distance from major glacial effluxes. For example, gravel at the base of the Derryoge exposure is similar to subaqueous outwash described by Rust and Romanelli (1975) formed close to a point input, whereas the erratics in the overlying stratified diamict sequence point to a mass flow origin from a westerly source. The extent of the diamict apron across the Mourne Plain is attributed to rain-out and resedimentation seaward of successive tidewater grounding positions (Fig. 6.25). This depositional model not only explains the sedimentary variability and absence of glaciotectonics in the coastal exposures of south county Down, but is also consistent with the erratic distribution and sectional geometry.

Phase 2 (Subaerial channelling)
The steep (10–20°) flanks of each mesa are not consistent with the mass flow origin of the diamicts which was essentially cohesionless during deposition. In addition the flat-lying diamict beds within the mesas are truncated abruptly along their flanks implying that there was an original diamict spread which was later cut by channels (Figs. 6.25, 6.28). Because the channel axes are graded well below modern sea level it is likely that the channels were cut by meltwaters draining south from the mountains during a fall in relative sea level. It is possible for quasicontinuous turbidity currents to cut channels on valley fan systems (Hay et al., 1983) but four arguments suggest a subaerial origin. First it must be recognised that isostatic uplift resulting in emergence was well underway even before this area was finally deglaciated. Second, the cut (into coarse-grained diamict) and fill (mud) nature of the channels denotes

marked erosion followed by quiet water marine deposition. These two events are distinct geological episodes requiring quite different energy levels and formative conditions. Third, if channels are subaqueous then there should be deltas deposited by these streams near their contemporary sea level. None have been observed. Four, a subaqueous origin is excluded because there is no *a priori* reason why subaqueous processes would initially be erosional, followed by a switch to processes that result in nearly instantaneous infilling of the channels with mud, particularly on low angle slopes and at shallow water depths (<30 m). In theory the formation of subaerial erosional channels is consistent with the timing of early postglacial isostatic uplift which is confirmed by radiocarbon dating of the marine mud within the channels.

Phase 3 (Marine transgression)

Marine mud infilling the channels contains an *in situ* marine microfauna dominated by *Elphidium clavatum*, which is dated to ~19,000 years ago (Fig. 6.28). The laminated sand/mud couplets are similar in texture,

thickness, cyclicity and variable laminae characteristics to rhythmites formed at least 0.5 km from ice fronts in Alaska (Mackiewicz et al., 1984). They are termed 'cyclopels' and are thought to result from the interaction between overflows/underflows and tidal currents in a glaciomarine/marine environment. The continuous extent, streaked form and sharpness of the very fine laminae within the mud may be due to sorting of silt and clay by oscillatory currents. On the flanks of the mesas the muds are interbedded with mass flow of diamict and gravelly sediment from unstable slopes. In some cases these packages of sediment which formed during slope adjustment show downslope transitions from cohesive mass flows to more stratified and gravelly beds with increasing distance from source (Fig. 6.26).

The marine limit at Kilkeel is 30 m above present sea level and formed ~20,000 years ago (Clark et al., 2004). At this time global sea level was 130–140 m lower than present (Yokoyama et al., 2000) indicating that the marine limit at Kilkeel records a net isostatic uplift of 160–170 m. The presence of channels cut into the deglacial diamict shows that the coast had risen

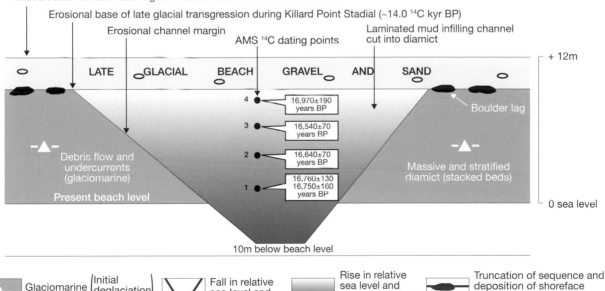

Fig. 6.28 Cartoon illustrating the facies geometry, facies types, sites sampled for AMS ^{14}C dating and the main deglacial events at Kilkeel Steps, County Down (Clark et al., 2004). The channel was cut into glaciomarine diamict during a fall in relative sea level when isostatic uplift occurred immediately after early deglaciation. Four similar channels are cut into glaciomarine diamict between Kilkeel and Cranfield Point (McCabe, 1986). Monospecific samples of the foraminiferan *Elphidium clavatum* in pristine condition from marine mud infilling these channels are dated to 19 cal ka BP and record a global rise in sea level. The *in situ* marine microfauna also includes delicate specimens of the ostracod *Polycope* sp. with valves still joined at the hinge line. *By kind permission of GSNI.*

at least 30 m by the time they had formed. The mud infilling the channels therefore records a sea level rise to ~10 m above sea level. Rapid deposition of the mud is indicated by the similarity of five radiocarbon dated foraminifera samples collected from the base to the top of one channel fill providing a weighted mean of 19,111 years BP for the event. These data therefore confirm a rapid rise in eustatic sea level of at least 10 m at 19,000 years BP, that was previously inferred from the Bonaparte Gulf (Yokoyama et al., 2000). The global impacts and climate changes associated with this far-field event are discussed in Clark et al., (2004).

Phase 4 (Emergence)

The horizontal, planar surface truncating all deposits along 6 km of coast between Kilkeel and Cranfield Point is overlain by ~4 m of horizontally-bedded pebbly gravel and sand (Fig. 6.28). Gravel beds show degrees of both size and shape sorting. These deposits are part of an extensive marine terrace leading up to a fossil notch at 19 m OD. The lateral extent of the truncation below the beach gravel, the sedimentology of the overlying gravel and sand and their relationship to the fossil notch show that they are part of the regional raised beach system. The presence of ice wedge pseudomorphs cutting across primary bedding confirms a late glacial age. However, this raised beach system can be traced as far south as Cranfield Point where it merges with the terminal outwash of the moraine indicating contemporaneity. Furthermore, the beach system does not occur within these moraine limits, again indicating contemporaneity. The Cranfield moraine and the shoreline is dated to ~17 cal ka BP and shows that there is a considerable hiatus between deposition of the marine muds in the channels and subsequent marine truncation (Fig. 6.28).

Although the overall facies signature along the coast of south County Down is one of isostatic emergence, it was not either a steady, unidirectional process or one entirely dominated by local controls. Essentially it was the process of early and rapid deglaciation that led to a high sediment preservation potential in a local depocentre while the land area was still deeply isostatically depressed. Although rapid uplift occurred early, the land surface was sufficiently depressed, permitting a far-field global meltwater signal to be preserved when sea level rose. An hiatus within the sequence around Kilkeel shows that records of environmental change are missing locally for a few thousand years but elsewhere within Carlingford and Dundalk Bays these breaks are recorded by significant ice sheet readvances. It is therefore critical to note that depositional sequences rarely contain a facies sequence that accurately records a full deglacial cycle. In northern Britain most sections on exposed coasts simply consist of a deglacial diamict truncated by late-glacial marine erosion and overlain by emergent beach gravel. The exposures in County Down and indeed along the coast of eastern Ireland in general show that complex interplays of land and sea level changes are contained within the stratigraphic record.

7

Terrestrial Deglaciation

Background and factors influencing deglaciation

Standard views on the pattern and nature of deglaciation in Ireland are generally based on synchronous ice sheet advance to ice sheet limits somewhere on the inner continental shelf followed by ice retreat towards the coast. Geophysical models depict a monotonic ice sheet retreat back towards centres of ice sheet dispersal accompanied by estimates of ice-marginal positions during retreat. Over the last decade the validity of these static models has been undermined by at least three major structural advances on the nature and anatomy of a relatively small ice sheet situated on lowlands adjacent to the climatically sensitive northeastern Atlantic. First, the sedimentary successions formed around the periphery of the retreating ice sheet as it contracted from the continental shelf can best be evaluated in terms of marine and glaciomarine deposition. The critical inference is that isostatic depression exceeded eustatic fall in sea level. Second, the range of climatic proxies available from North Atlantic records now show that ice age events responded to climate signals on millennial and lesser timescales. Therefore the field evidence would be expected to contain records for at least the major climate signals emanating from sea surface conditions and temperatures of the North Atlantic. Third, it is now possible to present an age-constrained deglacial stratigraphy which tracks the course of events punctuating deglaciation.

For the first time it is now possible to date moraines which mark significant ice sheet limits and area covered by ice at successive deglaciation positions using both AMS ^{14}C dating of marine transgressions and estimations on the timing of deglaciation by surface exposure dates (Fig. 7.1). These advances in dating techniques allow correlations to be made between moraines from widely spaced ice sheet sectors and reconstructions of the ice sheet systems operating at a particular time

slot. Together these advances offer an improved methodology which can be used to explore more precisely the nature and patterns of deglaciation.

Holistic reconstructions of the events during the last deglaciation are therefore based on the fact that ice sheets in Ireland were extremely mobile over the last 30,000 years and oscillated possibly on millennial timescales. Formerly the main patterns of moraines across the country were interpreted in terms of a fairly static ice sheet wasting back towards poorly defined centres of ice sheet dispersal. AMS ^{14}C dates from sites in the north Irish Sea Basin (Port, Cooley Point, Rathcor, Linns, Cranfield Point, Kilkeel, Killard Point and Rough Island), north County Donegal (Corvish) and north County Mayo (Glenulra, Belderg) provide a much more complex deglacial stratigraphy. This includes three major deglacial periods within the termination which ended the last major cold phase. In the west a regional deglacial event occurred ~28 cal ka BP, early deglaciation of the entire ice sheet occurred shortly after 22 cal ka BP and final deglaciation of remaining ice sheet sectors in the north, east and west occurred after 15.6 cal ka BP. Each major deglacial period is supported by an age-constrained stratigraphy and different terrestrial records of ice sheet retreat across separate sectors of countryside (Fig. 4.10). The standard idea that the ice sheet was unresponsive to climate drivers is not supported by a wide range of field evidence. Deglaciation events included periods of widespread and catastrophic ice wastage, regional changes in ice sheet configuration, ice sheet readvances at a variety of scales, patterns of phased ice wastage and moraine formation around the final centres of ice sheet dispersal. Finally, each of the three major periods of deglaciation were associated with somewhat different ice sheet systems comprising different centres of ice sheet dispersal, ice flows and limiting moraines.

Deglacial records cannot be interpreted in terms of climate change alone. At a smaller than ice sheet scale

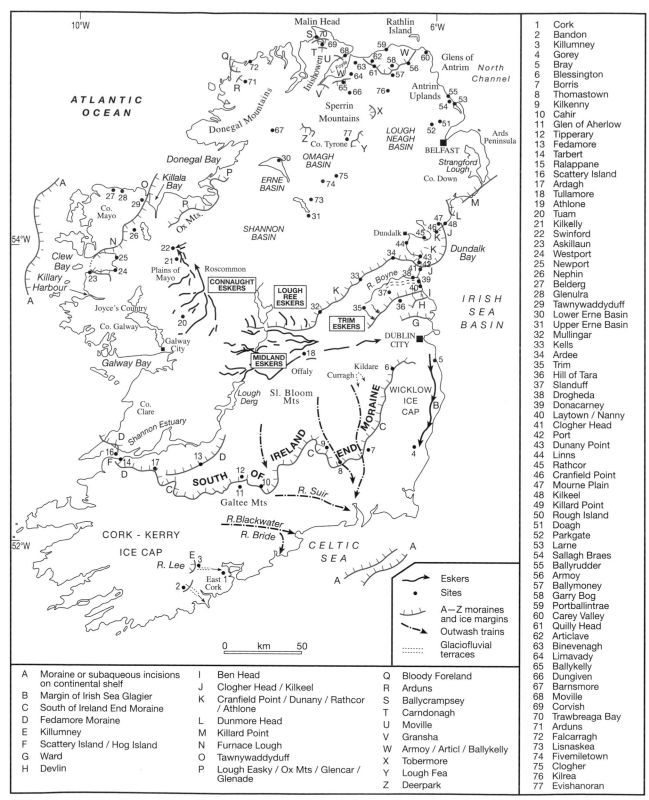

Fig. 7.1 Moraines, landform patterns and locations of deposits used to reconstruct patterns of deglaciation in Ireland.

1	Cork
2	Bandon
3	Killumney
4	Gorey
5	Bray
6	Blessington
7	Borris
8	Thomastown
9	Kilkenny
10	Cahir
11	Glen of Aherlow
12	Tipperary
13	Fedamore
14	Tarbert
15	Ralappane
16	Scattery Island
17	Ardagh
18	Tullamore
19	Athlone
20	Tuam
21	Kilkelly
22	Swinford
23	Askillaun
24	Westport
25	Newport
26	Nephin
27	Belderg
28	Glenulra
29	Tawnywaddyduff
30	Lower Erne Basin
31	Upper Erne Basin
32	Mullingar
33	Kells
34	Ardee
35	Trim
36	Hill of Tara
37	Slanduff
38	Drogheda
39	Donacarney
40	Laytown / Nanny
41	Clogher Head
42	Port
43	Dunany Point
44	Linns
45	Rathcor
46	Cranfield Point
47	Mourne Plain
48	Kilkeel
49	Killard Point
50	Rough Island
51	Doagh
52	Parkgate
53	Larne
54	Sallagh Braes
55	Ballyrudder
56	Armoy
57	Ballymoney
58	Garry Bog
59	Portballintrae
60	Carey Valley
61	Quilly Head
62	Articlave
63	Binevenagh
64	Limavady
65	Ballykelly
66	Dungiven
67	Barnsmore
68	Moville
69	Corvish
70	Trawbreaga Bay
71	Arduns
72	Falcarragh
73	Lisnaskea
74	Fivemiletown
75	Clogher
76	Kilrea
77	Evishanoran

A	Moraine or subaqueous incisions on continental shelf	I	Ben Head	Q	Bloody Foreland
B	Margin of Irish Sea Glagier	J	Clogher Head / Kilkeel	R	Arduns
C	South of Ireland End Moraine	K	Cranfield Point / Dunany / Rathcor / Athlone	S	Ballycrampsey
D	Fedamore Moraine	L	Dunmore Head	T	Carndonagh
E	Killumney	M	Killard Point	U	Moville
F	Scattery Island / Hog Island	N	Furnace Lough	V	Gransha
G	Ward	O	Tawnywaddyduff	W	Armoy / Articl / Ballykelly
H	Devlin	P	Lough Easky / Ox Mts / Glencar / Glenade	X	Tobermore
				Y	Lough Fea
				Z	Deerpark

Eskers
Sites
A–Z moraines and ice margins
Outwash trains
Glaciofluvial terraces

many local factors influenced the pattern of deglaciation through all of the stratigraphically defined periods of deglaciation. Central to understanding deglaciation is the fact that the Irish ice sheet was based on the lowlands and consisted of multiple centres of ice sheet dispersal whose emphasis changed or shifted during the course of deglaciation. Variable cross-cutting striae on the summits of Carboniferous outliers together with changes in bedform orientations within lowland basins provide primary evidence for massive ice sheet reorganisation during the course of deglaciation (McCabe, 1969; Synge, 1969). At times when the ice was at its maximum extent the ice flowlines were longest and therefore susceptible to very rapid change driven either by climate change or massive ice loss associated with deep isostatic depression and marine downdraw (Eyles and McCabe, 1989a). Resulting facies sequences will either be sedimentologically complex, reflecting rapidly changing environmental conditions, or have a very low preservation potential. However, if rapid retreat of the ice margin occurs it may suddenly re-equilibrate when a topographic high or pinning point is reached. An ice-contact spread or ridge can therefore form at this equilibrium position which is not climatically controlled. Interpretations on the precise meaning of ice-contact stratified mud, sand, diamict and gravel sequences can vary between glacio-marine, marine, and glaciolacustrine. Therefore the importance or levels of confidence given to a particular exposure can be debatable. However, given the fact that thick ice (>800 m) was present in the peripheral ice sheet sectors of the southwest and around the Wicklow Mountains when deglaciation began (Ballantyne et al., 2006, 2007), the resulting isostatic depression must have led to ice sheet–marine interactions. In addition if interpretations on the pattern and direction of deglaciation are based on individual exposures it is important to identify precisely where the exposure is located and any palaeoflows inferred from the internal facies geometry.

Patterns of deglaciation across swathes of countryside have been traditionally reconstructed using 'lines of moraine' (Fig. 7.2). One regional example, the Armoy moraine of north County Antrim, can be used with confidence because its outer ridges are fairly continuous and obvious landscape features (Stephens et al., 1975; Creighton, 1974). However, the North Donegal moraine as mapped by Stephens and Synge (1965) has been used in countless ice sheet reconstructions even though its basis is a series of unconnected and isolated gravelly deposits situated within a topographically diverse area. A very useful tool has been to map the pattern of moraines which are fed by subglacial esker ridges and meltwater channels (Fig. 7.3). These couplets together

Fig. 7.2 Ice-marginal gravelly ridges and meltwater channels along the eastern margin of the Slievenamuck ridge, County Tipperary. These features were considered once to be part of the South of Ireland End Moraine.

Fig. 7.3 Flat-topped, cross-valley ridge at Cam Lough, County Tyrone composed of coalesced ice-marginal deltas. The ice-contact part of the ridge is marked by kettle holes filled with water. Note the small winding esker ridge in the left of the photograph which acted as a subglacial meltwater feeder channel leading into the ice-proximal face of the delta.

with patterns of ice-contact ridges provide local evidence used to reconstruct larger scale patterns. In a similar way the duplet of a moraine ridge grading into an outwash spread will provide data on the direction of deglaciation. The general pattern of outwash terraces within a valley provides similar information. When all known deglacial phenomena are placed on maps and general patterns of ice wastage inferred, the deglacial systems cannot be resolved until major moraines and ice limits within the overall pattern are dated and then correlated (Fig. 7.1). Because deglaciation is naturally time-transgressive, dating of ice sheet limits and estimates of ice volumes remaining during successive phases of deglaciation are necessary to evaluate the crustal response to ice sheet loading and the timing of isostatic uplifts.

Evidence for the three main periods of deglaciation identified from stratigraphic and dating evidence is preserved for three main reasons. First, centres of ice dispersal early in the deglacial cycle shifted and result-ant changes in ice sheet configuration preserved some areas of sediment offlap, especially in north County Mayo. Second, early, widespread and rapid deglaciation over perhaps a few thousand years resulted in patterns of sediment offlap northwards from some maximum ice sheet limits. Third, the last major ice sheet readvance known as the Drumlin Readvance (Synge, 1968, 1969)

only affected northern and western parts of the island and did not overrun the midlands or the south of the island (Fig. 1.4). It should be noted that in some areas, especially in the southwest, there is very little dating evidence on which to reconstruct detailed patterns of deglaciation. The remainder of this account focuses on major stratigraphic, morphological and sediment-ological accounts which support three major periods of deglaciation during the last glacial termination. More detailed descriptions of the stratigraphic parts of these records are given in the relevant chapters describing the complex environmental interactions which characterise the last glacial cycle, ice sheet readvances and sea level history. Detailed descriptions of the eskers of Ireland are omitted from this chapter and presented later in order to provide information on the significance of this landform distribution.

Deglaciation prior to the LGM (~28 cal ka BP)

Traditionally the ice directional indicators and landforms along the west coast have been interpreted in terms of ice advance westwards onto the continental shelf during the Last Glacial Maximum (LGM, 23–19 cal ka BP) followed by monotonic ice retreat (Mitchell et al., 1973). In most instances the ages of the directional indicators used to reconstruct ice flows are largely unknown and regional patterns are simply grouped along the western seaboard. Inherent in this approach is that the ice sheet itself was fairly static, did not change configuration rapidly, did not respond to climate changes quickly and was glaciologically steady with no major contrasts in bed conditions. However, raised marine muds overlain by deltaic gravels from Glenulra, along the southern margin of Donegal Bay provide evidence for the build up of a thick ice sheet in western Ireland before 28 cal ka BP (McCabe et al., 2007a). Preservation of these deposits at heights approaching 80 m OD records the isostatic depression of ice sheet loading and adjacent striae record the northwestward ice advance onto the continental shelf (Fig. 7.4). The ice sheet must have covered extensive areas of the west prior to the traditional LGM because the Glenulra site was not overridden after deglaciation. The cause of deglaciation from the continental shelf so early in the last glaciation is not known for certain but it is clear that the ice sheet loading must have exceeded the contemporary fall in eustatic sea level. In the absence of any obvious signs of climate forcing it is concluded that isostatic depression followed by marine down draw was mainly responsible for local deglaciation of outer Donegal Bay at this time. Evidence for a similar ice sheet advance westwards along Clew Bay onto the continental shelf is documented by the streamlined (east–west) rock cores of the bedforms at the head of the bay between Westport and Newport (Fig. 4.11). These were later overridden and overprinted by ice moving from the south during drumlinisation around 16 cal ka BP.

Early deglaciation (<21 cal ka BP)

At its maximum the last lowland ice sheet moved onto the continental shelf off the south coast and into the Irish Sea Basin (Fig. 7.5). Offshore limits are not known with any certainty though it is likely that ice only moved tens of kilometres off the present coastline because of extended ice flow lines and the effects of ice wastage in the sea. Ice withdrawal inland from the inner Celtic Sea

Fig. 7.4 Glenulra valley, County Mayo with a basal valley infill of disorganised boulder gravel, sand and fine gravel interbedded with thick (~1 m) mud beds and overlain by gravelly foresets dipping downvalley (northwards). The latter are thought to record high relative sea level (~80 m) at 28 cal ka BP when ice withdrew from the continental shelf. Note figures for scale.

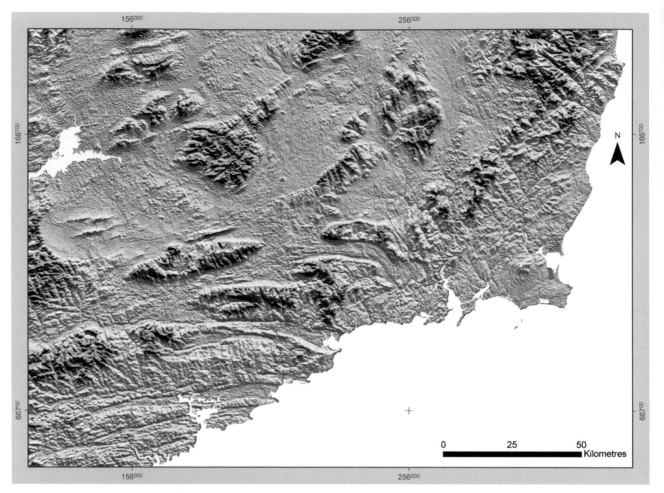

Fig. 7.5 Relief map of south eastern Ireland showing large scale ice grooving and large scale topographic moulding by ice flows moving directly offshore from the Irish midlands and the Cork/Kerry ice centre. Note that the erosional marks point to coeval southerly ice flows with no deflection of the ice flows from the Cork/Kerry area. The offshore ice sheet movement along the entire south coast (east of Cork) is confirmed by the regional pattern of glacial striae. There is an absence of ice flow indicators associated with an onshore ice movement either from the Celtic Sea or Irish Sea.

occurred around 22 cal ka BP (Bowen et al., 2002) and was followed by rapid deglaciation of the Irish Sea Basin which finished around 20 cal ka BP (Clark et al., 2004). Meltwater incisions on the floor of the Irish Sea Basin, ice-marginal/subglacial drainage systems flanking the eastern Wicklow Mountains and the internal geometry of morainal banks along the eastern coast together are consistent with meltwater drainage southward and vacation of the basin by tidewater ice margins (Fig. 6.4). The deglaciation of the basin was effected in a few thousand years (Eyles and McCabe, 1989a). Ballantyne et al., (2006) argue that the limited carriage of Wicklow granites means that at the LGM the Wicklows were encircled by thick (>700 m) ice from both the Irish Sea and midlands. Cosmogenic [10]Be exposure ages for rock outcrops on summits overridden by warm-based ice gave post-LGM ages of 19–18±1.2 ka BP (Ballantyne

et al., 2006) which are reasonably consistent with other estimates for early deglaciation.

The patterns of moraines, fluvioglacial features and stratigraphy from critical type areas provide evidence for important sedimentation events, directions of ice sheet decay following ice withdrawal from shelf areas and timing of deglacial events (Fig. 7.1). Early deglaciation is documented by events and landforms from five main sectors of the ice sheet:

i) *The Blessington delta complex*

Retreat of ice from the northwestern foothills of the Wicklow Mountains is marked by extensive deltaic sediments up to 60 m thick which were formed in a glacial lake impounded between the ice margins and the mountains (Farrington, 1957). The lake was bounded on its western side by the Slievethoul ridge stretching southwestwards from Brittas towards Toor in the south (Fig. 7.6). Breaches or cols in the ridge provided meltwater entrances into the lake from ice margins sited along the ridge (Cohen, 1979). The lake drained southwards through rock-cut channels at Toor Glen and around Hollywood (Fig. 7.7). This drainage system continues southward from Hollywood ending in gravel terraces around Whitestown (Farrington and Mitchell, 1973). Synge (1971, 1977, 1979b) recognised that it was possible to compare delta levels basinwide around glacial lake Blessington with ice sheet limits both from mountain ice and ice from the midlands. A series of lake levels between 309 and 180 m OD were reconstructed from shorelines and deltas and some of these were correlated with minor readvance moraines and positions of ice lobes (Fig. 7.8). The ice-marginal deltas are best seen at Brittas in the north, Blessington in the centre and Pollaphuca in the south (Fig. 7.9). Although delta morphology is uneven and channelled, lakeward slopes generally parallel dip directions of major gravelly foreset sequences which reach 40 m in height.

Detailed sedimentological observations by Philcox (2000) has now shown that delta formation is even more complex than the relatively simple model of falling lake

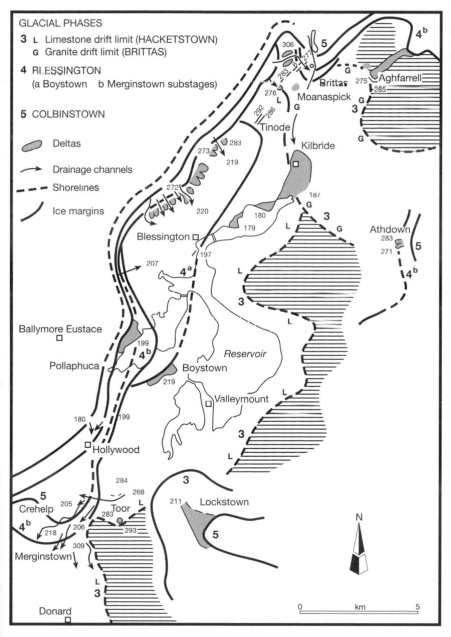

Fig. 7.6 The evolution of Glacial Lake Blessington, an ice-marginal lake associated with the maximal phases of the last ice sheet east of the Wicklow Mountains. Numbers refer to heights in metres of intakes or exits of meltwater channels, or delta surfaces (after Synge, 1979). *By kind permission QRA (Quaternary Research Association).*

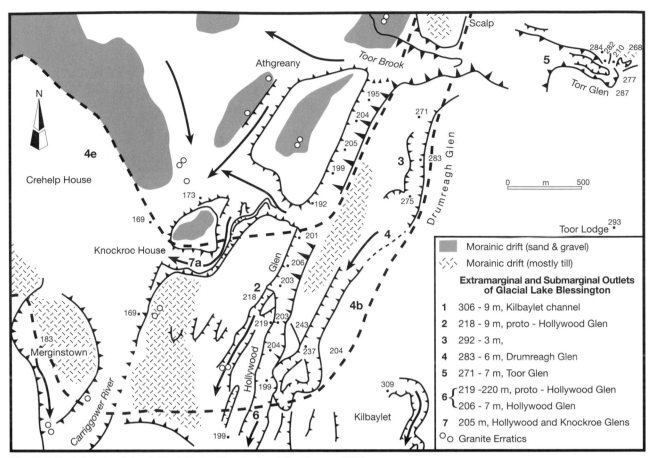

Fig. 7.7 Postulated outlets of Glacial Lake Blessington (after Synge, 1979). *By kind permission QRA (Quaternary Research Association).*

levels proposed earlier (Fig. 7.10). He divides the delta complex immediately north of Blessington into two zones separated by a narrow transitional zone. The delta zone consists mainly of foreset gravel beds up to 300 m in width which grade laterally into flat-lying toesets and bottomsets forming a typical Gilbert-type delta. The transition zone is around 100 m wide and is characterised by cross-cutting gravelly sediments, gravelly lenses within sandy sediment, thick beds of grit, erosion surfaces, cobble lags and slumped gravel beds often deformed by low-angle thrusting. Sedimentary variability within this facies sequence has been interpreted as subaqueous fan deposition. The northwest zone is characterised by rapid facies changes including boulder gravels, openwork pebbly gravel, poorly-laminated silts and sands and lenticular units of poorly-sorted gravel. Philcox (2000) concludes that the presence of extensional meltout structures within the deposits points to slumping as ice withdrew from the initial delta phase. The fact that a channel up to 40 m deep cuts through all three facies zones points to a fall in lake level. Channel

infills of diamict and laminated silt and clay probably record temporary falls in lake levels before another local readvance re-established deltaic formation at a lake level around 280 m near to the initial lake level. Subsequent ice lowering still allowed meltwater to cross the Slievethoul ridge incising the delta surface and spreading a diamict drape across parts of the delta surface. Philcox (2000) concludes that the evolution of the Blessington delta complex involves repeated falls and rises in lake level but caution is necessary before these are related directly to the intakes of the numerous and morphologically diverse spillways which drained the glacial lake (Fig. 7.7).

ii) *The South of Ireland end moraine*

Ever since Charlesworth (1928b) rediscovered Carville Lewis's Southern Irish End Moraine (1894) many authors have reused a line stretching from the north Dublin Mountains southwestwards to Borris, Thomastown, Cahir, Ballingarry and across the Shannon estuary near Tarbert (Fig. 7.1). Charlesworth's moraines were

Fig. 7.8 Summary and correlation of glacial lake levels and readvance moraines, Blessington area, County Wicklow (after Synge, 1979).

By kind permission QRA (Quaternary Research Association).

Levels of Glacial L. Blessington	Levels of outlets cut in bedrock		Glacial Limits — General ice	Glacial Limits — Local ice
306m. Delta, Slademore	(1) 309m. Kilbaylet		**Retreat westward** 3 - HACKETSTOWN	**Retreat S.E.** 3 - BRITTAS
219 Delta, Boystown	(2) 218	Hollywood Glen	**Readvance from N.W.** 4^{a1} - BLESSINGTON	— ? —
292-3 Shoreline, Tinode	(3) 293 ↓	Drumreagh Glen	**Readvance from W.N.W.**	— ? —
283-6 Shoreline, Tinode Deltas Slademore, Aghfarrell Delta, Athdown	(4) 283		4^{a2} - BLESSINGTON II Readvance from N.E. 4^{b} - III	**Advance in Liffey valley** 4c - ATHDOWN I
272-7 Delta, Blessington Delta, Moanaspick Delta, Athdown	(5) 283 ↓ 268	Toor Glen	**Readvance from N.W.** 4^{c} - BLESSINGTON IV	**Phase in Liffey valley** 4d - ATHDOWN II
211 Delta, Lockstown	(6) 218 ↓ 206	Hollywood Glen	**Readvance from W.N.W.** 4e - BLESSINGTON V	**Advance in King's R. valley**
205	(7) 205	Hollywood Glen - Carrigower Channel		
197-9 Delta, Pollaphuca Shoreline, Blessington	(8) 199	Hollywood Demesne	**Phase from W.N.W.** 5a - COLBINSTOWN I	— ? —
179-180 Delta, Kilbride	(9) 180	Hollywood Demesne	**Readvance from N.E** 5B - COLBINSTOWN II	— ? —

described as kettle moraines recorded by festoon-like sweeps along the margins of ice lobes marking the end moraine of the Newer Drift. The moraine reached a height of about 410 m OD on the northern flanks of the Dublin Mountains and continued without many breaks for almost 500 km to the west coast. Individual swathes of moraine were up to 5 km in breadth. The construction of the SIEM is based on diverse morphological forms from the lowlands and the notion that ice lobes controlled ice-marginal depositional patterns. Methodologically, Charlesworth believed that the presence of glacial cave mammalian faunas at Doneraile (County Cork), the fact that Cork/Kerry ice was separate from the main ice sheet and the morphological differences between freshness of the drifts north and south of the SIEM supported the proposed ice limit.

There is little doubt that substantial areas of sand and gravel topography exist across the lowlands south-west of the Wicklow Mountains, but these depositional nodes are not continuous and few ridge moraines occur. Most features mentioned by Charlesworth (1928b) are outwash and probably record general ice wastage northwards. There is little sedimentological, periglacial or dating evidence to support a major ice sheet limit across the south midlands. On the other hand large-scale bedform patterns identified from satellite imagery show that ice flows moved from the north and west across most of the southeastern coast into the north Celtic Sea towards the end of the last glaciation (Fig. 7.5). The initial withdrawal of the ice from its near offshore limits is recorded by erosional marks on the rock platform, overlying emergent facies sequences and glacigenic deposits in east County Cork (Fig. 7.11a,b,c) (McCabe and O'Cofaigh, 1996). Therefore the regional pattern of glacigenic events along the south coast shows that the surficial deposits to the north were formed

Fig. 7.9 Large-scale delta foresets, Blessington, County Wicklow.

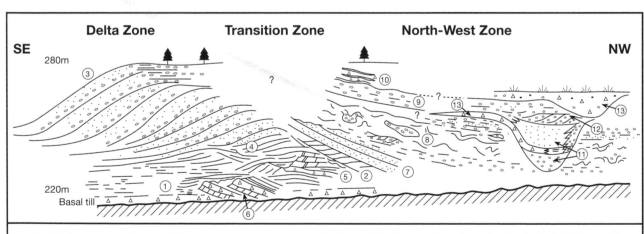

1. Horizontal sand with ripple drift
2. Ice-contact surface
3. Delta foresets <40 m high; topsets c.5m thick
4. Pre-delta foresets, covering smaller-scale sediment lenses in early fan
5. Ice-contact thrust complex
6. Proglacial thrust complex
7. Back-slope sediments in Transition Zone

8. Deerpark Facies: irregular beds ranging from laminated silt to boulder gravel, contorted
9. Top gravel: coarse boulder gravel
10. Diamict
11. Carnegie's Channel cuts across all zones filled with talus gravel and sand
12. Small-scale delta foresets, covered by boulder beds
13. Channel, filled with dirty gravel
14. Hudson Diamict (till) clay and silt

Fig. 7.10 Philcox's (2000) model of the main sedimentary and tectonic units in the gravel pits at Blessington. *By kind permission of Mike Philcox.*

Fig. 7.11a North to south subglacial erosional furrows cut into the surface of the platform at Simon's Cove, east County Cork. Note that the furrows cut across the west–east strike of the rock, are closely spaced, up to a few metres deep and can be cut into topographic highs on the platform. Potholes occur at intervals along the features.

Fig. 7.11b General arrangement of the beach facies overlying the rock platform, Simon's Cove. Note the diffuse lamination in the main granule to pebbly gravel facies, the presence of discrete horizontal beds and channel fills of small well-rounded and sorted cobbles and large angular, bed-parallel slabs of shale. These facies are typical of shallow marine settings subject to mass flow from local slopes, sorting and sediment downdraw by return marine currents. In most cases the boulder a-axes are subparallel to the shore and form vertically-stacked boulder pavements. Rule is 0.6 m long.

Fig. 7.11c Detail of gravelly facies described in Figure 7.11b. Rule is 0.5 m long.

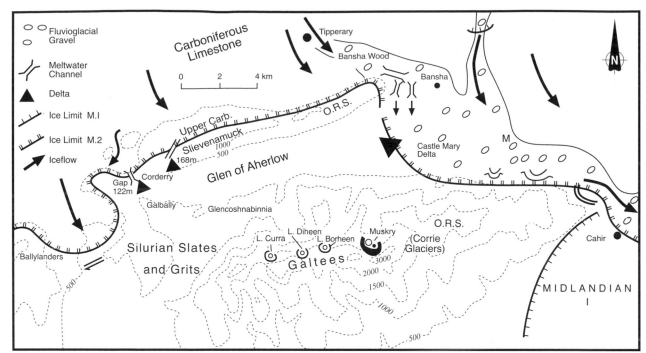

Fig. 7.12 Deglaciation features on the margins of the Slievenamuck range, Counties Tipperary and Limerick (after Synge, 1970a, b).

during recession of the ice margin. This inference is supported by the absence of organic deposits older than the typical late-glacial sequences in hollows either on the line of or to the south of the SIEM.

In Counties Tipperary and Limerick Synge (1970a,b) mapped clearly defined ice limits stretching west from the Slievenamuck Hills towards Galbally, Ballylanders and the Dromcolliher lowland (Fig. 7.12). From here the morainic ridges turn abruptly northward to Ardagh where they are truncated by an east–west moraine associated with a later ice sheet readvance known as the Fedamore Readvance (Synge, 1970a,b). The Ballylanders moraine was thought to represent the limit of the last glaciation based on the presence of fresh, unweathered boulder clay rich in Carboniferous limestone to the north of the moraine. To the south of the Ballylanders moraine Synge reported that the drifts are weathered and do not contain limestone clasts, only silicified limestone. However, there are no known organic deposits within the 'area of Older Drift' to the south which can be used to separate the two drift sheets. The ice limit is marked by thick (>40 m), ice-contact fluvioglacial gravels at eastern end of the Slievenamuck range and by transport of limestone rich tills onto the northern slopes of this ridge.

When each end of the Glen of Aherlow was blocked a glacially impounded lake developed between the Slievenamuck range and the northern slopes of the Galtee Mountains (Fig. 7.12). Delta moraines at Castle Mary and hummocky moraine at Ballylanders record the position of the ice fronts which blocked the glen. Along the eastern end of the Slievenamuck ridge a series of parallel meltwater channels cut in rock conducted subglacial meltwaters directly across the ridge (Fig. 7.2). Synge (1970a,b) recorded three lake levels at Corderry (168 m), Galbally (122 m) and Lisvarrinane (90 m) related to a decrease in ice thickness as deglaciation proceeded. Extensive glaciofluvial terraces were developed at this time in the areas immediately to the east at Golden and Bansha conducting outwash south along the Suir valley. Further retreat of the ice margin to the north is recorded by the topographically controlled development of morainic ridges along the Owenbeg valley draining east from the Knockalough Uplands and near Rear Cross.

The last ice movement in this general area is associated with a well-defined glacial system based on northeast–southwest trending drumlin ridges which occur north from Tarbert northwards along the Shannon covering most of east County Clare (Farrington, 1965; Finch and Synge, 1966). Synge (1970a,) noted that the drumlin bedforms were bounded by well developed morainic ridges on their southern margin about 10 km south of Limerick City on the south side of the estuary at

Fedamore (Fig. 7.1). The line of moraine is at right angles to drumlin long axes and occurs either as single ridges on lowland basins or as kame moraines banked against the northern slopes of the Old Red Sandstone ridge at Ballingarry. Where the moraine crosses the Shannon estuary the sediments in the estuary were deformed by compression into a distinct push moraine. Surviving parts of this moraine occur on the estuary margins at Ralappane and Aylevaroo though the most impressive glaciotectonic features form the core of Scattery and Hog Island in the estuary itself (Fig. 7.13a–d). This pattern shows that ice deformation extended for over 6 km across the widest part of the estuary. On Scattery Island push ridges up to 12 m high trend northnortheast to southsouthwest across the centre of the island. Exposures on the western margins of the ridges consist of a tectonic sequence (>10 m thick) of massive and laminated mud, poorly-sorted cobble gravel and matrix-supported pebbly gravel (Fig. 7.13a,c). At the base of the exposure beds of mud and muddy diamict (3–4 m thick) containing isolated pebbles and cobbles are cut by closely-spaced thrusts which strike NNE–SSW. In some cases thrust planes are concentrated along narrow bands of deformed or streaked out mud a few centimetres in thickness. Up to 4 m of poorly-sorted gravel contains thin (~20 cm) tectonic slices of mud and is characterised by a pervasive clast fabric parallel to major thrust planes (Fig. 7.13d). At the top of the exposure interbedded sand and pebbly gravel do not contain many shear structures though outsized clasts (~1 m long) are common (Fig. 7.13b). The entire tectonic sequence is marked by stacked, listric shear planes resembling an imbricate thrust system. The pattern of shears which rise to the west and the aligned cobbles was generated by ice pressure and proglacial thrusting from the east. This sense of shear is consistent with the ice flow direction and direction of drumlinisation along the estuary from an easterly direction. The type of sediments in the ridge include massive/laminated mud, mud with isolated cobbles and pebbles, disorganised boulder and cobble gravel and laminated sand. Facies of this type were probably deposited in water because of the presence of dropstones. This association strongly suggests that the ice sheet overran its own outwash and that the tectonic structures record an ice sheet readvance from the northeast. The timing of the event is not known for certain but it is likely to be the same age as drumlinisation around Galway Bay because the ice came from the same centre of ice dispersal in south Galway. If this is the case then the Fedamore moraine is younger than the moraine at Ballylanders, possibly forming around 17–16 cal ka BP.

iii) *The Midlands*

There are about five main morainic stages recognised along the eastern lowlands of counties Dublin and Meath (Fig. 7.1). Several minor east–west morainic ridges and meltwater channels record small halts in the ice margin as it receded across north County Dublin at Ward and Delvin (Hoare, 1975). The orientation of the ridges suggests that the ice margin maintained an east–west trend during initial deglaciation. The Delvin moraine complex is represented by glaciofluvial terraces and by a thrust complex buried by later outwash at the southern end of the coastal exposure at Gormanstown (McCabe, 1993).These moraines can be traced inland up to ten kilometres and are spaced at 6–10 kilometres apart. Farther ice recession north is marked by the prominent moraine at Ben Head which can be traced fairly continuously westwards to the northern flanks of the Hill of Tara where the trend changes to southwest where it joins the Galtrim moraine. Esker ridges aligned NW–SE join the ice-proximal face of the delta moraine and acted as subglacial feeder routeways (Synge, 1950). Exposures at Ben Head and inland at Cross show that diamict, gravel and sand comprising the crest of the moraine occur as a nested set of shear planes rising steeply to the south (Fig. 6.20) (McCabe, 1972). The east–west trend of the ice margin at this time is therefore associated with ice pressure from centres of dispersal and ice flow from the north and west. However, the southwesterly continuation of this ice margin cannot be identified directly by discrete lines of moraine across the Irish midlands in Kildare and Offaly. In this area as far west as the Shannon the east–west esker system across the midlands records rapid ice sheet retreat westwards towards a major centre of ice sheet dispersal in southern Galway and possibly northern Clare. The continuity of the sinuous esker ridge system (>100 km long) is punctuated by small subaqueous fans and hummocky, kettled topography crossing the subglacial ridges at right angles (Farrington, with F. M. Synge, 1970). It is thought that many of these crossings represent local halts facilitating deposition at successive ice sheet margins. The overall pattern is one of rapid ice wastage across the midlands towards an ice centre in the west.

At this point during early deglaciation the pattern of ice retreat changes from the general northward retreat from the south coast and the SIEM. The Ben Head/

Fig. 7.13a Laminated and massive mud overlain by poorly-sorted gravel from the western side of Scattery Island, Shannon estuary.

Fig. 7.13b Detail of the tectonic deformation in the gravelly facies sheared by ice push over the mud beds, Scattery Island.

Fig. 7.13c Section showing stacked tectonic units of mud and disorganised gravel, Scattery Island.

Fig. 7.13d Close-up showing total reorganisation of the gravel fabric during tectonic emplacement, Scattery Island.

Galtrim line records a minor readvance and stillstand as ice wasted north towards ice centres in the north. The close proximity of the eastern end of the midland esker system to the southwestern end of the Galtrim moraine near Enfield suggests that both started to develop at the same stage of deglaciation. This pivotal area therefore represents the suture separating ice pressures associated with ice retreat to the west and north and contrasting patterns and directions of ice retreat. It is emphasised that the relative continuity of the midland esker system is intimately associated with the decay of an ice system retreating westward to a single ice centre in Galway. It cannot be regarded as an interlobate formation deposited between northern and southern lobes or domes because ice of northern provenance had already retreated to the Boyne valley as esker sedimentation progressed. In addition the marked east–west pattern of the entire esker system simply records a consistent hydrologic gradient towards one ice centre in the west. An unresolved problem is the presence of subaqueous facies associations formed at successive ice margins during ice sheet retreat across the midlands (Warren and Ashley, 1994; Delaney, 2001). The origins of these ice-contact, proglacial lakes have not been satisfactorily resolved and may result from local topographic controls, high relative sea levels or misinterpretation of the sedimentary sequences.

Outwash from the Ben Head moraine forms a flat spread around Gormanstown and buries part of the earlier formed Delvin moraine. Outwash spreads along the river Nanny and at Mosney simply record further ice wastage northwards. However, a readvance of ice from the Irish Sea formed multiple tills along the south bank of the Boyne at Drogheda and a large moraine at Donacarney (McCabe, 1973). This readvance blocked meltwater escape eastwards down the lower Boyne valley. It is thought that meltwaters from the middle Boyne escaped along the large meltwater gorge at Slanduff and along the Nanny (McCabe, 1971). To the south of Donacarney an extensive spread of outwash occurs as far south as Laytown. Parts of this outwash and the Donacarney moraine have been eroded by the sea to a height of 16 m OD.

The patterns of moraines in the Boyne valley demonstrate that the ice margin withdrew northwestwards forming a suite of moraine and terrace couplets along the Boyne drainage system (Fig. 7.14). Four main terraces have been recognised including the highest or Drogheda terrace which slopes downstream from the

Sheephouse moraine with a gradient of around 6 m/kilometre. Lower terraces associated with the Proudfootstown moraine are similar in gradient. The most important terrace slopes downvalley from the Harlinstown moraine ending in a delta terrace at Drybridge where gravelly foresets dip downvalley. The foresets indicate that the lower Boyne was flooded by rising sea level to about 20 m OD. Sea level fell as the ice retreated farther to the northwest, forming moraines at Boyne House and Stackallen. Sometime later the Morningtown delta formed, recording a sea level fall to 14 m OD at the mouth of the estuary. Reconstructions of patterns of ice sheet retreat to the north during early deglaciation are difficult because a later ice sheet readvance drumlinised the landscapes from north County Louth across the north central lowlands as far south as Athlone around 16.5 cal ka BP.

iv) *The 19-kyr global meltwater pulse*

Although drumlinisation and ice sheet readvance dramatically reorganised the landscapes in northern and western counties there is excellent stratigraphic evidence preserved that shows that the coast of south County Down was ice-free during early deglaciation. Thick sequences of stratified diamict interbedded with gravel, sand and other laminated facies occur along the coast of south County Down between Kilkeel and Derryoge (McCabe, 1986). These facies are glaciomarine and accumulated in a depositional sink known as the Mourne Plain which was fed with sediment from decaying ice sheet margins. The facies pile accumulated well below the local marine limit of 30 m OD (McCabe and Dunlop, 2006). As theory predicts isostatic uplift occurred during deglaciation resulting in rapid uplift and deep subaerial channelling of the diamict apron (Clark et al., 2004). Fossiliferous marine muds dated to 19,000 years infill their channels and mark a significant rise in global sea level because regional isostatic uplift was well underway at this time (Fig. 6.28). The channels were graded to a relative sea level below present, indicating that the coast had emerged at least 30 m by the time they had formed. The presence of channels infilled with mud at Kilkeel provides a critical date for the end of the period of early deglaciation around 19,000 years ago. The relative absence of dropstones and coarse sediment in the laminated mud suggests that the ice margin had withdrawn at least 20 km inland and any coarse debris was contained in temporary sediment traps inland.

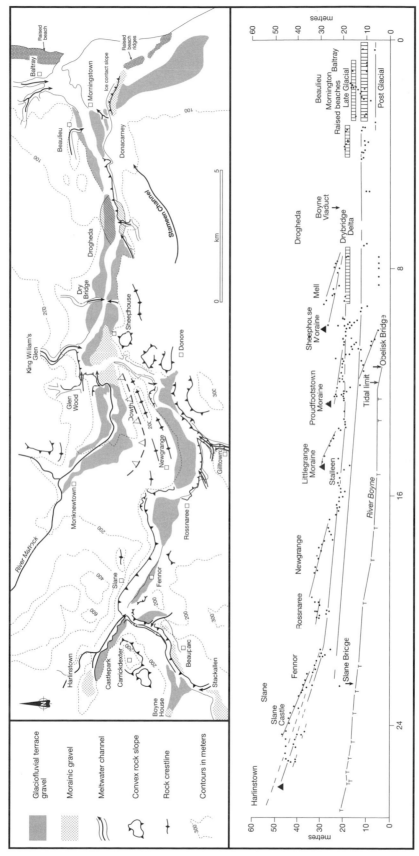

Fig. 7.14 Moraines, fluvioglacial terraces and sea levels formed during deglaciation of the lower Boyne valley, eastern Ireland. Note that the ice margin retreated generally westwards along the valley. This work was jointly undertaken by F. M. Synge and A. M. McCabe in May 1978.

The ice masses remaining after early deglaciation were located in northern and western Ireland. These were the areas where ice regenerated and developed into the Drumlin Ice Sheet which readvanced to maximum limits at 16.5 cal ka BP (Fig. 4.10). Although the precise limits of the remaining ice masses are unknown at the end of the period of early deglaciation, it is inferred that the large lowland basins stretching across from Lough Neagh southwestwards into the Omagh, Erne and Upper Shannon catchment contained remanie ice domes. These are the precise areas where continuous fields of pristine ribbed moraine are found. This bedform distribution probably represents a form of reorganisation at the base of the remaining ice sheet. It is significant that most ribbed moraine fields contain evidence for only one subsequent phase of drumlinisation. Regional coeval ice flow indicators show that drumlinisation was largely a synchronous event in northern and western sectors that postdated early deglaciation. Along the margins of Dundalk Bay the marine muds formed in areas vacated by ice during early deglaciation were subsequently overridden at the maximum of the Drumlin Readvance (McCabe et al., 2007b).

v) *Northwestern Donegal and Mayo*

There is little evidence in the north of the island which can be used to identify the precise areas where early deglaciation occurred because of later, widespread drumlinisation (Fig. 7.1). This ice sheet readvance covered most of the areas that were vacated during early deglaciation in the north and northwest. However, three sites situated on northwestern margins of the ice sheet immediately outside Drumlin Readvance limits provide critical evidence that early deglaciation occurred in northwestern sectors ice sheet, not just in the south and midlands. First, in northwestern Donegal on the western slopes of Bloody Foreland there is a series of subparallel morainic ridges consisting of large granitic boulders (1–2 m across) at the surface. Local pits a few metres deep often contain matrix-supported boulder facies. The moraines occur up to an altitude of 130 m OD and were emplaced at the eastern margin of an ice sheet which advanced onto the continental shelf. Clark et al., (2007b) reported *in situ* cosmogenic [10]Be ages from six boulders which provided an average age of 19.4±0.3 [10]Be ka. This age suggests that the ice flows coming from the western part of the Donegal ice cap vacated the site during early deglaciation. It is not certain whether or not the Bloody Foreland moraine marks the maximum vertical limit reached by the ice or a deglacial phase of surface lowering. Second, 80 km to the east in northern part of the Inishowen Peninsula fossiliferous marine muds at Corvish indicate that the area was deglaciated by 20.2±0.1 cal ka BP (McCabe and Clark, 2003). However, the extent of early deglaciation is unknown because the later readvance from the south overran the Corvish site into Trawbreaga Bay. Third, the two dates from north Donegal which document an interval of a retracted ice sheet margin back into the Donegal Mountains are in close agreement with age-constrained muds from Belderg on the south coast of Donegal Bay. These raised muds are dated to 19.6±0.1 cal ka BP, formed in an iceberg zone, and show that a glacier margin situated in the inner bay was the main source of ice-rafted detritus. The presence of the muds records ice-free conditions and possibly a retreat or destabilisation of an ice margin in Donegal Bay. Collectively, these ages date the retreat of ice from the continental shelf towards the coastline of northwestern Ireland and are in good agreement with ages constraining early deglaciation of the Irish Sea Basin. A critical inference is that the entire ice sheet experienced widespread deglaciation at about the same time. A conservative estimate is that the ice sheet lost more than two thirds of its area and mass during early deglaciation (McCabe et al., 1998).

Deglaciation from drumlin readvance limits (<16.5 cal ka BP)

The subglacial landscapes of northern and western sectors of the island owe their main characteristics, especially the topographic grain of the countryside, to drumlinisation during the last widespread glacial readvance termed the Killard Point Stadial (McCabe and Clark, 1998) (Fig. 1.4). The main centres of ice sheet dispersal developed across the great lowland basins of the north stretching southwest into Joyce's Country and County Galway. The ice sheet limits originally proposed by Synge (1968) were based on morphological grounds. These have been progressively refined by new dating methods, though a central theme to all reconstructions is that drumlinisation effected sediment transport towards ice sheet limits forming ice-marginal moraines. Many ice margins ended at tidewater and overstepped areas vacated by ice during early deglaciation (McCabe and Clark, 1998). Dating of individual moraines is therefore a method used to identify the timing of drumlinisation from different ice sheet sectors.

The Killard Point moraine fronts the drumlin swarms of County Down and was deposited as marine outwash dated to ~16.5 cal ka BP. This ice limit occurs around the margins of Dundalk Bay and occurred slightly after 16.9 cal ka BP. A slightly earlier readvance occurred ≤18.5 reached Clogher Head and is part of the same ice sheet system (McCabe et al., 2007b). These ice limits continue southwest across country to Athlone where the Lough Ree esker system transported meltwater southwards. The Furnace Lough moraine on the north side of Clew Bay was deposited by ice which drumlinised the lowlands north and northwestwards from the centre of ice sheet dispersal in Joyce's Country. On the south side of the bay the glaciomarine spread at Askillaun marks a similar limit to the ice lobe in Clew Bay. Final deposition of this moraine occurred around 15.6 ± 0.4 [10]Be (Clark et al., 2007a). The continuation of this moraine northwards on the west side of Killala Bay is known as Tawnywaddyduff moraine and can be traced across the floor of Donegal Bay (Fig. 1.7). The southern margin of the ice lobe in Donegal Bay is marked by a regionally significant moraine at about 300 m OD on the northern flanks of the Ox Mountains which was finally formed at 15.6 ± 0.5 [10]Be ka (Clark et al., 2007a). Finally, in Inishowen the last ice sheet readvance overran marine muds at Corvish which date the readvance to ≤16.7 cal ka BP. These dates using two different methods (AMS ^{14}C on marine microfaunas and [10]Be surface exposure ages) from widely spaced ice sheet limits demonstrate an ice sheet readvance which reached its maximum around 16.5 cal ka BP and possibly remained there for about one thousand years.

Information is available on a variety of levels on the pattern and nature of ice sheet retreat from the age-constrained outer limits of the drumlinising ice sheet. At least four distinct types of ice decay are recognised from different sectors of the drumlin ice sheet:

i) Stagnation zone retreat occurred in southeastern Ulster, the north central lowlands and Donegal. This inference is based on the fact that no significant or continuous lines of moraine have been located between the ice limits and associated centres of ice dispersal. For example, there are no extensive areas of sand/gravel or frontal moraines between the Killard Point/Athlone ice limit and ice sheet centres to the north. The absence of significant evidence for morainic halts indicates that ice retreat was rapid with little time for frontal deposition. In east County Down a marine transgression dated to 15.5 cal ka BP shows that glacier ice had completely disappeared from northern Strangford Lough by this time. In the Donegal Mountains there are no well-defined patterns of cross-valley ridges which could be used to reconstruct a phased ice sheet withdrawal. It is concluded that rapid ice wastage occurred following ice advance to the morainic limits on the northern and western periphery of the mountain ice cap at Carndonagh and Arduns. However, it should be noted that deglaciation events were quite different to the east of the mountains.

ii) In the west deglaciation records the wastage of the ice that moved northwards from Joyce's Country into Clew Bay and Killala Bay. There are no discrete moraines marking stillstands and phased retreat of the ice margin though Coxon and Browne (1991) noted that the presence of kames, deltas and undifferentiated sand and gravel deposits may represent retreat of ice back into Roscommon. The great esker system stretching from south of Tuam, 70 km northwards to Kilkelly across the Plains of Mayo records a major meltwater system (Fig. 7.1). The gently curving esker system forms an arc that bends gradually to the northwest which subparallels changes in the local trend of striae, streamlined hills and glaciated lake basins. This close association demonstrates that the mainly subglacial system was associated with the decay of a single ice mass. Areas of hummocky topography adjacent to and partly superimposed on the esker ridges indicate a general pattern of ice wastage to the south which is also suggested by the pattern of northward converging ridges. Although the eskers are not associated with either stagnation zone retreat or marked backwasting features, their pattern suggests that the ice source to the south remained fairly active during ice sheet decay.

iii) The lowlands to the north of the Sperrin mountains were supplied from southeastern Donegal, the Omagh Basin and the Lough Neagh Basin with ice that moved across the north coast during drumlinisation (Fig. 7.15). When this ice withdrew from the Malin Sea into Lough Foyle hummocky moraine formed during a stillstand in the lower Faughan valley at Gransha (Fig. 7.15). Ice retreat

Fig. 7.15 Generalised ice flows, glacial landforms and evidence for deglaciation in the Roe and Faughan valleys, Foyle estuary, Northern Ireland. The ice limit at Ballykelly is probably of the same general age as the Armoy moraine farther east in County Antrim. Note that the prominent deltas in the valleys record ice wastage southwards towards the Sperrin valleys while meltwater outlets to the north were blocked by the ice at Ballykelly (from McCabe et al., 1998).

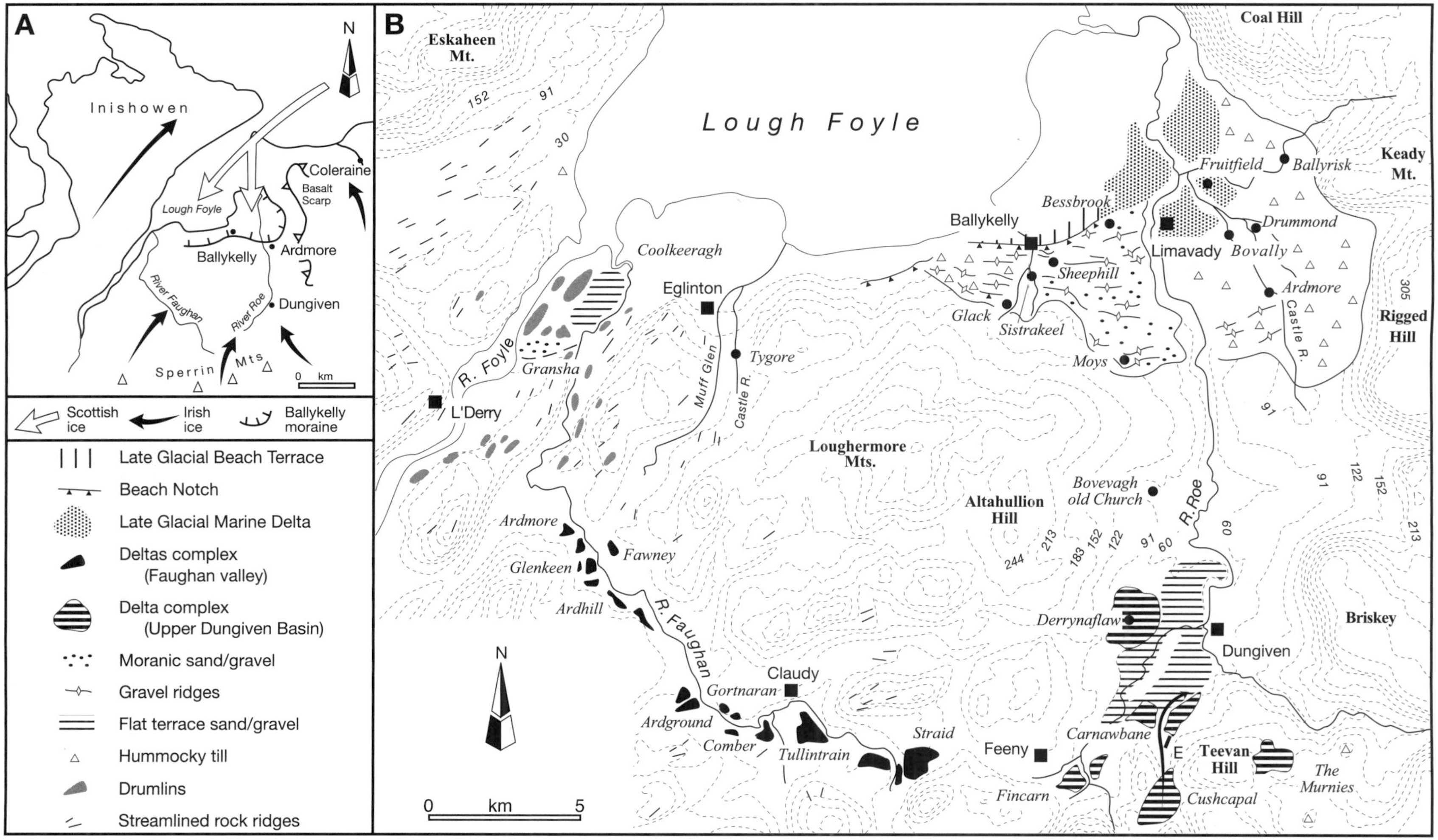

farther south for 25 km along the Faughan valley is marked by successive ice-contact deltas recording ponded water up to 120 m OD. A similar pattern of high level Gilbert-type deltas occurs to the east in the upper Roe valley with water planes up to 240 m OD (Fig. 7.16). In both valleys the heights of the delta surfaces increase upvalley. These deltas are particularly well developed around Dungiven and consistently contain large-scale foresets dipping to the north and northeast (Fig. 7.16b). Because both valleys open northwards into Lough Foyle it is likely that ponding of a large body of open water in both valleys was caused by the same agency. The well-defined morainic topography around Ballykelly on the southern shore of Lough Foyle is associated with a major ice sheet which blocked northward drainage along both valleys (McCabe et al., 1998). The fact that the delta levels increase upvalley suggests that the lake levels increased as ice wasted southwards along the valleys. This field relationship demonstrates that the ice blocking the northern valley exits was advancing and thickening as the inland ice sheet margin retreated southwards and formed deltas.

The Ballykelly moraine is probably contemporaneous with the Armoy moraine to the east of Binevenagh Mountain on the Bann/Bush lowlands (Fig. 7.17). The morainic belt is convex southwards and consists of a nested series of linear ridges with discrete crestlines aligned subparallel to the overall convex trend of the feature. Ridges are up to 40 m high forming an almost continuous landscape feature between Armoy, Ballymoney, Quilly Head and Articlave. At Ballymoney, Shaw and Carter (1980) described a section through one of the main

Fig. 7.16a Flat-topped extraglacial delta at the Murnies, Dungiven, County Londonderry. The delta was fed by the Altnaheglish channel from an ice margin that had vacated the Roe depression and was located on the north flanks of the Sperrin Mountains. Delta surfaces and large-scale foreset beds record two former lake levels at 230 and 243 m OD.

Fig. 7.16b Gravel foresets from the Murnies delta dipping northeastwards the lake impounded in the upper Roe valley around Dungiven.

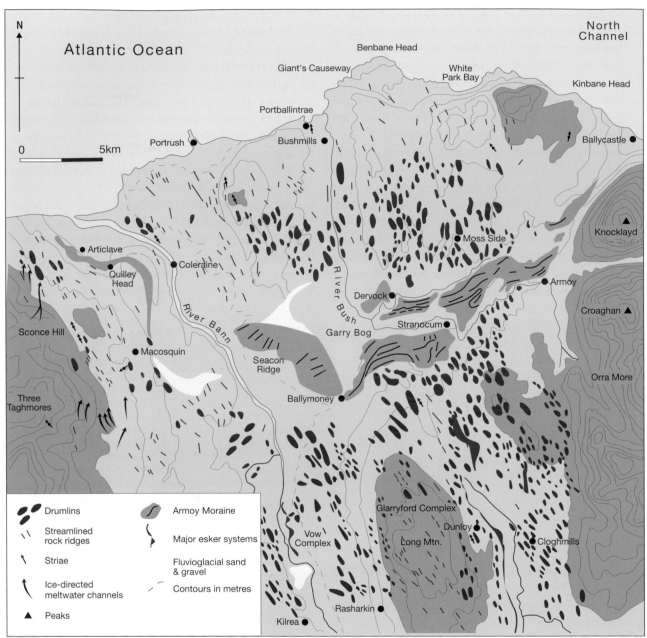

Fig. 7.17 The Armoy moraine, north County Antrim and the patterns of subglacial flow-parallel bedforms. The moraine is convex southwards and was formed by a southwesterly ice readvance from western Scotland into lake sediments deposited as the Irish ice masses withdrew southwards (Shaw and Carter, 1980). It postdates the bedform patterns which maintain a similar trend on both sides of the moraine. The bedforms record ice sheet flow from the Lough Neagh basin northwards onto the inner continental shelf (McCabe et al., 1998). *By kind permission of GSNI.*

ridges consisting of interbedded diamict, laminated silt and mud, coarse gravel and massive sand with a total thickness of >60 m. Beds were steeply inclined to the northwest and closely related to ice thrusting. The interbedded sediments are typical of proglacial lake deposition with ice pressure causing thickening of the sediments and vertical duplication of the original stratigraphy. On the basis of the large-scale glaciotectonic structures, erratics of northeastern provenance, the presence of an ice-toe depression immediately to the north of the ridges now occupied by Garry Bog and the convex bulge in ridge morphology southward, it is argued that ridge building was caused by an advance of Scottish ice onto the lowlands of north Antrim. The readvance of Scottish ice occurred as Irish ice withdrew southwards towards the centre of ice dispersal in the Lough Neagh basin. Rhythmites associated with density underflows drape subaqueous gravelly spreads deposited from esker tunnel mouths during ice sheet wastage southwards. This ice sheet readvance occurred late in the general deglaciation but did not destroy the drumlin bedforms across north County Antrim (Fig. 7.17). It is possible that ice erosion was minimised by the presence of extensive areas of water-saturated fine-grained sediment. In essence the ice cannibalised its own outwash as it moved southwestwards across the lake bed before forming the outer moraine ridges by compression and sectional shortening.

During the Armoy readvance the North Channel was choked with thick ice because drumlin orientations along the Glens of Antrim were fashioned by ice flows moving westwards (inland) up the glens almost into the Lough Neagh basin. The hummocky topography and eskers between Doagh and Parkgate mark one ice sheet limit associated with ice moving inland along the Glenwhirry valley. This ice readvance probably helped to maintain ponded water in the Lough Neagh basin along retreating ice sheet margins. The Armoy readvance was contemporaneous with the readvance of ice into the Glens of Antrim as originally postulated by Dwerryhouse (1923). A section at Ballyrudder described by Praeger (1892/3) to the north of Larne records the basal till of this readvance overlying shelly gravels and laminated red clays. Shell dates from the gravels were infinite, suggesting that they are derived from earlier deposits on the margins of the North Channel. It is likely that the landslips at Sallagh Braes, 5 km northeast of Larne, were ice-moulded during this readvance. If this is the case it is possible that the initial period of rotational slumping occurred during early deglaciation when the greatest amount of ice unloading occurred. After Scottish ice vacated the north Antrim coast a large marine delta formed at Limavady and shallow wave-influenced rhythmites are preserved in the horseshoe bay at Portballintrae (Fig. 7.18) (McCabe et al., 1994).

iv) As the Irish ice retreated southwards through and around the eastern and western flanks of the Sperrin Mountains a distinctive suite of deglacial landforms developed. Originally Dwerryhouse (1923) and Charlesworth (1924) identified a series of ice-

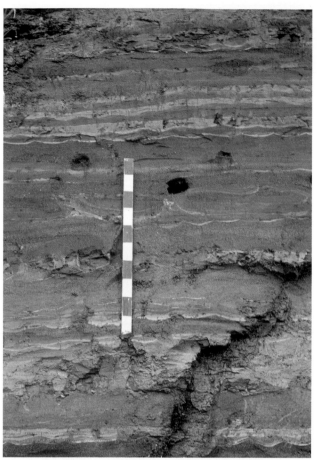

Fig. 7.18 Raised, shallow marine rhythmically-bedded sand and mud couplets formed during storm events at Portballintrae, County Antrim. These facies are characterised by ripple trains draped by mud which extend for tens of metres (McCabe et al., 1994). The rhythmites are part of an emergent facies sequence comprising glaciomarine diamict, shallow-water rhythmites and upper shoreface gravels. The rhythmites have a low preservation potential and are generally missing from poorly-preserved coastal exposures.containing only diamict truncated by wave action and overlain by upper beachface sand and gravel.

dammed lakes in the Sperrins using mainly morphological criteria. Later, Colhoun (1970, 1972) used the absence of lake shorelines, minor occurrences of deltas and lack of lake bottomsets to reject this general model and replace it by a fluvioglacial one. Dardis (1982, 1985a, 1985b, 1986) re-examined the sedimentology of the Tyrone gravel complex which is bordered in the east by the western Lough Neagh drumlin field and the central Tyrone drumlin field to the west (Figs. 7.19, 20, 21). This position along the junction between two major ice sheet flows suggests that the complex developed along this deglacial suture (Fig. 7.21). The landform assemblages which

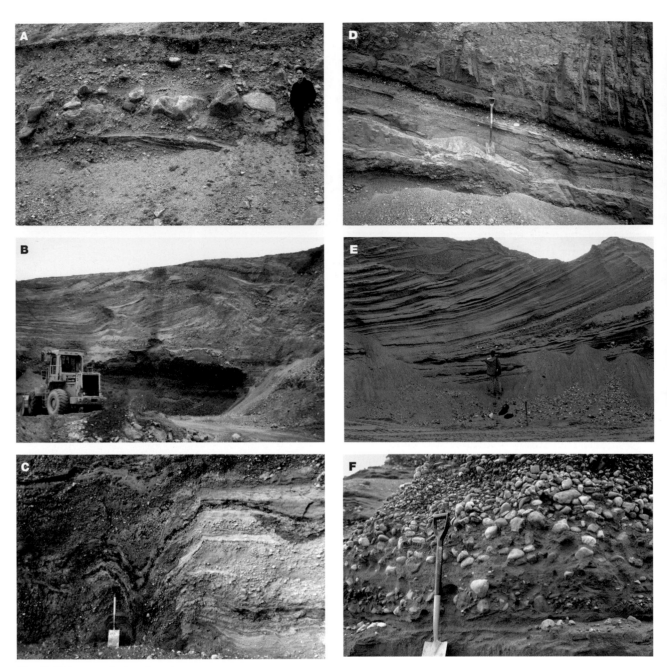

Fig. 7.19 Ice-marginal sediments:

a. Coarse-grained, poorly-sorted gravel sequences formed by debris flows and high energy flows, Ballypriest, County Tyrone. Note the extreme variability in grain size associated with pulsed flows.

b. Cross-valley ridge composed of ice-sheared subaqueous sands and gravels, Ballinderry river valley. Sense of shear from right to left.

c. Outwash gravel containing pull-apart structures developed when supporting blocks of ice melted out after deglaciation.

d. Outwash gravel and sand beds forming the limb of a syncline formed when blocks of ice buried by outwash slowly melted out, Castleroe, County Londonderry. Note the ripple trends along the rising limb.

e. Concave foresets from a delta moraine, Knockaleery, Ballinderrry river valley.

f. Bimodal gravel consisting of cobbles set in sand and granules, Muntubber cross-valley delta moraine, County Tyrone. The cobbles are formed by longitudinal sorting in a glaciofluvial stream prior to delivery at the brinkpoint of a delta foreset sequence and buried by a sandy return current coming up the delta foreset slope.

Fig. 7.20a *(above)* The ice-contact slopes and kettles within the delta moraine at Gortin, County Tyrone looking north. The delta formed in a glacial lake in the Owenkillew valley and was fed by subglacial meltwaters from the main ice margin to the south. This lake developed when the western exits of the Owenkillew and the Glenelly valleys were blocked by ice moving north from the Omagh basin.

Fig. 7.20b *(left)* The Butterlope Glen looking north. This meltwater channel acted as an overflow from the large glacial lake impounded in the Glenelly valley to the south. The intake of the channel is marked by a deeply-eroded scabland topography marked by a variety of erosional fluvioglacial bedforms.

occur within the complex include kame terraces, glaciolacustrine deltas, feeding beaded eskers, and irregular hummocky gravel topography (Fig. 7.21) (Dardis, 1980). Dardis (1985a, b) mapped flat-topped deltas, lake outlets, shoreline remnants and outwash terraces in south-central Ulster and concluded that glaciolacustrine sediments were widespread across the area (Fig. 7.19). He rationalised the pattern of deglaciation by examining the heights of delta tops, locations of cross-valley moraines and the feeding esker systems, palaeocurrent directions and evaluation of proximal to distal facies relationships in cross-valley moraines. On this basis ten major glacial lake phases were reconstructed in south-central

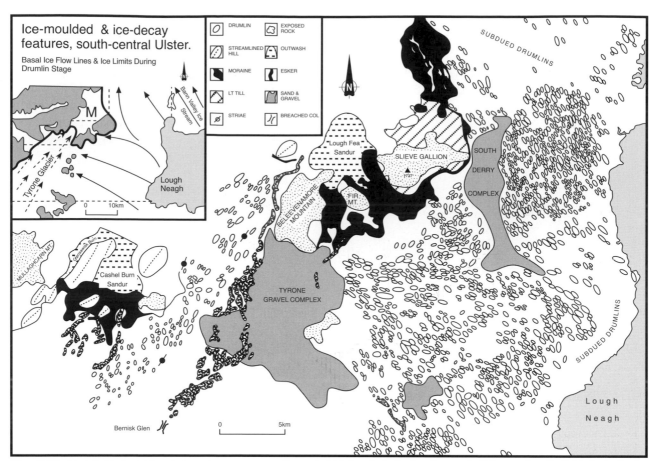

Fig. 7.21 Ice-moulded and ice-decay phenomena, south-central Ulster. A- indicates the central Tyrone drumlin field; B- indicates western Lough drumlin field; C-indicates Lough Fea-Fir mountain moraines; D- indicates the Evishanoran esker; M-is the Moyola valley. Inset shows flow lines during final drumlinisation (from Dardis, 1986). *By kind permission QRA (Quaternary Research Association).*

A Newtownstewart
B Glengink
C Gortin
D Cashel Burn
E Killucan
F Evishanoran
G Evishatrask
H Sultanbane
I Sluggen
J Cappagh
K Edenfore
L Lurganmore
M Todd's Leap
N Garvaghy
O Knockaleery
P Annahavil
Q Carmeen
R Cookstown
S Coalisland

Fig. 7.22 Major ice retreat stages in south-central Ulster including esker ridges and meltwater flow (from Dardis, 1986).

Ulster in relation to major retreat phases extending from Newtownstewart in the west to Cookstown in the east (Figs. 7.22, 7.23). In general the ice sheet margins receded from the uplands into the lowlands towards centres of ice sheet dispersal in the Lough Neagh, Omagh and Erne basins (Fig. 7.21). In the upper Foyle catchment ice receded westwards along the river Derg towards the remnants of the Donegal ice cap around Barnesmore. The presence of ice sheet thrusting of delta sediments and cross-valley moraines suggests that the ice remained active (Fig. 7.19b). However, where an ice lobe became separated topographically from the main ice sheet as in the central Sperrins, active ice gave way to ice stagnation. This process may explain Colhoun's (1970, 1972) emphasis on ice stagnation especially along some east–west Sperrin valleys (Fig. 7.24, 7.25).

Fig. 7.23 An example of one major lake system and associated ice sheet margin during the Cashel Burn-Gortin stage, County Tyrone (from Dardis, 1986).

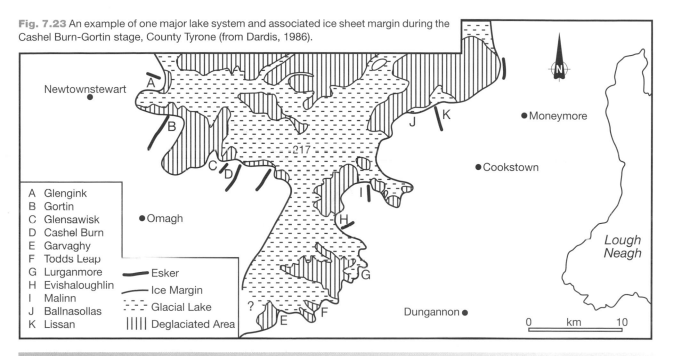

A Glengink
B Gortin
C Glensawisk
D Cashel Burn
E Garvaghy
F Todds Leap
G Lurganmore
H Evishaloughlin
I Malinn
J Ballnasollas
K Lissan

— Esker
— Ice Margin
---- Glacial Lake
|||| Deglaciated Area

Fig. 7.24a Hummocky moraines near Pomeroy, County Tyrone.

Fig. 7.24b Hummocky mounds of sand and gravel deposited across the Glenelly valley at Goles Bridge as the ice margin wasted westwards from the central Sperrin range.

Fig. 7.25a Outwash gravel at Comber delta, Claudy, County Londonderry, containing cross-cutting ice-wedge pseudomorphs.

Fig. 7.25b Coarse-grained outwash in the Douglas Bridge fluvioglacial terrace, Foyle outwash system. The outwash is very poorly sorted and was deposited kilometres from the nearest ice margin. It may represent a catastrophic event such as the collapse of an ice dam or rapid drainage of an impounded lake. This interpretation is consistent with the presence of erosional terraces along the Foyle system unrelated to distinct ice sheet margins.

The last glacial termination

The three major periods of ice sheet deglaciation illustrate the complexity of events during the last termination. Ice sheet readvances and major periods of ice sheet wastage during deglaciation suggesting that a range of major drivers influenced the glacial system. The first deglaciation on the western seaboard was related to isostatic depression and ice wastage into the Atlantic Ocean. However, the deposits formed at this time at Glenulra on the southern margin of Donegal Bay were not subsequently overridden by ice during the traditional LGM (22–25 cal ka BP). Therefore ice sheet configurations were complex and are not known in any great detail during the course of the glaciation. More importantly what were the mechanisms which caused early deglaciation (<22 cal ka BP) when most of the ice sheet disappeared to a few centres in northern and western sectors? Clark et al. (2004) suggested that because the ice sheet was located in a region which is particularly sensitive to changes in North Atlantic climates, it is possible that this early deglaciation occurred in response to an increase in the Atlantic meridional overturning and the associated warming of the North Atlantic. This began around 22,000 years ago (Grootes et al., 1993; Bard et al., 2000). However, initial deglaciation may have been triggered by relative sea level rise associated with isostatic depression and ice sheet downdraw. Later ice sheet regeneration during the Killard Point Stadial and drumlinisation across the northern and western sectors occurred during the *Oldest Dryas* and bracket Heinrich event 1 (McCabe et al., 2007b). Ice sheet regeneration must have involved a more positive mass balance as a result of a more vigorous hydrological cycle. Finally, it is not known for certain why ice from western Scotland readvanced into north Antrim at a time when the Irish ice sheet margin had decayed southwards. It is possible that the lowland ice sheet was more sensitive than a highland one to relatively small temperature and moisture changes.

8

The Eskers of Ireland

Introduction

An Irish author, Charles Smith, was in 1774 perhaps the earliest writer to liken the appearance of a drumlin swarm to the surface of a basket of eggs, and it is fitting that the earliest known description of an esker in British literature should also have come from an Irish observer, Richard Prior in 1699 (Davies, 1970). Prior's description of the ridge extending from Timahoe to Port Laoise and Mountmellick included comments on trend, length, breadth, steepness of slopes, sinuosity and sediment texture. The geomorphic attributes of the ridges were rationalised within the prevailing intellectual environment of the time which supported the idea of the Naochian Deluge. The widespread occurrence of gravelly ridges across swathes of the Irish lowlands has prompted more detailed comments on their origins and geomorphic significance for over one hundred and fifty years. The Rev. Maxwell Close in 1865 distinguished flow-parallel, streamlined drumlins composed of unsorted boulder clay from elongated and often discontinuous esker ridges formed of stratified or water-sorted drift.

Victorian naturalists such as Kinahan (1864) recognised the linear esker patterns, variability in internal structure and topographic setting. However, interpretation was dominated by reference to one of the prevailing geomorphic models of the time which envisaged Ireland as an archipelago of islands submerged by the sea. Therefore one early classification of esker ridges, based on tides and currents flowing among these islands, recognised fringe eskers around high ground, barrier eskers stretching between two areas of high ground and shoal eskers similar to present day shoals with low, undulating hills. Observation of local topographic detail in the form of blunt and tapering patterns of drift ridges were related to the inferred directions and topographic deflections of regional marine currents. A related example uses the idea that icebergs drift to particular localities, melt and deposit angular clasts and erratic blocks on esker slopes.

Although this classification appears weak it emphasised field observations and for the first time recognised complexity, topographic position and structure in eskers as important features to be explained by later work. More importantly, Kinahan (1864) tried to group field observations into a working model of sea current patterns over the entire island, and in doing so initiated an early systems approach to geomorphic explanation. Indeed he hypothesised that esker form changed or evolved as the system variables (shallowing and erosion) altered during marine emergence.

W. J. Sollas

All subsequent research on the distribution of Irish eskers is based on the classical paper by Sollas (1896) which included a map of the 'esker systems' of Ireland (Fig. 2.1). Sollas's object was to test the writings of Jukes and the marine origins of eskers using published maps of the Geological Survey supplemented by field observations. A marine origin was immediately rejected because no marine shells were found when fieldwork began. He initially realised that these gravelly ridges were not being formed by the sea at the present day (modern analogue method). Subsequent questions posed centred around how a ridge of loose gravelly debris could be deposited by water and result in steeply sloping sides, have little spatial and directional relationship to modern streams, why internal structures dipped parallel to ridge sides, why did ridges unite and enclose enclosed depressions and why should ridges branch and meander? He concluded that esker systems presented a remarkable resemblance to a map of a river system including linear, meandering, branch converging or diverging, loop and knot elements. Perhaps the most significant conclusion was that eskers are casts of glacial streams deposited within ice walls during melting of an ice sheet. This hypothesis could explain many features of eskers including gaps in a linear system related to breaching efficacy of floods, formation on lowlands (<150 m) where ice remained longest, steep sides related to formation within narrow ice-walled channels, the run (convergence or divergence direction) of an entire esker system and ice flow/pressure, deformation and faulting of esker deposits in association with buried ice, settling of large erratics on top of eskers from ice melt late in the deglaciation process and the coincidence of ice flow directional indicators with the trend of the two great esker systems in Ireland (Fig. 1.4).

Sollas refined his general model by several field observations and inferences. He noted that the eskers of greatest dimensions were related to the thickest ice which provided the greater 'fall' of water and therefore the largest tunnels. He suggested that tunnel fill by water-worn debris was followed by channel abandonment and formation of a subjacent new tunnel. Associated with this idea was the observation that the linear postglacial lakelets and bogs flanking eskers represent esker moats which result from the scouring effects of meltwater. Because the moats are subparallel to and closely associated with the gross esker morphology, Sollas argued that they may in some cases be the first stage in the formation of an esker. In essence the small, isolated elongated postglacial bogs and larger integrated bog networks and meshes may be used as a surrogate for ice flow direction. A very perceptive comment was that ground moraine (till) is converted by degrees into a water-worn material by being forcibly pressed into nearby tunnels. Finally, he argued that the wide areas of bogs and the absence of elongation of the Midland system is suggestive of sluggish ice flow in an immense pool of ice to the west that was drained by esker channels transporting Connemara marble and Galway granite eastwards.

General characteristics of esker systems

At the outset it should be realised that although the literature records small esker-like ridges in the process of formation at contemporary ice sheet margins, the extensive esker systems formed beneath the large Pleistocene ice sheets are much larger in scale. Whether or not a modern analogue exists for these fossil lineations snaking across many kilometres of countryside is questionable. It is therefore not surprising that the literature abounds with alternative interpretations of both diversity in esker morphology and sedimentology.

In Ireland the main esker systems are generally regarded as sinuous ridges of stratified sands and gravels, up to 100 km long, recording meltwater drainage of the ice sheet during final deglaciation of the lowlands (Fig. 7.1). Much smaller esker ridges occur in close association with ice-contact fluvioglacial accumulations which formed during temporary halts or re-equilibrations of the ice sheet margins (Fig. 8.1). Although eskers may form by subglacial water flow at the base of active ice, formation of the actual arborescent network of drainage channels must predate the final sedimentation events which constitute the final esker ridge. The channels follow hydraulic-potential lows on the glacier bed and trend normally in the same direction as the ice-surface slope because the latter largely controls hydraulic equipotentials (Clark and Walder, 1994). Commonly, eskers have steep sides, single crestlines and typically are a few tens of metres in height (Fig. 8.2). McDonald and Vincent (1972) suggest that sedimentation is built up over time from the channel floor upwards as successive conduits are created by melting of subjacent ice walls. However, many single crested forms are associated with

Fig. 8.1 The Munville esker near Lisnaskea, County Fermanagh (see Fig. 8.15 for location). The feature is a cast of a subglacial tunnel showing a sinuous planform, steep sides and a rising crestline towards the viewer. The esker conducted water from the north and increases in height southwards towards the ice margin at Doneen Hill.

Fig. 8.2 Winding esker ridge at Cloonacannana north of Swinford, County Mayo. Note the steep, ice-contact slopes and the small knolls studded along the ridge crest

sheets of sand and fine gravel forming flat-lying topography subparallel to and extending hundreds of metres from the esker flanks. This close field relationship suggests that the sedimentation events that formed the main ridge core are different from those which deposited the flanking flatter topography. Possibly repeated meltwater pulses spread laterally from the main esker conduits during phases of high hydrological pressure. Sedimentation of coarse debris producing steep-sided, sharp-crested eskers is consistent with high-energy deposition within R-tunnels although lower energy flows can produce low energy cross-bedded gravel and sand. Facies associations within eskers can therefore include a wide range of facies related to different energy levels ranging from high density heterogeneous suspensions (chaotic boulder gravel) to parallel laminated sand (low energy density flow) (Fig. 8.3).

The eskers which form within long R-tunnels are generally recognised by the presence of sinuous almost continuous ridges extending across country for kilometres. Many sectors of the two great esker systems of Ireland have sectors of this type in the central midlands and on the Plains of Mayo (Figs. 8.4, 8.5). Both systems are developed across Carboniferous limestones and not crystalline bedrock with the variable till cover tending to be thin. The morphological characteristics of esker complexes vary even along the run of an individual system. Long meandering sectors formed within R-channels formerly extended back into the ice sheet. More morphologically diverse sectors seem to be associated with sedimentation nearer the ice margin recording environments including short tunnel, re-entrant, marginal and fan/delta depositional settings. Resultant topographies can include coalesced ice-marginal deposits forming a ridge perpendicular to the subglacial feeder channels, which could be interpreted as a moraine with hummocky topography. In such cases the ridge records a local or short-lived halt providing a record of the position and perhaps configuration of the ice sheet margin at a particular point during

A

B

C

D

Fig. 8.3a Disorganised boulder gravel deposited in a closed esker conduit, Enniskerry.

Fig. 8.3b Planar and cross-laminated sand and pebble gravel from a subaqueous fan, Rooskagh esker, County Roscommon.

Fig. 8.3c Subaqueous outwash fan showing rhythmically bedded laminated and rippled sand, Rooskagh esker, County Roscommon.

Fig. 8.3d Arched bedding from the Glarryford esker, County Antrim. This example is indicative of the withdrawal of lateral ice support rather than primary flow in a conduit.

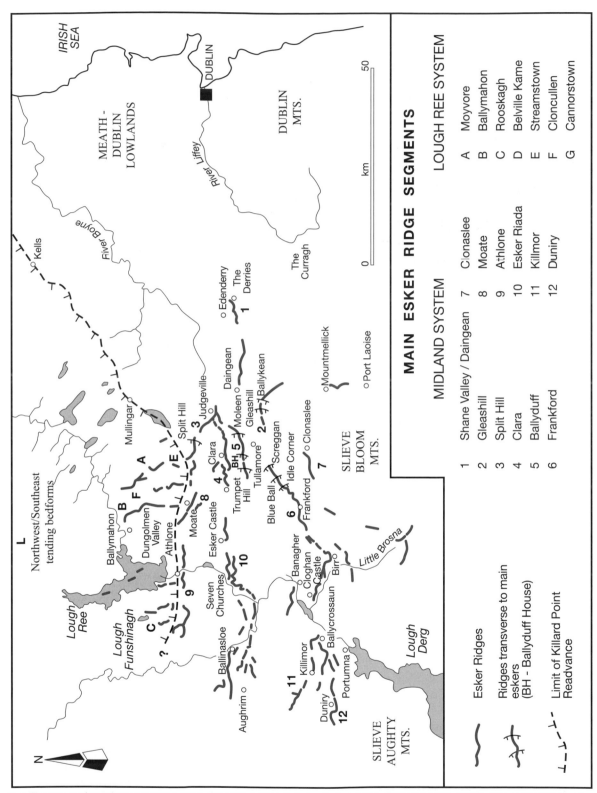

Fig. 8.4 The two major esker systems across the Irish midlands (after Sollas 1896, Delaney 2001, 2002). The earlier Midland system records ice retreat westwards towards a centre of ice sheet dispersal in County Galway. The later or Lough Ree system is at right angles to the Midland System and conducted meltwater southwards towards an ice front situated between Athlone and Mullingar. The Midland Esker System formed during early deglaciation around 20–19 cal ka BP and the Lough Ree Esker System formed at 16.5 cal ka BP shortly after the maximum of the Killard Point Stadial.

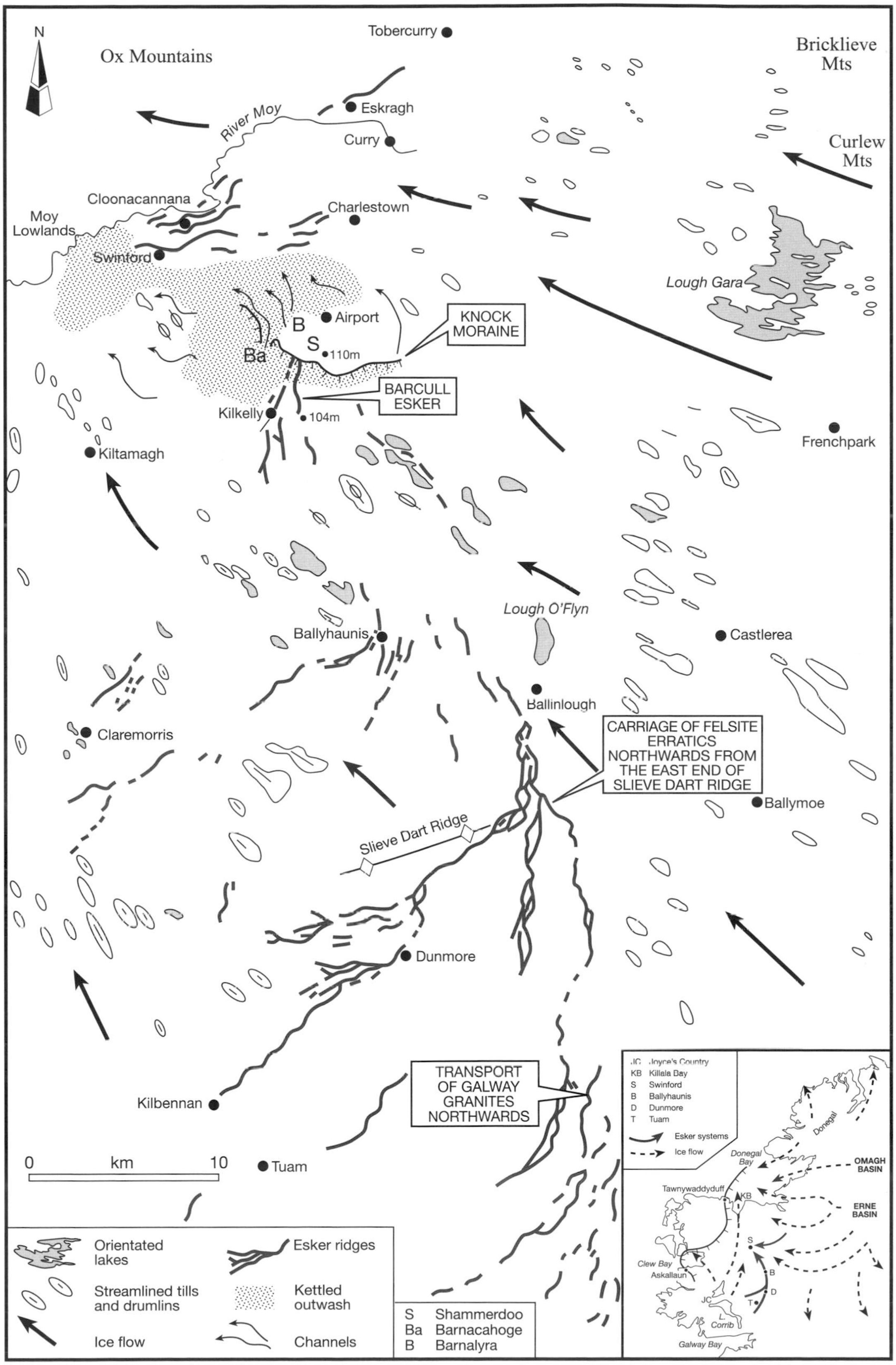

N
Ox Mountains
Tobercurry
Bricklieve Mts
Eskragh
River Moy
Curry
Curlew Mts
Cloonacannana
Charlestown
Moy Lowlands
Swinford
Lough Gara
Airport
B
S
Ba
110m
KNOCK MORAINE
BARCULL ESKER
Kilkelly
104m
Frenchpark
Kiltamagh
Lough O'Flyn
Castlerea
Ballyhaunis
Ballinlough
Claremorris
CARRIAGE OF FELSITE ERRATICS NORTHWARDS FROM THE EAST END OF SLIEVE DART RIDGE
Ballymoe
Slieve Dart Ridge
Dunmore
TRANSPORT OF GALWAY GRANITES NORTHWARDS
Kilbennan
0 km 10
Tuam
Orientated lakes
Esker ridges
Streamlined tills and drumlins
Kettled outwash
Ice flow
Channels
S Shammerdoo
Ba Barnacahoge
B Barnalyra
KB Killala Bay
S Swinford
B Ballyhaunis
D Dunmore
T Tuam
Esker systems
Ice flow
Donegal
Donegal Bay
OMAGH BASIN
Tawnywaddyduff
ERNE BASIN
Clew Bay
Askallaun
JC
L. Corrib
Galway Bay

Fig. 8.5 The Connaught Esker System on the Plains of Mayo (partly based on Sollas 1896 and McCabe 1985).

deglaciation. Ridges formed by sequential deposition near to or at the ice margin from a common tunnel can resemble a string of beads recording diachronous events as the ice margin recedes. Individual beads can be conical, elongate or mesa-shaped depending on the initial shape of the depositional node, permanency of adjacent ice walls, available sediment load, presence or absence of standing water and subsequent erosion as local deglaciation progresses. Generally these forms are regarded as diachronous along the depositional system with conspicuous steep-sided ridges or cones separated by topographic lows depicting successive, ice-contact depocentres at a receding ice sheet margin. The precise nature of the ice-marginal depositional environment may be reconstructed by the facies assemblages present and whether or not subglacial discharge ended in standing water (Fig. 8.3).

Not all beaded eskers are necessarily associated with diachronous deposition at successive tunnel mouths. Some R-channels with variable cross-sections comprising restricted and narrow tunnel sectors linking broader, more open cavities can result in a beaded morphology because in order to maintain water flow along the system, water must flow faster in the narrower reaches. Therefore, erosion is more likely to dominate in the narrower reaches during full pipe-flow, and deposition where flow expansion occurs in lower energy, broad cavities. It is also recognised that elongated esker systems can either alternate with meltwater channels along a slope, or their paths can be preserved up a reverse slope. In both cases ice walls are necessary factors involved in formation. However, the latter situation requires either a hydraulic head in a subglacial situation or time-transgressive sedimentation during ice-marginal retreat (Fig. 8.6).

There is no doubt that eskers in Ireland formed during deglaciation because they are commonly found in channels eroded into till or rest on subglacial till. This general situation occurs across the lowlands of central Ireland, but in the northern counties eskers are found superimposed on either drumlins or transverse ribbed moraines (Knight and McCabe, 1997a). It is also fairly certain that even our largest esker systems extending tens of kilometres across the lowlands only represent a snapshot of the entire glacial drainage system. Theory suggests that the main sources of melting would have been concentrated near to ice sheet margins and there is little field evidence for groundwater supply to the basal melting zones and near-surface melting. For these reasons alone it is likely that the large, apparently continuous esker systems across the midlands and on the Plains of Mayo are time-transgressive and linked to ice-marginal recession. The main eskers record surface melting over a relatively narrow marginal zone where the meltwater reached the ice/sediment/rock interface.

The two main esker systems formed at quite different times during deglaciation. The east to west system across the midlands from The Derries towards Athlone formed as the ice margin withdrew westwards after the early deglaciation of the Irish Sea basin shortly after 22 cal ka BP. The pattern and general run of the esker ridges clearly shows that the ice margin withdrew rapidly towards a major centre of ice sheet dispersal in the west of the island. The major ice sheet readvances which followed around 17–16 cal. ka BP reached almost to Athlone town and filled the large embayments along the western coast (Fig. 7.1). Eskers with a north to south orientation located on either side of Lough Ree are almost superimposed upon the earlier east–west trending midland eskers at Athlone (Fig. 8.4). They formed during the Killard Point readvance because they are flow-parallel to the ice directional indicators of this readvance (McCabe et al., 1998). The Connaught system stretching north towards Swinford across the Plains of Mayo is also subparallel to the ice flow indicators related to ice flow northwards from a major centre of ice sheet dispersal in County Galway in Joyce's country (Fig. 8.5). To the north the ice flow passed Killala Bay forming a prominent submarine moraine across Donegal Bay. The Connaught eskers record ice wastage southwards and meltwater drainage northwards from this ice limit (Fig. 8.5).

Although esker ridges often provide spectacular landscape elements they may provide insights into the nature and patterns of deglaciation. Before listing some of these attributes it is worth being reminded that their potential for providing critical data on ice sheet history is reduced because there may be a lack of suitable exposures for environmental reconstruction, and it is generally recognised that the preserved esker systems represent less than 10 percent of the former drainage system; the rest was in the ice itself. Eskers provide information on patterns, phases and ages of deglaciation, positions of former ice sheet margins, whether or not the ice was active or stagnant, the subglacial flux of sediments, combinations of processes operating along ice sheet margins, shapes of former subglacial conduits, the general slope of the ice surface and hydraulic gradients and whether or not the ice margin ended in standing

Fig. 8.6 Air photograph of the Evishanoran esker, County Tyrone. The esker ridge trends SSW to NNE and conducted meltwater northwards across the main Cookstown/Omagh road near Teebane Cross. The feature is composite in origin, consisting of a northern subglacial channel fill towards the Davagh Forest area which trends uphill, a middle portion at Teebane Cross where a cross-valley delta is present and a southern portion again showing morphological and sedimentological characteristics typical of a subglacial feeder channel. Note the deeply kettled topography immediately south of the road and the anastomosing pattern of ridges verging northwards. The main esker conducted meltwater northwards over a vertical height of 200 m into the ice-marginal glaciofluvial deposits around Tullybrick and the Lough Fea Platform. The ridge segment is 4.5 km long on this photograph. Top of photograph is north.

water. Each of the major esker systems recognised in Ireland are uniquely situated both in a topographic and stratigraphic sense to provide some answers to these parameters of ice sheet history. In order to avoid confusion and to add structure to the sections below, the term 'moraine' is applied loosely to linear complexes deposited parallel to the ice margin and perpendicular to the general drainage within the ice sheet which is recorded by tunnel fills or eskers.

The Midland Esker System

This esker system extends across the Carboniferous limestones of the Central Plain from The Derries westwards to the Shannon river, a distance of about 80 km (Fig. 8.4). In the east the feature begins around 85 m OD, rises towards 100 m OD around Daingean and falls to 70–80 m OD around Athlone. It crosses the shallow watershed near the esker stream, Mount Lucus. Clearly, the ridge planiform pattern resembles a fossilised, branched river pattern that conducted water eastwards with lower order tributaries beginning at the Shannon and joining eastwards to form fewer higher order tributaries around Tullamore (Fig. 8.4). At localities such as Judgeville and Daingean the system has been reduced to one major ridge which continues and merges into a large spread of sediment at The Derries. This analogy to an integrated river system is more apparent than real because the scale and magnitude of the ridges does not change or appreciably increase eastwards as might be expected with increased discharge. Therefore segments of the system will be described from east to west in order to identify the origins, nature, linkages and operation of the system. Although the dominant ridges are linear with a significant east–west component there are other significant ridges and spreads of sediment at right angles to the overall system.

At the eastern end of the system an esker ridge known as the Shane Valley esker merges with an extensive area of flattish, bog covered ground. This landscape duplet consists of a single, sinuous subglacial feeder channel and an area of outwash, though it is unknown whether or not the debris discharge was deposited into a lacustrine environment. Westwards fairly continuous esker ridges occur at Daingean consisting of steep-sided, ice-contact ridges with occasional breaks. Available exposures suggest that the ridges up to 12 m high are composed mainly of disorganised boulder gravel though finer, cross-bedded facies are also present.

Around Tullamore the character of the system changes with many more separate ridges spaced 2–3 km apart, larger ridge heights up to 30 m and a marked increase in gravelly ridges transverse to the general esker run. Nevertheless, between Judgeville and Tullamore the main features are narrow sinuous ridges of symmetrical cross-section, typical of true, subglacial eskers (Fig. 8.7a) (Farrington, 1970). The Ballyduff esker, 2 km north of Tullamore, is the most southerly and most prominent esker ridge segment in the dendritic system of ridges which converge eastward to Judgeville. The Clara esker occurs in the middle of the system and the most northerly ridge is the Split Hill esker. The overall dendritic pattern of the three main ridges and their convergence towards an area of flat outwash around Judgeville demonstrates that they conducted water eastwards. The steep, ice-contact slopes of the ridges strongly suggest that they are mainly subglacial in origin, an inference which is supported by the prominent cores of poorly-sorted, amalgamated beds of boulder gravel. The latter contain boulders up to 2 m across.

Although parts of the Ballyduff ridge are symmetrical in cross-section, other parts are wide (~0.4 km) and depart from the standard steep-sided esker form (Fig. 8.8). Esker embankments occur along the flanks of the main esker together with kettle holes, suggesting that processes other than tunnel sedimentation occurred. Northwest of Ballyduff House Farrington (with F. M. Synge, 1970) recorded a small, ice-contact esker ridge superimposed across the main Ballyduff ridge. Other features including broadening of narrow esker segments and flat top segments together with asymmetrical slopes clearly require additional explanation. The most logical explanation is that the main part of the ridge is clearly subglacial, mainly high-energy tunnel sedimentation, and that later depositional events altered the form of the ridge. Four lines of evidence support this interpretation. First, thickening of the eskers is associated with kettle holes and fluvioglacial deposition around detached ice blocks as at Moleen Hill (Farrington, 1970). Second, the boulder gravel core is overlain directly by ripple drift cross-lamination interbedded with cross-bedded pebbly gravel. Warren and Ashley (1994) interpret these facies in terms of subaqueous outwash. The fact that the outwash is pitted with kettles demonstrates that deposition occurred after subglacial ice walls collapsed and the main ice sheet margin began to contract westwards possibly in association with ponded water. Third, the preservation of arched bedding in narrow parts of the

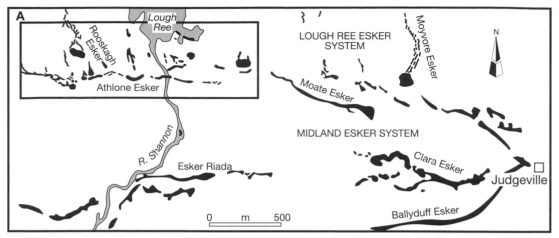

Fig. 8.7a Field relationships between the Midland and Lough Ree Esker Systems (after Delaney, 2002).

By permission of Elsevier.

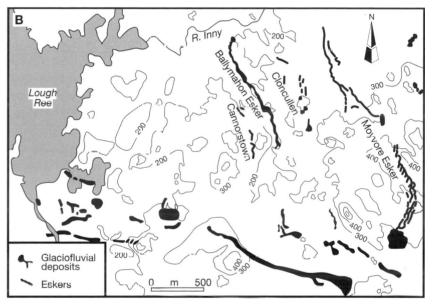

Fig. 8.7b The Ballymahon, Cloncullen and Moyvore eskers of the Lough Ree Esker System (after Delaney, 2001, 2002).

By kind permission of Cathy Delaney.

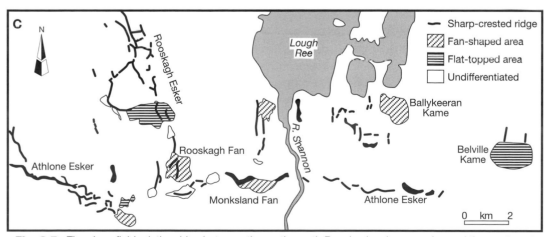

Fig. 8.7c The close field relationships between the north–south Rooskagh esker complex and the west–east Athlone esker (after Delaney, 2001, 2002).

By permission of Elsevier.

ridge point either to slumping on ridge margins after ice support disappeared or that the arched bedding is a primary sedimentary feature of tunnel sedimentation or pseudoanticlinal macroforms with crest convergent fabrics (Brennand, 1994). Fourth, well-defined gravelly hillocks and ridges occur at right angles (NE–SW) to the main esker trend at localities including Trumpet Hill and are considered moraines because of their transverse relationship to the core esker ridge pattern (Fig. 8.8) (Farrington, 1970). Transverse elements therefore record phased ice-marginal retreat westwards and not southward towards the hypothetical ice dome inferred by Warren and Ashley (1994).

The esker topography in this sector of the system is therefore composite, recording at least a subglacial followed by a proglacial event as deglaciation progressed. This inference is supported by the highly variable, pitted morphology of the Clara esker which merges locally with hummocky topography and obscures the classic ridge features (Smyth, 1997). Even though there are breaks along the length of individual esker ridges the dendritic pattern comprising the Moate, Split hill, Clara and Ballyduff ridges conducted meltwater eastwards towards the Judgeville convergence. However, evidence for the presence of an extensive lake bordering the decaying ice margin is not conclusive because the marginal gravelly spreads are not draped or interbedded with fine-grained rhythmites. The east–west fabric and trend of the dendritic pattern around Ballyduff is continued and maintained westwards for a farther 50 km across the Shannon to near Aughrim (Sollas, 1896). Esker elements of this sector include the Athlone esker and the Esker Riada both providing important highways across and above the alluvial plain bordering the Shannon (Fig. 8.7a).

Farrington (1970) describes a small but prominent esker ridge about 3 km southeast of Tullamore at Geashill (Fig. 8.8). It is 13 km long, up to 6 m high, 38 km long and is punctuated by gaps. This small feature records the position of subglacial channels which acted as feeder channels and a sediment supply to the two ice-marginal morainic ridges which cross the esker ridge at right angles near Geashill village and Ballykeen. The NE–SW trend of the moraines suggests that the phased withdrawal of the local ice margin was westwards. To the west the line of the Geashill esker ridge dies out for 4–5 km but the main thread of probable meltwater routing continues in the form of another prominent esker ridge known as the Frankford esker (Fig. 8.8). This complex extends southwest from Screggan almost to Birr where the ridge changes to a northwesterly direction paralleling the Little Brosna River, before ending near Cloghan Castle (Fig. 8.4). Farrington (1970) considered the esker to be the cast of a subglacial stream because the ridge is narrow, sinuous and continuous, it is bounded by steep ice-contact slopes, symmetrical in cross section and consists of coarse, disorganised

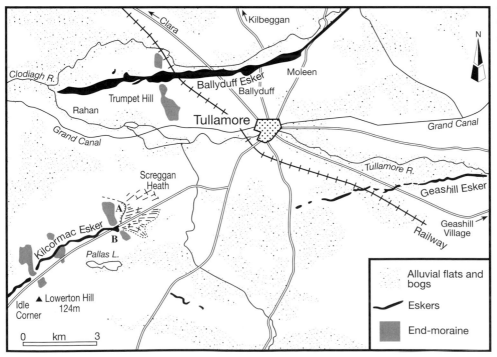

Fig. 8.8 Esker ridges near Tullamore, County Offaly (after Farrington, 1970). The Kilcormac Esker was formerly known as the Frankfurt Esker.

gravelly facies deposited by frictional freezing of high-energy flows. The ridge ends in a flat spread of finer-grade gravel and sand known as Screggan Heath. This field association of a subglacial feeder tunnel ending at a major spread (4 km²) of subaerial outwash that is bounded by an ice-contact slope to the west and slopes gently northeastwards shows that the ice margin was retreating southwestwards and westwards towards an ice mass west of the Shannon river. The fact that at least four NW–SE belts of hummocky morainic gravel topography are superimposed across the run of the narrow esker shows that there was a phased retreat of the local ice margin westwards and southwestwards to the north of the Slieve Bloom mountains.

On the western side of the Shannon the dendritic esker system north of Portumna extends for another 13 km northwestwards towards Killimor and Duniry on the northeastern flanks of the Slieve Aughty Mountains (Fig. 8.4). The Portumna segment consists of about four main braches which coalesce eastwards and end on the western bank of the Shannon at Ballycrossaun. This point of convergence or meltwater node does not link up precisely with the western end of the Frankford esker, but is displaced southward by 2–3 km. However, the displacement is more apparent than real because a wide erosional breach now occupied by the present Shannon river and associated alluvial flats separates the remaining ends of both esker systems which could be easily joined by a gentle curve which is within the scale of the local ridge sinuosity. This interpretation extending the line of the Midland eskers across the Shannon is strongly supported farther north where the Athlone esker and Esker Riada are breached by the Shannon but are clearly linked to the Moate and Clara eskers of the main Midland chains to the east of the Shannon.

The Lough Ree Esker System

The esker ridges and associated kames which occur on both margins of Lough Ree are distinct from the eskers of the Midland system even though their southern segments occur within touching distance of the Midland eskers (Fig. 8.7). The most northerly branch of the Midland esker system consists of the Split Hill, Moate and Athlone esker association which crosses the Shannon at Athlone and records a west to east meltwater flow and a phased ice retreat westward (Fig. 8.9). Immediately to the north are three major esker ridges, trending NNW–SSE at Moyvore, Ballymahon and Rooskagh (Delany,

1997, 2001, 2002). This esker pattern is parallel to the local subglacial bedform pattern and is at right angles to the persistent west to east hydraulic gradient of the earlier Midland system (Figs. 8.7, 8.9). Regional patterns of striae and subglacial bedforms show that the last ice sheet advance in this area occurred between 17 and 16 cal ka BP during the Killard Point Stadial (McCabe et al., 2005, 2007b). Drumlinisation at this time occurred as far south as Mullingar where the major lakes and large-scale rock lineaments are aligned parallel to the last ice sheet movement from the northwest (Fig 8.4). The streamlined islands and gross morphology of Lough Rea itself form part of the same coeval ice sheet flow lines which etched the last subglacial footprints across the north central lowland from Dundalk Bay west to Galway Bay. The eskers around Lough Ree extend north into the drumlinised topography of the main Irish drumlin belt and therefore mark the southern limit of the readvance.

The Ballymahon esker is 12 km long, trends NW–SE, is punctuated by short breaks especially south of wider segments and is bordered by hummocks, gravelly topography and other discontinuous esker ridges at Cannorstown and Cloncullen (Delaney, 2001) (Fig. 8.7b). Ridge morphology varies from narrow and sharp-crested which locally rapidly increase in height and width to wider, flat-topped segments (Fig. 8.10). Some straight segments may be up to 350 m in length and consist of poorly-sorted, amalgamated beds of boulder gravel. In contrast the wider, flat-topped to undulating segments consist of parallel-laminated, ripple-laminated and massive sand interbedded with cross-stratified pebble gravel (Delaney, 1997). It is possible to interpret these field associations with deposition of coarse sediment within tunnel segments and the fine-grained fan and spread deposits representing deposition at the ice margin in very shallow water. In one case Delaney (2001) recorded cross-bedded gravel overlying the disorganised gravelly ridge core and suggested that the crossbeds represented part of a Gilbert-type delta formed in a temporary lake in a re-entrant along the ice margin. It is tempting to relate the segmented nature of the esker to phased ice retreat to the north accompanied by time-transgressive deposition of gravelly, undulating topography as the ice margin backwasted. The continuity of the ridge, the linear arrangement of its individual ridge segments and its persistent discharge regime to the southeast along the axis of the Dungolman valley possibly records rapid and progressive ice-marginal decay from the ice maximum. The Moyvore esker is

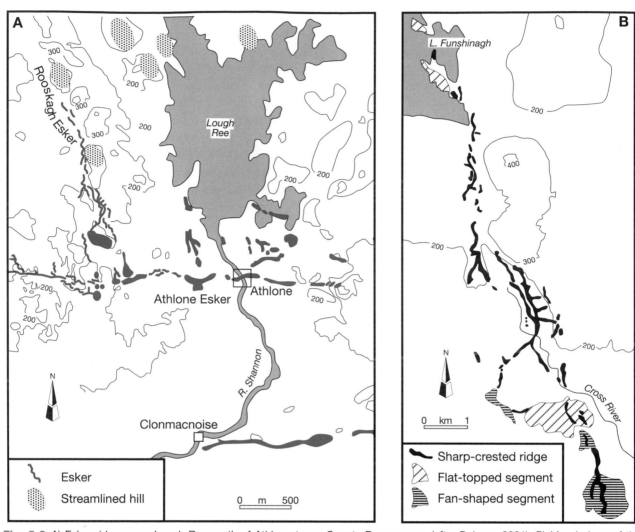

Fig. 8.9 A) Esker ridges near Lough Ree north of Athlone town, County Roscommon (after Delaney 2001). B) Morphology of the Rooskagh esker (after Delaney 2001).

subparallel to the Ballymahon esker, is 16 km long and ends in a prominent isolated kame at Streamstown (Fig. 8.7b). It essentially has a similar morphology to the Ballymahon esker though more complex subparallel ridges are present. It conducted subglacial meltwater southwards and formed in the same deglacial phase as the other north–south eskers around Lough Ree. The relative absence of integrated esker systems farther north records accelerated ice wastage and stagnation zone retreat (McCabe and Clark, 1998).

The Rooskagh esker trends NW–SE from the eastern side of Lough Funshinagh and continues south for 10 km along the Cross River, 3 km west of Lough Ree (Figs. 8.7c, 8.9a, b). The southern part of the esker together with adjacent small esker fragments ends in fans or kames immediately to the north of the Athlone

esker which marks the westerly limit of the Midland system. The Rooksagh esker consists of over thirty segments or beads separated by discontinuities. It also includes parallel esker ridge development, flat-topped segments and narrow cross-ridges trending across the main ridge. Delaney (2001) recognises three morphological forms including sharp-crested ridges up to 20 m high, flat-topped and steep-sided areas up to 15 m high and fan-shaped areas (Fig. 8.9b). Small esker-like ridges cross the surface of the Rooksagh fan and sharp-crested ridges join the steep ice-contact northern face of the Belville kame. Sharp-crested ridges consist of a wide range of gravelly sediments up to boulder grade which are sometimes organised into anticlinal bedding transverse to the esker ridge. Delaney (2001) concludes that these coarse-grained sediments including bimodal

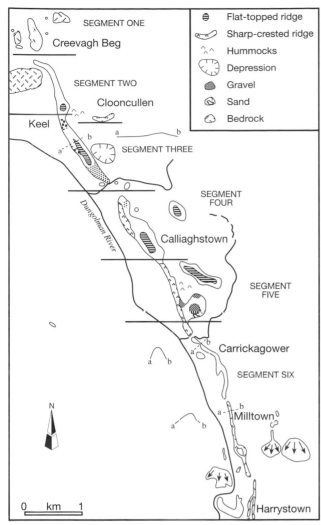

Fig. 8.10 Morphological map of the Ballymahon esker east of Lough Ree (after Delaney 1997). *By kind permission QRA (Quaternary Research Association).*

The legend for the map reads:

SEGMENT ONE — Creevagh Beg
SEGMENT TWO — Clooncullen
Keel
SEGMENT THREE
SEGMENT FOUR — Calliaghstown
Dungolman River
SEGMENT FIVE
Carrickagower
SEGMENT SIX
Milltown
Harrystown

Flat-topped ridge
Sharp-crested ridge
Hummocks
Depression
Gravel
Sand
Bedrock

0   km   1

N

gravel were deposited mainly under full pipe conditions (McDonald and Vincent, 1972). In the fans Delaney (2001, 2002) found horizontally-bedded and planar cross-bedded sediments which were interpreted as aggrading gravel-bed barforms (Fig. 8.7b). Some more distal fan sequences contain interbedded packages of climbing-ripple cross-laminated sand, massive sand and thin beds of diamict (Fig. 8.3c). It was concluded that these facies represent a combination of under-flows and suspension deposition typical of subaqueous outwash fans (Rust, 1977). In the Belville kame fine sand, silt and clay are arranged in rhythmic, fining, upward packages interbedded with large-scale cross-sets of pebble gravel. This relationship may suggest a bottomset to foreset transition typical of Gilbert-type deltas (Delaney, 2002). The position of the main fans

and the kames suggest that they were formed in ice-marginal re-entrants which were supplied directly with sediment from high-energy subglacial conduits. The flat-topped delta fragments such as Belville kame seem to record the presence of a lake up to around 92m OD though Delaney (2002) suggests it was ephemeral and its outflow related to local cols. The location and pattern of eskers in this area cannot be used to reconstruct the presence of synchronous ice domes to the north and south because the Athlone/Midland system predates the Lough Ree system by about five thousand years and both systems are geomorphologically discrete.

The Trim Esker System

The Trim esker system is significant because it conducted meltwater to a well-defined ice limit which ended in a proven glacial lake (Fig. 8.11). This ice limit is known as the Galtrim moraine (Synge, 1950) which can be traced from Enfield northeastwards around the northern flanks of Tara Hill and east to the coast at Ben Head, a distance of 55km (McCabe, 1973). At Ben Head the morainic sediments are sheared over earlier sediments by ice readvance from the north. The position of the Galtrim across the lowlands of County Meath and its southwesterly continuation to Enfield suggests that it represents a slightly earlier phase of deglaciation immediately before ice withdrawal across the Irish midlands and deposition of the midland esker system. Typically the moraine is a broad, flat-topped ridge up to 16m high, bounded on its northwestern face by steep ice-contact scarps over-looking marshy depressions. The latter often contain isolated, symmetrical and conical shaped kames up to 10m high representing the infills of depressions or per-forations in the ice surface. The marshy depressions are tongue-basins and represent the main frontal positions reached by the former ice margin. Usually the position of the ice margin is represented by a single ridge, though towards the southwestern end of the complex multiple ridges occur at Rahinstown and Cloncurry. Morpho-logically the ridge is a delta-moraine consisting of coa-lesced delta fragments showing gravelly beds (foresets) dipping southward. The upper kettled surfaces of the moraine at Galtrim, Drumard, Ballinrig, Rahinstown and Cloncurry occur at about 33 m OD supporting the presence of an ice-dammed lake ponded against the rising slopes of the Summerhill escarpment (Fig. 8.11). This level approximates to the presumed intakes of the overflow channels which drained the ponded water

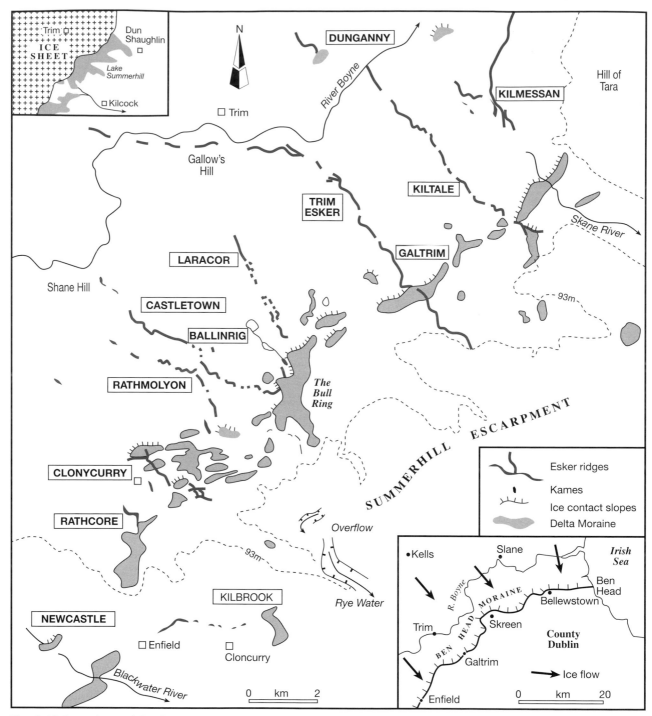

Fig. 8.11 General map of the glacial features around Trim, County Meath (redrawn from Synge, 1950).

southeastwards along the Rye valley. Lacustrine clays are preserved on the floor of the former lake between Summerhill and Batterjohn.

Twelve subparallel esker ridges trending north-west–southeast join the northwestern slopes of Galtrim moraine at right angles and appear to have acted as melt-water feeder channels to the landform (Synge, 1950). The ridges are generally single and morphologically similar with sharp crests, swales and knobs, winding outlines, symmetrical cross-sections, up and down long profiles and beaded sections. The largest eskers tend to be fairly continuous across country and are characterised by beaded sections. Synge (1950) remarked that each bead or bulge along otherwise narrow esker ridges represents a partially formed delta laid down near the mouth of a subglacial stream. Narrower esker segments represented tunnel sediments though Synge (1950) commented that morphology was not easily correlated to the calibre or structure of sediment within the ridge. This model implies that the ice-marginal lake persisted as the ice margin withdrew northwestwards until meltwater could escape freely eastwards along the Boyne river. The Trim esker is the largest (14.5 km) and best preserved drainage channel within the system. Synge (1950) observed that the Trim esker (single ridge) continues across the ice-contact face onto the surface of the delta-moraine as two subparallel ridges composed of coarse-grained gravel. He suggested that these deposits collapsed from an overhanging ice cliff though it is likely that this process would result in smudging of the esker ridge morphology. It is more likely that the ice sheet margin readvanced locally over the top of the moraine because of decreased ice wastage and protection as the morainic shoal increased in height. Bifurcation at this point would also have been a consequence of blocking of the main esker channel by debris and diversion around detached ice blocks. The readvance of the ice margin is also recorded by nested sets of shears on the ice contact slopes of the moraine farther east at Cross and Ben Head. It is therefore not always safe to conclude that the sediments overlying the core, boulder gravel facies represent subaqueous fan deposition.

The Connaught Esker System

This extensive system formed after the final phase of drumlinisation in the west when ice withdrew from the Tawnywaddyduff ice limit in Donegal Bay which occurred during the Killard Point Stadial after 16.5 cal ka BP (Fig. 8.5). The eskers date to around 15 cal ka BP when deglaciation was well underway in the east. The general pattern of esker ridges is related to two major ice flows which coalesced to form the glacier which moved northwards into Killala Bay along the western flank of the Ox mountains (Fig. 8.5). One ice flow advanced north northeastwards from Joyce's Country, Lough Corrib and south Galway and the other swung in a marked arc from the Erne and Leitrim Basins towards the Ox Mountains. Because all of the eskers show a general convergence towards the River Moy lowlands they are considered to be of the same age and part of the same deglacial system. Two major esker patterns occur, termed the Charlestown–Swinford and Tuam–Kilkelly systems. The outwash from both converge with a westerly palaeoflow though meltwaters especially from the south have extensively eroded earlier outwash forming both deep and smaller interlinked channels similar in places to scabland morphology. Neither esker system was deposited in interlobate settings and both are essentially the products of drainage below discrete ice masses of different provenances.

The smaller system extends from Tobercurry southwestwards to Charlestown and Swinford, a distance of 20 km. The main ridges are up to 4 km in length, 10–15 m high, 30–50 m across and are bounded by steep sides. Occasional exposures consist of edge-rounded boulders set in a finer gravelly matrix suggesting high energy flows in closed conduits (Fig. 8.12). In the west these ridges terminate at Swinford where they are fronted by kettled spreads of gravelly outwash. This association and the general esker convergence westwards shows that ice was retreating eastwards along the Moy valley. The general absence of cross-esker moraines suggests that ice retreat was rapid, characterised by stagnation zone retreat. However, bulges and high kame-like features punctuate esker symmetry especially around Cloonacannana, a few kilometres northeast of Swinford (Fig. 8.2).

The larger system occurs mainly on the area of slightly higher ground (~100 m) immediately east of the Plains of Mayo. Esker ridges are traceable as a huge arc which bends gradually from SW–NE to SE–NW over a distance of about 70 km. The height of the system rises from about 40 to 100 m OD from south to north. The marked change in orientation of the esker system suggests that it formed time-transgressively. The northern portion formed first as the ice front withdrew southeastwards towards the north central lowlands. After this, the hydraulic gradient was controlled by the remaining ice dispersal centre located in Joyce's Country, Lough Corrib and south County Galway driving major meltwater flows to the northeast. Most of the esker system is parallel to the regional ice flows to the northwest with the exception of the ridges to the south of Dunmore which may to some extent be controlled by the topographic barrier of the Slieve Dart ridge. General northward meltwater flow is shown not only by the ridge convergence to the north but by the erosion and marked carriage of felsites north from the eastern flank of the Slieve Dart ridge (Fig.

Fig. 8.12 Boulder gravel with minimal matrix of pebble gravel, Eskragh near Tobercurry. The boulder gravels are disorganised and were deposited en masse by frictional freezing before sorting and grading could develop.

8.5). There is no doubt that the sedimentology of the esker fills is complex. In places around Dunmore and Ballyhaunis sinuous ridges up to a few kilometres long composed of boulder gravel lead into beaded portions containing large-scale foresets or rippled sands. These segments suggest tunnel sediments accumulated within feeder channels that led into deltas deposited into standing water (McCabe, 1985). However, most ridge segments are relatively long (3–5 km), sinuous, narrow, sharp-crested and composed of coarse, boulder gravel suggesting that many are typical tunnel fills. There is little evidence to suggest that meltwater flow was constrained by two ice masses anywhere within the system or that the eskers around Ballyhaunis are some sort of marginal deposit as suggested by Warren and Ashley (1994).

The northern part of the Tuam-Kilkelly system is important for four main reasons (Fig. 8.5). First, it records essentially the same flow pattern as the esker ridges farther south though there is a slight break in continuity to the north of Ballyhaunis. Overall the meltwater was driven against a reverse bed gradient northwards. Second, several esker ridges end directly at the ice-proximal face of a major moraine at Knock which skirts the bedrock hill of Shammerdoo. The Barcull esker is well developed, about 2 km long and rises in altitude by about 60 m as it climbs the southeastern slopes of Shammerdoo to join the moraine. The large eskers at Kilkelly also join the morainic topography which is particularly hummocky at Lough Nambrackkeagh

and Shammerbaun. Third, a large (>40 km²) area of hummocky, pitted outwash up to 40 m thick records general meltwater escape to the west and northwest showing that thick ice remained to the east. The moraine and outwash couplet may record a temporary re-equilibration of the ice sheet margin along local slopes after rapid withdrawal from the inner con-tinental shelf. Fourth, all of the deposits on lower ground have been deeply dissected by meltwater channels and furrows. In places the topography resembles scablands with drift replaced by erosional forms. The meltwater floods from the south did not obliterate the esker ridges farther north at Swinford, suggesting that ice still protected these ridges.

The Ulster Esker System

i) *Introduction*

The large esker systems stretching across the plains of central and western Ireland record large-scale patterns of ice wastage characterised by tunnel deposition and phased ice-marginal halts in ice recession. All three major systems document ice wastage related to ice sheet margins fed from separate sources of ice dispersal. In contrast, although esker ridges are numerous in Ulster they do not occur as extensive integrated meltwater systems because they formed as the last ice sheet wasted southwards and westward across and around the Sperrin Mountains (Fig. 8.13).

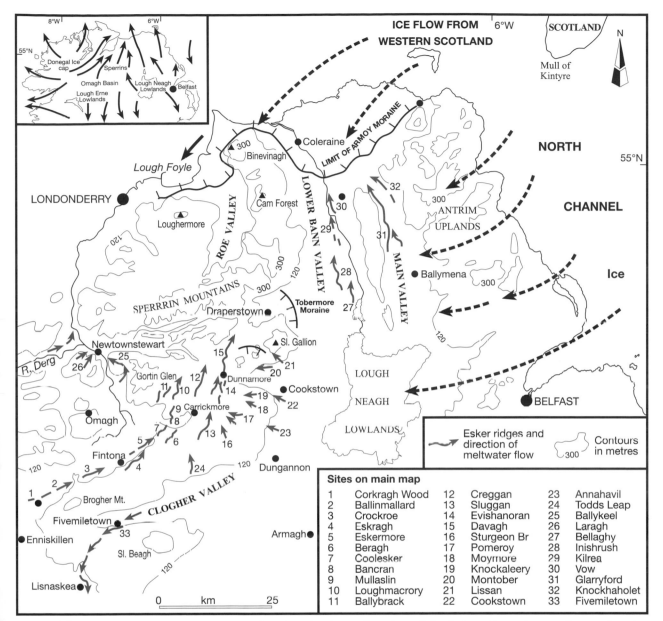

Fig. 8.13 The Ulster Esker System showing general northward meltwater flows towards ice limits on the eastern, southern and western flanks of the Sperrin mountains. Other esker systems conducted meltwater northwards along the Bann valley and southwestwards along the Clogher Valley. Inset shows generalised ice flows (based on Dardis, 1982).

Depositional patterns and meltwater routing therefore were dependent not only on topographic diversity and major structural lineaments within the mountain block but on the regional slopes of the composite ice sheet which was fed from three ice sources in Lough Neagh, the Omagh Basin and southeastern Donegal. All of the eskers in the north of the island occur within Killard Point Stadial ice sheet limits and most occur peripheral to the final centres of ice sheet dispersal in central Ulster. Four main areas will be considered.

ii) *The Bann Eskers*

The ice sheet margin in the north began to contract southwards as the last ice sheet readvance from western Scotland reached the Armoy moraine in north County Antrim. An extensive lake floored by thick sequences of fine-grained rhythmites developed between the ice sheets along the Bann valley. Rapid ice-marginal recession southwards towards the Lough Neagh Basin is recorded by two south–north eskers at Glarryford and Vow (Fig. 8.13). The Glarryford esker is a single lakes-ridge, about 10 km long, 10–15 m high, sinuous, steep-sided and is located along the axis of the southward draining Main river. However, the esker sediments were transported northwards against the local topographic gradient. The general NNW–SSE trend of the esker is parallel to the elongated drumlins which border the esker ridge. The esker ridge at Vow is parallel to the Glarryford esker but lies to the west of the streamlined ridge of Long mountain. It is parallel to local drumlin long axes and to the other small eskers feeding the delta moraines at Kilrea. Although it is a relatively small esker it provides evidence for the existence of an ice-dammed lake along the Bann valley as ice wasted southwards. The northern terminus of the esker ridge ends in a subaqueous spread consisting of coarse-grained tunnel mouth gravelly facies which are interbedded with thick (~1 m) sequences of laminated silts and clays formed by density underflows. Deposition was so rapid that coarse-grained gravels were immediately draped by rhythmites when sedimentation paused, resulting in a classical sequence of core outwash gravel draped by thick (~1 m) packages of rhythmites as the lake extended southwards. This deglacial sequence is complemented by the regional pattern of ice-directed meltwater channels on the flanks of the adjacent uplands of The Three Taghmores and the slopes of Slieve Gallion farther south. The meltwater drainage patterns therefore record ice sheet withdrawal southwards towards a major or final centre of ice sheet dispersal in the Lough Neagh Basin.

iii) *The Tyrone Eskers*

There are literally hundreds of small, short (~1–2 km) esker ridges located around the southern foothills and flanks of the Sperrin mountains (Dardis, 1982). In general they are located in structural depressions, broad upland surfaces and in valleys and are closely associated with extensive spreads of ice-contact ridges, deltas and spreads of fluvioglacial sand and gravel (Fig. 8.13). These areas of stratified sand and gravel topography can be found in a broad band in the mainly west to east valleys which dominate the uplands between Newtownstewart and Cookstown, for up to 25 km south of the main Sperrin range (Fig. 8.13). This pattern of small eskers at first sight seems unimportant, but as a group they point to location of ice sheet limits and also provide local evidence for ice sheet slope and therefore hydraulic gradient during the final stages of deglaciation. Both the limiting areas of gravelly topography and the trend of the feeding esker ridges show that the ice was still active

in core areas until a very late phase of deglaciation, and that ice fronts withdrew east from the south Sperrin Hills towards the Lough Neagh depression, southwest to the Omagh and Erne basins and west towards the Lough Derg area of south Donegal.

A critical inference here is that lowland centres of ice sheet dispersal were the drivers of ice sheet flow and earlier phases of drumlinisation, and that the main Sperrin range did not act as a centre of ice sheet dispersal. In a more general sense it was the growth of lowland ice sheets which provided the main ice flows effecting geomorphic change. In essence the peripheral mountain groups of the island are far too small to support the Irish ice sheet. The esker evidence from County Tyrone supports the idea that relatively small areas of active ice record final ice sheet decay towards the main centres of ice sheet dispersal during deglaciation. This inference is supported by the fact that most esker ridges are orientated parallel to drumlins, reflect structure-controlled glacial drainage and formed after drumlinisation.

The patterns of individual esker chains in Tyrone provide information at a series of different levels (Dardis, 1982, 1985a, 1986). The short (~1 km) eskers on the southeastern flanks of the uplands between Moneymore, Cappagh and Todd's Leap are all linked to ice margins ending in glacially impounded lakes where the hydrostatic head was driven by the ice dome situated in the Lough Neagh basin. The ice dome margins back-wasted eastwards towards the low ground around Lough Neagh. A more impressive pattern of discontinuous esker ridges can be traced from northeast to south west along the Draperstown-Fintona corridor for a distance of 60 km (Fig. 8.13). Ridges are long (1-3 km), sinuous, steep-sided and sharp-crested features parallel to local drumlin orientations. Individual ridges are not joined directly but tend to end in fluvioglacial complexes such as the Murrins. Nevertheless they record a prominent and persistent pattern of subglacial meltwater flow towards the northeast. Clearly the ice surface sloped northeastwards and was associated with a powerful Tyrone glacier which may have been sourced by ice from Lower Lough Erne. The small (~5 m high, <1 km long) esker ridges in the Derg, Baronscourt and Strule valleys are closely associated with ice-marginal deltaic gravel ridges on the valley floors and record meltwater flow from the south and southwest. The presence of discrete subglacial-esker and delta moraine duplets in each valley marking an ice sheet halt shows that the ice was still active and geomorphologically active. These ice-marginal positions represent the final phase of ice-marginal wastage around the centres of ice dispersal because only patchy, low-relief hummocky moraine occurs in the area immediately to the south.

iv) *The Evishanoran Esker*

The most notable esker chain in south-central Ulster is the Evishanoran system which lies at the eastern end of the Omagh-Draperstown corridor. It extends north northeastwards from Evishanoran Mountain to Davagh Forest, over a distance of 12 km (Fig. 8.6) (Dardis, 1982). The esker begins near Bernisk Glen 20 km south of Davagh Forest transporting erratics northwards across Evishanoran Mountain and across the east–west Ballinderry valley before rising uphill and ending as the Davagh eskers. The northern part of the Davagh eskers consists of steep-sided ridges with exposures of laminated clays deposited in close association with coarse-grained, disorganised gravel and sand (Dardis, 1982). This part of the system may have been characteristic of tunnel mouth deposition in ponded water. The main ridge north from Dunnamore consists mainly of coarse-grained boulder gravel deposited in a closed tunnel system that climbed uphill about 200 m from the axis of the Ballinderry valley to Davagh Forest (Fig. 8.14). However, exposures on the south side of the Ballinderry valley show that the 'ridge' has a more complex origin. The basal gravels record south to north erratic transport that is parallel to most ice sheet directional indicators when ice entirely covered the upland area. As deglaciation proceeded the ice sheet thinned and topographic control of ice sheet margins increased. At this stage ice pressure from the east produced an ice margin in the valley at Evishanoran which formed a prominent cross-valley moraine on top of the earlier esker sediments. The cross-valley ridge or the 'southern part of the Evishanoran esker' is therefore related to deltaic sedimentation along the western limits of the Lough Neagh drumlin field (Dardis, 1982). This explanation, which recognises two different depositional phases, readily explains the debate in the Geological Magazine between Gregory (1925) and Charlesworth (1924, 1926) on whether the Evishanoran Esker was a moraine or a subglacial esker. Both workers failed to identify sedimentological changes along the feature in relation to the position of supporting ice walls or to differentiate between ice-contact slopes and primary slopes controlled by progradation of delta foresets.

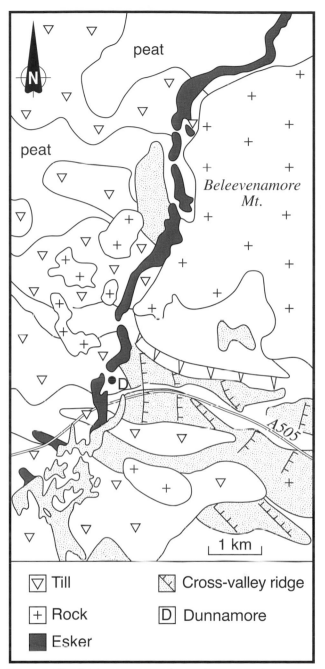

Fig. 8.14 The Evishanoran esker, County Tyrone near Dunnamore (after Dardis, 1982).

v) *The Clogher Valley Eskers*

This system occurs along the central axis of the valley and is 25 km long (Fig. 8.15). It begins as a series of steep-sided, subparallel ridges around 10 m high at Andrew's Wood and continues in a southwesterly direction to Colebrooke where segments coalesce and demonstrate westerly meltwater flow. Ridges climb over drumlins at Pulpit Hill and show a general southwesterly

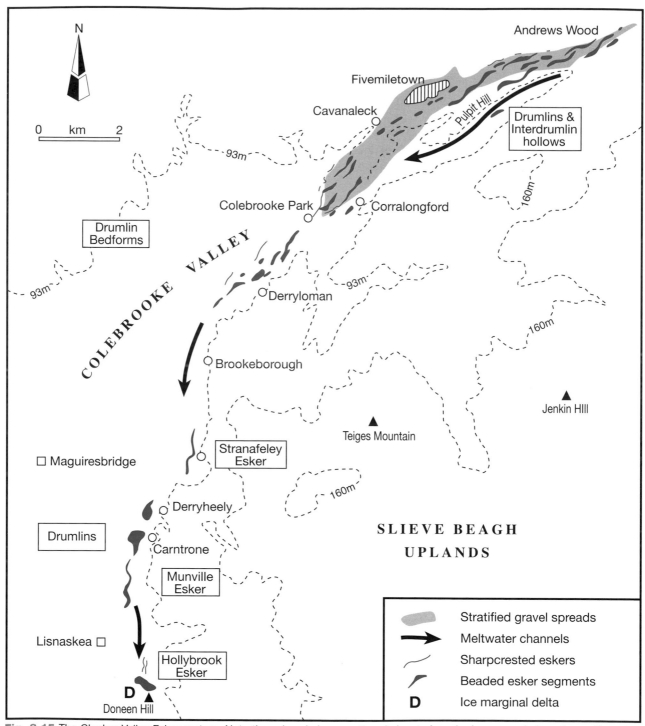

Fig. 8.15 The Clogher Valley Esker system. Note the esker drainage pattern is located on the low ground along the northwestern margin of Teiges Mountain leading to an ice margin at Doneen hill.

convergence. The entire complex around Fivemiletown is closely related to a kettled complex of sand and gravel and in places is draped by later deposits (Fig. 8.16a). The form of the ridges together with cores of boulder gravel suggest a subglacial origin. Southwards the esker

run continues along the flanks of Slieve Beagh towards Lisnaskea both as gravelly spreads, meltwater incisions and sharp-crested ridges. At Munville the feature snakes and rises ontop of rising ground at Croghan but is immediately replaced on its southern side by a

Fig. 8.16a Slumping of esker sediments along the flanks of the Pulpit Hill esker, Fivemiletown, County Tyrone.

Fig. 8.16b Detail of slumping and faulting on the margins of an esker ridge, County Tyrone.

marked meltwater gash at O'Brien's Cross (Fig. 8.1). The maximum extent of the feature is marked by two small, parallel esker ridges which feed into the ice-proximal face of the Delta moraine, known as Doneen. Different levels within the delta may relate to former higher lake levels on the margins of Upper Lough Erne. Although the system is probably time-transgressive, it records deglaciation of Upper Lough Erne while the Clogher valley was still blocked by ice. This pattern shows that

a significant ice sheet existed in the Cloghfin/Clogher/ Ballygawley area which was probably the southwesterly arm of the Lough Neagh ice centre. This area of active ice was one of the last centres to melt towards the end of deglaciation and funnelled ice along the topographic low as far as the margin of Upper Lough Erne at Lisnaskea. The eskers record the final decay of this ice lobe (Figs. 8.15, 8.16a, b).

9

Ice Sheet Readvances around the North Irish Sea Basin

Introduction

Drumlinisation and the moraines fringing the drumlin fields were the basis for Synge's (1969) hypothesis of a Drumlin Readvance Ice Sheet across the north Irish lowlands (Fig. 1.4). Readvances during deglaciation were also identified from the grey, clay-rich till/diamicts around the margins of Dundalk Bay and in south County Down. McCabe and Hoare (1978) argued that these deposits containing large amounts of silt and clay (80–90%) and moderate amounts of matrix carbonate could not be generated by direct glacial erosion of local, non-calcareous Silurian bedrock. They demonstrated that the fine-grained, glacigenic diamicts and tills were scavenged by readvancing ice from muds on the floor of Dundalk Bay (Fig. 1.3b). The widespread extent of the muds suggested a marine origin and that high relative sea level closely followed retreat of ice from the Irish Sea Basin (Fig. 9.1).

Following early and probably catastrophic deglaciation of the Irish Sea Basin (>20 cal kyr BP) the BIIS lost over two thirds of its mass and was reduced to a series of inland ice domes across the north Irish lowlands (Fig. 4.10). This ice sheet configuration was distinctly different from earlier, thicker ice masses which drained onto the continental shelf. Directional indicators (McCabe, 1969) record successive shifts in these new centres of ice sheet dispersal which regenerated and readvanced as far as coastal margins but not onto the continental shelf. Undoubtedly the invigorated ice sheet drumlinised the substrate by fast ice flow and ice streams which cross-cut earlier transverse, ribbed moraine. Limits of ice-marginal readvance are marked by outwash and moraines composed of debris delivered to the ice fronts from widespread subglacial sediment transfer during drumlinisation. The glacigenic deposits formed at these ice fronts are commonly associated with raised marine muds containing *in situ* monospecific microfaunas dominated by the foram *E. clavatum* (85–95%). These faunas represent opportunistic biocoenoses recorded from Arctic–subarctic areas recently vacated by glacier ice (Fig. 1.5) (Hald et al., 1994). Therefore muds below, within and above glacigenic units generated during glacial readvances provide a method to constrain the age of successive sea level and glacial events and a robust AMS[14]C chronology (Fig. 5.4).

Ice sheet readvances during the latter stages of the last glacial termination are specifically linked to ice sheet reorganisation following a major early deglaciation of the BIIS. Because widespread readvances occurred after a significant deglaciation, they provide critical information on ice sheet history, relative sea levels, amounts of isostatic depression and on the driving forces within the North Atlantic climate system. Reconstruction of the anatomy of ice readvances are based on a range of inter-related observations which include: the overall ice sheet system comprising fast ice flow from the Irish lowlands, drumlinisation of the lowlands and subglacial transfer of detritus to tidewater ice sheet margins, deposition of terminal outwash in the sea forming morainal banks, moraines marking the former configuration of marine ice lobes, position of late-glacial raised beaches and subaqueous outwash immediately beyond ice sheet limits and a chronology constrained by AMS ^{14}C dating of fossiliferous marine muds.

Two major ice-marginal readvances have been identified from the coastal exposures in east central Ireland and the larger scale erosional/transportational system inland (Fig. 9.2). The earlier ice readvance reached as far south as Clogher Head and the younger readvance was restricted to Dundalk Bay. Contemporary limits of this readvance occur at the mouth of Carlingford Lough, along cols in the Carlingford Mountains at Windy Gap and farther northeast along the coast of County Down to Killard Point (Fig. 9.3). Two non-glacial exposures at Cooley Point and Rough Island provide data on marine

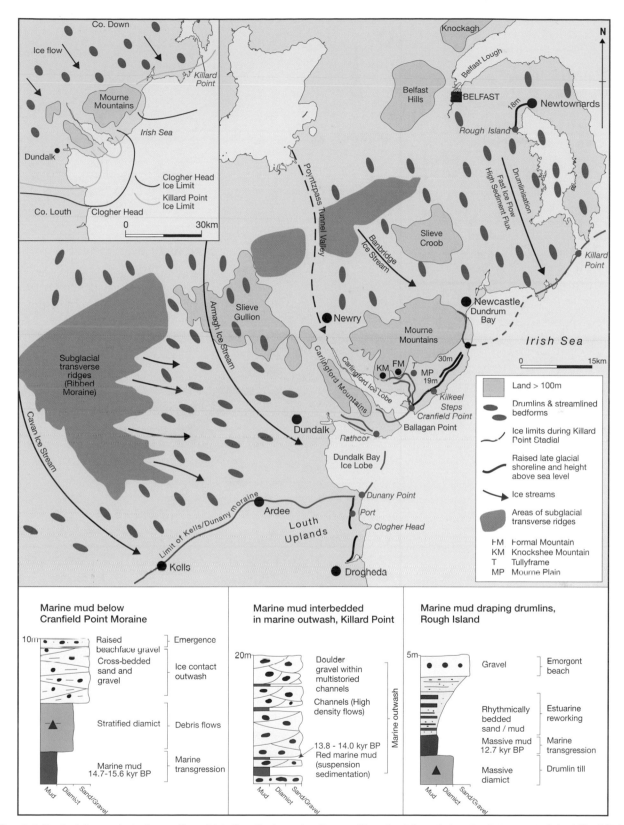

Fig. 9.1 Deglacial records in the northern Irish Sea Basin and patterns of ice sheet flow during the latter part of the Killard Point Stadial. Location of critical sections together with beds of dated marine muds are shown in relation to ice sheet limits. Ages are given in radiocarbon years (after McCabe and Dunlop, 2006). *By kind permission of GSNI.*

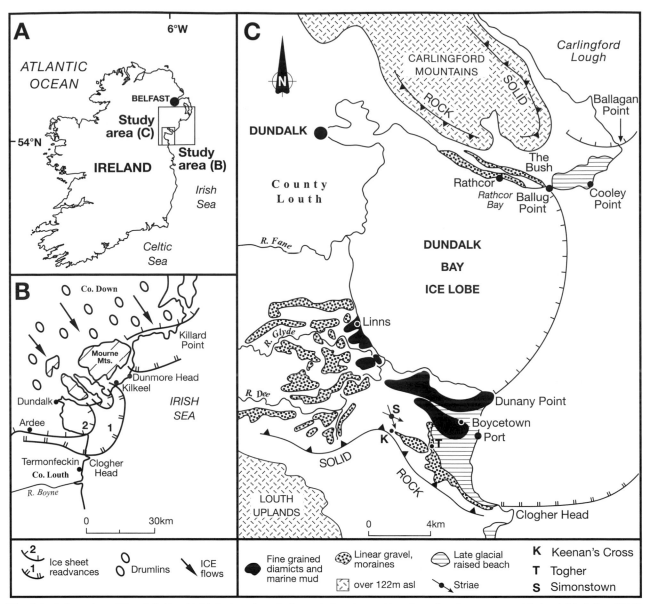

Fig. 9.2 A. Location of Dundalk Bay. B. Major ice sheet limits and directions of ice sheet flow in east-central Ireland (McCabe, 1986). C. Detail of the pattern of moraines around Dundalk Bay. Note that the Lateglacial raised beach and shoreface deposits record sea level that was contemporaneous with ice lobe maxima because high relative sea level is not recorded within the ice lobe limits at Rathcor and Dunany Point (after McCabe et al., 2007b).

mud chronology and sea level trends during and after the ice sheet readvances. Much of the present coastal morphology between south County Down and Clogher Head is related to erosion of thick drift sequences which are a direct result of subglacial sediment transfer to tidewater ice margins when ice sheet readvances were at their maximum. For example, the preglacial cliffline in the Cooley Point and Ballagan areas would have been 2–3 km inland and similar estimates are likely for the Mourne Plain.

The Clogher Head Readvance

On the southern margin of Dundalk Bay morainic ridges occur between Keenan's Cross and Clogher Head, a distance of 8 km (Fig. 9.2). The linear ridges trend north-west to south-east, are up to 25 m in height and are bounded by steep slopes to the east. To the west the ridges feather out against rising rock slopes. The eastern slopes are interpreted as ice contact because the internal push structures within the ridges at Togher are associated with ice pressure westward (McCabe and Hoare, 1978).

Fig. 9.3 Boulders and ridges of the moraine at Cloughmore, Carlingford Peninsula. The moraine formed when ice moving through Carlingford ice advanced southwards through the Windy Gap col into upper basin of the Big river. The Cloughmore megaerratic is dated to 17.0±0.6 ka BP (Bowen et al., 2002) which is similar in age to the AMS ^{14}C dates obtained from marine muds overridden by ice advance during the ice sheet into Dundalk Bay and Killard Point.

For example, poorly-sorted cobble and boulder gravel has been glaciotectonically disturbed by ice pressure from the east. On the eastern side of the ridge at Togher, 3 m of diamict that exhibits westward rising shear planes overlies at least 10 m of boulder gravel disturbed by glacitectonics. The location of the moraine and its internal structure show that it records high-energy fluvioglacial deposition near the frontal margin of a large ice lobe which reached as far south as Clogher Head.

Immediately distal to the moraine at Clogher Head indicators of former higher relative sea levels including swash gullies and raised foreshore gravels occur as far south as the Boyne estuary. North from Keenan's Cross and Togher the Clogher Head moraine is truncated abruptly by large ridges at Boycetown and Dunany Point. These ridges are composed mainly of muddy diamict and are associated with later tidewater sedimentation from a more restricted ice lobe position (McCabe et al., 1987) (Fig. 9.2c). Striae at Simonstown record the south-easterly movement of ice to the Clogher Head ice limit whereas the orientation and pronounced southward bulge of the Dunany ridge records ice pressure directly from the north. Inland from Keenan's Cross the southern margin of a gravelly morainic complex has been mapped for 16 km to Ardee (McCabe, 1973) and has been interpreted as the limit of Synge's (1970a) drumlin readvance (Fig. 9.4). The moraines continue southwestwards from Ardee as a belt of ice-marginal ridges that are ~1 km across.

Coastal erosion a few kilometres north of the Clogher Head moraine at Port, has exposed up to 3 m of diamict overlain by 2.5 m of sand and gravel (Fig. 9.5). The diamict is compact and matrix supported with matrix texture ranging from sand to mud. About 60 percent of clasts (pebble-boulder) are glacially facetted and most (~85%) are derived from local Silurian siltstones, though a significant fraction represents igneous erratics of northern provenance. The diamict is interpreted as a basal till because the clast fabric is parallel to stacked subparallel shear planes 5–20 cm apart that rise southwards at 5–10° and can be traced along the entire (40 m) exposure. In addition the variability within the matrix is usually abrupt suggesting that deposition involved discrete packages of either sandy or muddy sediment which were only partially mixed and attenuated. At one point along the section a steeply inclined intrusion of grey fossiliferous mud 1.5 m high and up to 0.5 m across cuts across the tectonic stratification of the till (Fig. 9.6). These relationships suggest that the ice overrode and scavenged marine mud during advance into Dundalk Bay, and the mud dyke was injected into the till from below as a result of ice loading subsequent to the shearing of the till. Four AMS ^{14}C dates between 16.0 and 15.2 kyr BP were obtained from monospecific samples of *Elphidium clavatum* taken from different levels in the vertical mud dykes in the till at Port (Fig. 9.6).

The tectonic structures in the till at Port are truncated and overlain by a pronounced, horizontal boulder lag along the entire section. Boulders vary between 10 and 35 cm long, or within the same size range as those in the till below. The sharp truncation and boulder lag are attributed to marine erosion and sorting during the late

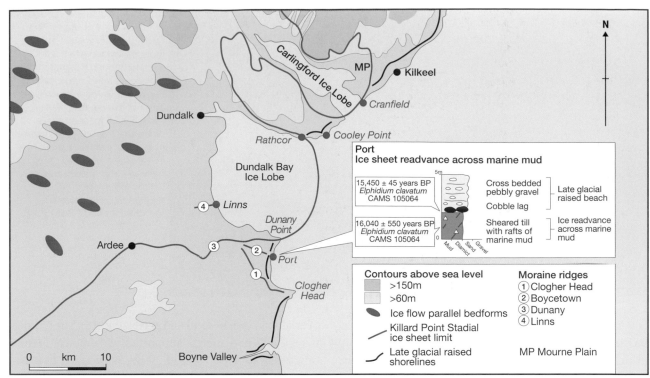

Fig. 9.4 Location, two limiting AMS ¹⁴C dates from marine mud and graphic log from Port, County Louth. The marine mud has been incorporated into basal till during ice advance south to Clogher Head. These dates help to constrain an ice sheet readvance into the Irish Sea Basin that was more extensive than the ice advance to Rathcor/Dunany Point at the end of the Killard Point Stadial.

By kind permission of GSNI.

Labels within Fig. 9.4:

Carlingford Ice Lobe · MP · Kilkeel · Cranfield · Dundalk · Rathcor · Cooley Point · Dundalk Bay Ice Lobe · Linns · Dunany Point · Ardee · Port · Clogher Head · Boyne Valley

Port
Ice sheet readvance across marine mud

15,450 ± 45 years BP
Elphidium clavatum
CAMS 105064

16,040 ± 550 years BP
Elphidium clavatum
CAMS 105064

Cross bedded pebbly gravel — Late glacial raised beach
Cobble lag
Sheared till with rafts of marine mud — Ice readvance across marine mud

Mud · Diamict · Sand · Gravel

Contours above sea level
>150m
>60m
Ice flow parallel bedforms
Killard Point Stadial ice sheet limit
Late glacial raised shorelines

Moraine ridges
① Clogher Head
② Boycetown
③ Dunany
④ Linns

MP Mourne Plain

0   km   10

N

Fig. 9.5 Exposure of basal till at Port, County Louth overlain by raised beach gravel. To the left of the spade (0.9 m long) is a pale coloured vertical dyke of fossiliferous marine mud scavenged into the basal till. The till has been truncated and winnowed by marine erosion leaving a horizontal cobble/boulder lag. The overlying planar gravel and sand are part of the late-glacial raised beach system which occurs immediately south of the Dunany Point ice limit (after McCabe et al., 2007b).

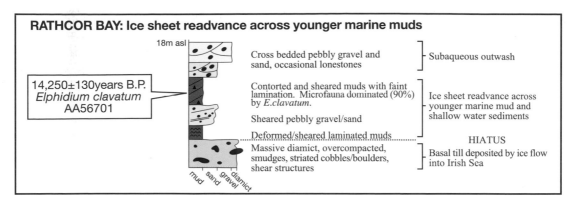

RATHCOR BAY: Ice sheet readvance across younger marine muds

18m asl

14,250±130years B.P.
Elphidium clavatum
AA56701

Cross bedded pebbly gravel and sand, occasional lonestones — Subaqueous outwash

Contorted and sheared muds with faint lamination. Microfauna dominated (90%) by *E.clavatum*.

Sheared pebbly gravel/sand

Deformed/sheared laminated muds

Massive diamict, overcompacted, smudges, striated cobbles/boulders, shear structures

Ice sheet readvance across younger marine mud and shallow water sediments

HIATUS

Basal till deposited by ice flow into Irish Sea

mud sand gravel diamict

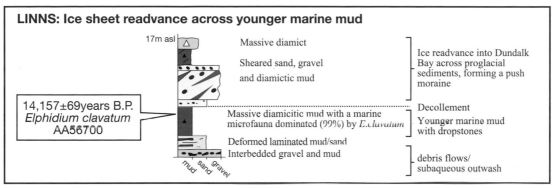

LINNS: Ice sheet readvance across younger marine mud

17m asl

14,157±69years B.P.
Elphidium clavatum
AA56700

Massive diamict

Sheared sand, gravel

and diamictic mud

Massive diamicitic mud with a marine microfauna dominated (99%) by *E.clavatum*.

Deformed laminated mud/sand
Interbedded gravel and mud

Ice readvance into Dundalk Bay across proglacial sediments, forming a push moraine

Decollement
Younger marine mud with dropstones

debris flows/ subaqueous outwash

mud sand gravel

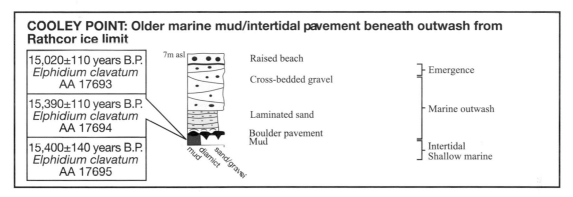

COOLEY POINT: Older marine mud/intertidal pavement beneath outwash from Rathcor ice limit

15,020±110 years B.P.
Elphidium clavatum
AA 17693

7m asl

15,390±110 years B.P.
Elphidium clavatum
AA 17694

15,400±140 years B.P.
Elphidium clavatum
AA 17695

Raised beach

Cross-bedded gravel

Laminated sand

Boulder pavement
Mud

Emergence

Marine outwash

Intertidal
Shallow marine

mud diamict sand/gravel

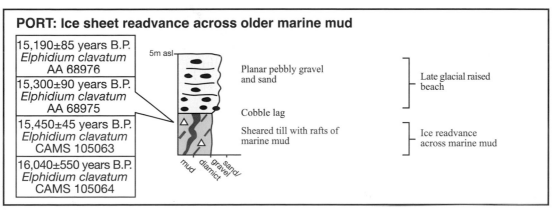

PORT: Ice sheet readvance across older marine mud

15,190±85 years B.P.
Elphidium clavatum
AA 68976

5m asl

15,300±90 years B.P.
Elphidium clavatum
AA 68975

15,450±45 years B.P.
Elphidium clavatum
CAMS 105063

16,040±550 years B.P.
Elphidium clavatum
CAMS 105064

Planar pebbly gravel and sand

Cobble lag

Sheared till with rafts of marine mud

Late glacial raised beach

Ice readvance across marine mud

mud diamict sand/ gravel

Fig. 9.6 Schematic logs from critical sedimentary successions exposed around the margins of Dundalk Bay together with the stratigraphic position of the nine AMS ^{14}C dates used to constrain the age of ice sheet readvances. Full sedimentary descriptions in McCabe et al., 1987, McCabe and Haynes, 1996 and McCabe et al., 2005, 2007b. The AMS ^{14}C ages are from monospecific samples of *Elphidium clavatum* and are corrected for a 400 yr reservoir age.

glacial transgression after ice vacated the site. This interpretation is supported by the overlying parallel-bedded, planar pebble gravel and sand which is characterised by both size sorting and shape sorting that are typical of foreshore gravels. Beds are often emphasised by single lines of pebbles, some of which are imbricated seawards. These foreshore sediments cover an area of ~4 km² between the Clogher Head and Dunany Point moraine ridges (Fig. 9.4). During the erosional phase of this transgression the eastern (seaward) part of the Boycetown ridge was eroded and replaced by the plain of raised beach gravel around Port. The presence of ice wedge pseudomorphs in the beach gravel confirms that the transgression was late glacial and not post glacial in age.

Although the southern limit of this ice readvance is age-constrained directly the northern limit of the large ice lobe is less well known. The northern limit is probably recorded by morainic ridges immediately west of Kilkeel, south County Down. These ridges between Mourne Abbey and Derryoge trend north–south, contain gravelly sediments, some of which is arranged as eastward dipping foresets at ~25 m OD, contain enclosed depressions and are up to 30 m high and a few hundred metres across. It is argued that the morainic complex can be correlated with the Clogher Head moraine because both occur a few kilometres outside or distal from the set of dated moraines formed during the less extensive Killard Point Readvance which reached Dunany Point and Cranfield Point (Fig. 9.2). It is also significant that the southern parts of the moraines at Mourne Abbey have been eroded and truncated by the late-glacial transgression which occurred during the later Killard Point readvance (Fig. 9.1). These moraines are therefore not preserved in the coastal exposures around Kilkeel where raised beach gravels truncate the section (Fig. 6.28). It is also thought that much of the Carlingford Peninsula was not covered by ice at this time. Associated ice limits are marked by gravel ridges at levels around 40–50 m OD between Rampark and The Bush on the southern flanks of the peninsula. The structure of gravels at The Bush are typical of deposition in ponded water between the ice sheet margin and the rising slopes of Barnavave. These ridges and kettled topography occur at much higher altitudes than the moraines formed during the later Killard Point ice sheet readvance to Rathcor near the present coast (Fig. 9.2).

The Killard Point Readvance

In eastern Ireland this readvance can be differentiated from the earlier readvance to Clogher Head because the Killard Point moraines record a less extensive ice readvance across marine muds younger than those overrun during the Clogher Head Readvance (Fig 9.6). The type site and type area for this readvance occurs at Killard Point, east County Down where most of the attributes necessary to reconstruct the anatomy of a significant readvance are present (Fig. 9.7). These include the ice flow lines from centres of ice dispersal and subglacial sediment transfer leading to ice-marginal moraines and outwash interbedded and underlain by *in situ* marine mud (Fig. 5.4). Clearly a non-glacial marine interval is recorded between the early deglaciation of this coast and the later ice sheet readvance to Killard Point. The moraine systems recording the former ice margin can be traced from Killard Point around the flanks of the Mourne Mountains, on the margins of Carlingford Lough and Dundalk Bay and southwestwards across the Irish midlands to Mullingar, a distance of 170 km (Fig. 9.1). The moraines, which are tightly age-constrained to a few hundred years from four key sites, record the time when ice was at its maximum (~14 ¹⁴C kyr BP) (Fig. 9.6). Contemporaneity along this morainic complex is also evident from the regional colinear iceflow lines leading to the reconstructed ice sheet limit and the presence of raised late-glacial shoreline facies immediately outside the ice limits (Fig. 9.1). The moraines represent maximum ice limits rather than decay features because the upper stratigraphy in the immediate offshore zone is fine-grained mud and does not include ice contact or outwash sediments. Variations in sediment geometry and depositional processes along the ice limit are typical of ice contact deposition at tidewater which resulted in a range of landforms from outwash spreads, constructional ridges to ice-pushed forms. Intricate reconstructions of ice limits were necessary in areas where topographic highs forced the ice margin to split into distinct lobes in Carlingford Lough and Dundalk Bay (Fig. 9.4).

The anatomy of the Killard Point Readvance, the extent of the limiting moraine, knowledge of its origins and its age late in the deglacial cycle make it the most significant moraine known in the British Isles. It is one of the few age-constrained moraines which not only record the sensitivity of the relatively small BIIS (<2 m sea level equivalent), but its millennial timescale can be

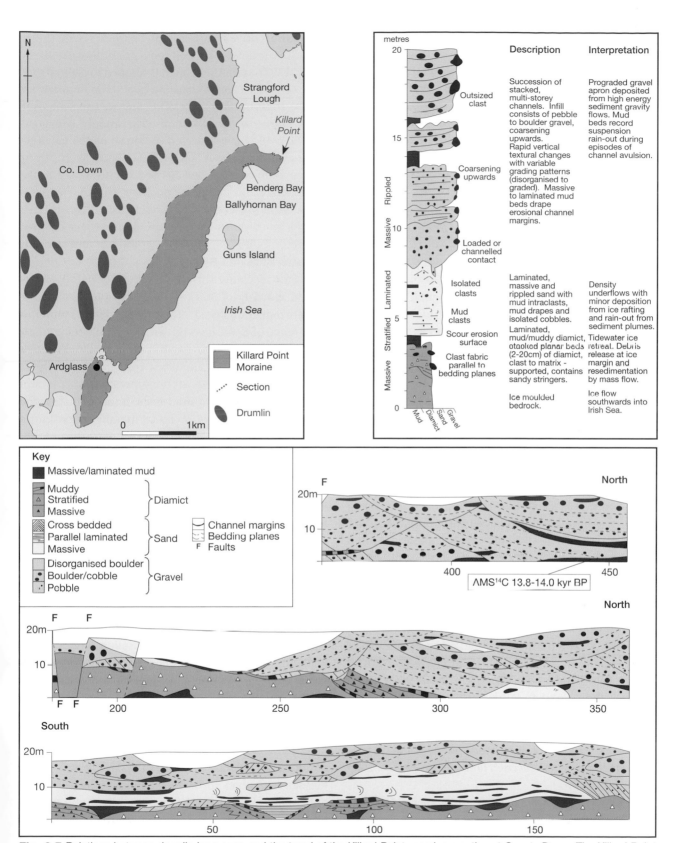

Fig. 9.7 Relations between drumlin long axes and the trend of the Killard Point moraine, southeast County Down. The Killard Point moraine occurs immediately outside the drumlin swarm and is perpendicular to the drumlin long axes. The three lithofacies associations exposed in Benderg Bay, Killard Point, and a schematic log illustrate the sedimentology and the sites sampled for radiocarbon dating.
By kind permission of GSNI.

evaluated against other abrupt and large fluctuations in the North Atlantic. The key field sites which occur along the line of the moraine will be evaluated from the type site at Killard Point southwestwards to Dundalk Bay. Critical sections related to contemporary sea level changes at several localities are included together with evidence for timing of stagnation zone retreat following the readvance.

Killard Point

A NE–SW zone of hummocky topography about 9 km long by 1 km wide occurs between Killard Point and Ardglass along the coastal fringe of County Down (Fig. 9.7) (McCabe et al., 1984). Sediments in this zone are at least 20 m thick and overlie an irregular rock base. Although the land surface varies from ridges to kettled topography with a local relief of 10–15 m, the rather subdued topography is the result of erosion or smoothing during high RSL which is identified on the basis of later raised beach forms, washing limits and regional drapes of red marine mud. The stepped profiles of Gun's Island immediately offshore records sea level fall from the late-glacial marine limit. The Killard Point Moraine is located 1–2 km in front of the drumlin swarms of east County Down whose long axes are perpendicular to the orientation of the moraine (Fig. 9.7). This close field association shows that both are part of the same integrated glacial system comprising southeasterly ice flow with associated drumlinisation, substrate erosion and sediment transfer to the ice margin around Killard Point. Drumlin morphologies record the last ice flow across County Down from the Lough Neagh centre of ice dispersal (Fig. 9.2b). Dating of marine microfaunas from mud within the succession at Killard Point place the maximum of this ice advance to ~14 ^{14}C kyr BP (McCabe and Clark, 1998). The original extent of the moraine cannot have been more than 1km offshore because the immediate offshore facies are dominated by mud. The ice readvance associated with formation of the moraine has been termed the Killard Point Stadial and dates, for the first time, the major oscillation of the BIIS which preceded the Loch Lomond Stadial (McCabe and Clark, 1998).

At Killard Point the glacigenic facies succession rests directly on ice-sculpted Lower Palaeozoic bedrock which was moulded by plastic ice flow to the southeast (Fig. 9.7). Erosional marks consist of parallel, generally straight grooves (~20 m long by ~0.5 m deep) cut across the entire convex surfaces of bedrock risers (1–2 m high). The grooves are sculpted across the strike of the bedrock with smoothed, open U-shaped to dish-shaped cross sections. Glacial striae up to a 5 mm in depth are parallel to the groove orientation and are superimposed on the larger erosional marks. The intense subglacial erosion could have occurred during an earlier ice sheet advance into the Irish Sea Basin but more likely during an active phase of subglacial sediment transfer to the ice margin a short distance offshore. Three facies associations are present and are distinguished on the basis of their position in the section, internal and external geometry, bounding surfaces, grain size and bedding characteristics (Fig. 9.7).

On the glaciated surface the 4m thick diamict association consists of stacked, massive and planar beds that are matrix and clast-to-matrix supported, contain a wide range of grain sizes (granules to boulders) and are often separated by stringers of coarse sand or granules (Fig. 9.8a). Diamict lithofacies vary from massive to crudely stratified and are interbedded with minor sand, pebbly mud and mud beds. In general stratified diamict infills hollows in more massive deposits, or occurs as thin beds (5–10cm) stacked along the margins of irregular mounds of massive diamict (Fig. 9.8a). The a–b planes of many cobbles and small boulders lie parallel to bedding planes in the stratified diamict. Some larger clasts project above the lower contacts of succeeding beds. Shallow scours (1–2m across) infilled with interbedded red mud, sand, gravel and diamict beds containing outsized clasts (<40cm across) occur within the upper part of the diamict association or are cut into its upper surface (Fig. 9.8b). Small-scale fold structures and localised contorted bedding are found within these facies. The degree of interbedding and stratification of the diamict association, and features including soft-sediment mud clasts and slump folds, are in keeping with a mass flow origin following debris release from a grounded tidewater ice margin. Textural variability of the mass flows is dependent on the rate and type of debris release from the ice margin, distances travelled and degree of winnowing (Lowe, 1982). A sand facies association infills irregularities and drapes the upper diamict surface for about 150 m. Up to 3 m of parallel laminated and graded sand contain discontinuous mud laminae, water escape structures, isolated cobble clasts, rip-up soft-sediment clasts, small folds and flame structures (Fig. 9.8b). Dominant transport mechanisms include sediment gravity flows of low to

Fig. 9.8a Massive diamict grading into stratified diamict with clast fabrics parallel to bedding planes. The base of the succession rests on striated and grooved bedrock, Killard Point, County Down. Larger clasts show clear evidence of freighting by thin debris flows.

Fig. 9.8b Detail of the sand association containing lone-stones, mud clasts and flame structures overlying stratified diamict, Killard Point. Note that the stratified diamict contains resedimented beds of marine red mud.

intermediate viscosity driven by density contrasts (Middleton and Hampton, 1976). The upper surface of the sand association is channelised and marked locally by soft-sediment gravel dykes. The overlying gravel facies association comprises fourteen stacked and infilled channels (50–100 m across, 10–15 m deep) identified on the basis of well-defined margins, erosional junctions between adjacent infills and sedimentary contrasts between adjacent infills (Figs. 9.9a, b). Channel infills are dominated (80%) by variable boulder gravel facies with subordinate sand and pebbly gravel facies. Gravel infills show a wide range of internal geometries (amalgamated to catenary infills with beds parallel to erosional channel margins) and clast sizes. Individual beds vary

Fig. 9.9a Catenary gravel infills within multistoried and cross-cutting channels, Killard Point, County Down. The channels were cut and infilled by powerful meltwater discharges into a shallow marine environment.

Fig. 9.9b Disorganised cobble gravel facies infilling the centre of a major channel, Killard Point. The lack of structure in the gravel suggests deposition from a high-density flow by frictional freezing.

Fig. 9.9c Marine red mud containing Arctic microfauna interbedded with diamict beds (debris flows) forming beds of pebbly mud, Killard Point.

greatly in thickness (0.2–1 m) and are mainly massive. Up to 20 percent of beds show crude forms of distributional and coarse-tail grading together with boulder clusters. About 80 percent of beds are clast supported. Overall, the gravel facies coarsens upwards and inferred channel axes indicate southward progradation of the channel system. The nested channels record repeated cut and fill and represent a distributary-type system subject to avulsion and changes in fluid/sediment input. The pulsed nature of the system is suggested by strong erosional channel margins, abrupt bed contacts, abrupt clast size changes and the presence of mud drapes lining erosional channel margins and within channel fills (Fig. 9.10). The variable grading patterns, together with evidence (outsized clasts and boulder clustering) for clast freighting (Postma *et al.*, 1988), suggest deposition by mass flows in which there were strong grain interactions and high dispersive pressures (Hein, 1982). Sediment geometries with disorganised to variably graded beds are very similar to the gravel lithofacies successions described by Walker (1975, 1984) from resedimented, deep-water conglomerates. It is suggested that the gravel lithofacies present are similar to the evolutionary continuum recognised from high-density sediment gravity flows.

Beds, laminae and clasts of red mud occur throughout the succession (Fig. 9.8b, 9.9c). The thickest and most continuous bed of laminated and massive mud is 1.5 m in thick and occurs in the gravel association where it lines channel margins, especially at the northern end of the exposure (Fig. 9.7). The presence of mud lining erosive channel floors and margins shows that the phases of cut and fill did not always closely follow one another. In the larger channels the presence of red pebbly mudstone along channel flanks is probably associated with collapse of gravelly banks into mud beds, mass flow and textural inversion. Beds of massive red mud at the northern end of the exposure contain well-preserved marine microfaunas dominated by *E. clavatum* (90%) whereas foraminifera in the laminated mud present at higher levels in the same channel are abraded and less abundant. Two monospecific samples of *E. clavatum* are dated to ~14 ^{14}C kyr BP (McCabe and Clark, 1998).

These data from Killard Point are critical to any interpretation of events within the ISB because it is one of the few sites where a zone of terminal outwash, dated from an *in situ* marine microfauna, is the product of subglacial transfer of debris during drumlinisation to a tidewater ice margin (Fig. 5.4). This integrated glacial system, which is correlated with similar events in the northern part of the basin and in other sites in the western British Isles, is important within an amphi-North Atlantic context for the following reasons.

a) It provides a terrestrial record of millennial-scale ice sheet oscillations around 14 ^{14}C kyr BP which is correlated with Heinrich event 1 (McCabe *et al.*, 1998).

b) It allows terrestrial records from a small ice sheet situated in the climatically sensitive northeastern Atlantic to be compared with other hemispheric climate proxies.

c) It is an excellent example of a raised morainal bank which did not evolve into a Gilbert-type delta. Sedimentologically the three facies associations record a prograding system with progressive transfer of coarse sediment by high energy mechanisms along channels into deeper water (Fig. 9.10).

d) The raised morainal bank shows that the ice margin ended at tidewater, showing that deglaciation was accompanied by high RSLs. This sector of the Irish Sea Basin must have been deeply isostatically depressed because global eustatic sea level was low

at this time (McCabe, 1997). This evidence clearly questions the numerical model predictions of Lambeck (1995, 1996) which do not record high RSLs at this time and therefore minimized the role played by deep isostatic depression during deglaciation. The current numerical models are deterministic and fail to consider the well-dated field evidence. Instead they are based on an amalgam of generalised data including LGM ice flowlines, ice flowlines of mixed ages and on monotonic ice-marginal retreat. The current numerical models (e.g. Lambeck and Purcel, 2001) therefore cannot aid palaeoenvironmental reconstructions around the basin because post-glacial uplift data cannot be easily projected back into the deglacial period when rapid environmental change dominated the palaeogeography of the Irish Sea basin.

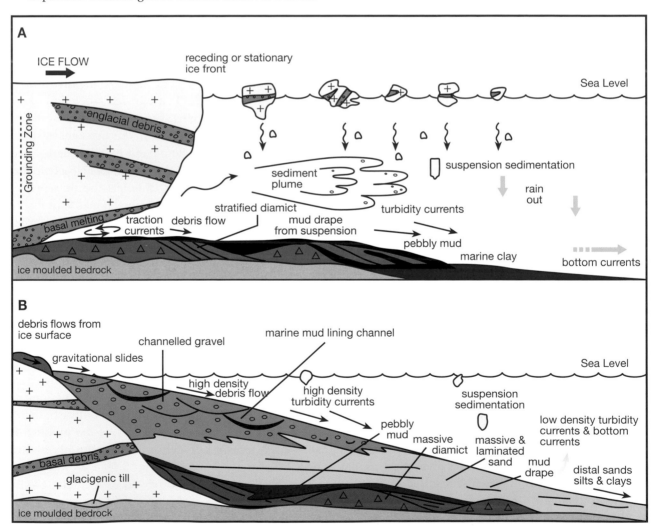

Fig. 9.10 Depositional model to account for the origins of the three lithofacies exposed at Killard Point (after McCabe et al., 1984).

Bloody Bridge and Dunmore Head

Topographically the granites of the Mourne Mountains (150 km²) form a dissected upland region rising to 500–700 m OD while the Silurian country rock forms low ground (<50 m OD) on the northern, western and southern flanks (Fig. 9.1). The mountains are cut by glacially-overdeepened valleys, some of which are fed from corries whilst others record ice flow through the mountain mass from lowland centres of ice dispersal to the north and west (Fig. 9.11). There is no real evidence to suggest that an independent ice cap developed in the mountains though many corries and erosional valley forms have been recorded (Stephens *et al.*, 1975). The presence of tors on mountain summits cannot be used directly as evidence that these summits were unglaciated because their location and form is best explained by post-glacial weathering along rebound joints (Figs. 9.12a–c).

Because of their location to the east of the main ice dispersal centres, the Mourne Mountains must have influenced ice sheet configuration, especially when ice margins thinned and readvanced during the last glacial termination (Fig. 9.1). The glacigenic deposits along their steep eastern margin between Bloody Bridge and Dunmore Head are associated with deposition from an ice lobe which moved southwards as far as Dunmore Head (Fig. 5.11). These deposits record an upper drift limit which begins at ~150 m OD near Newcastle and falls steadily in altitude southwards to around 35 m OD at Dunmore Head. This level marks the cultivation limit on the mountain and is replaced by screes at higher elevations.

The morphology of the Dunmore Head ridge is largely a result of meltwater channelling, but the generally hummocky lateral moraine (local relief ~15m) contains kettleholes and appears to be intact. Small exposures show that the moraine consists of poorly-sorted gravel interbedded with thin (0.3 m), discontinuous beds of diamict. There are rapid lateral and vertical changes of gravel facies from granules to boulders which generally infill shallow scours. Southwards the hummocky topography grades into a gravel spread which extends from Glassdrumman south to Mullartown Point and Annalong village (Fig. 6.22). The gravel spread has been notched at 30 m, 19 m and 9 m OD by late-glacial marine erosion. The highest notch can be seen 100 m west of the main road north and south of Annalong. The lower

Fig. 9.11 View eastwards through the eastern Mourne Mountains showing rounded summits and cols modified by ice flow through the mountains from the lowlands to the north.

Fig. 9.12a Granite tors, about 7 m high, on the summit of Slieve Binnian, Mourne Mountains. Tors are residual landforms resulting from post-glacial weathering of massive granite along rebound joints that developed after glacial erosion and removal of the ice and rock overburden. *By kind permission of GSNI.*

Fig. 9.12b Upslope view of a stone banked boulder lobe on the southern side of Slieve Donard at about 700 m above sea level. (Cunningham and Wilson, 2004). Person is 1.8 m in height. These landforms are probably periglacial and are associated with frost climates, frost weathering of rock, downslope movement of boulders, clast sorting, and lobe migration. *By kind permission of GSNI.*

Fig. 9.12c Cross-view of a boulder lobe. Person is 1.8 m high and is positioned at the junction between the riser (left) and the tread (right). These features postdate the last glacial maximum and were active during the coldest phases of the late-glacial. In some cases they represent detritus detached by mechanical weathering from upland summits and moved downslope while the tors remained as residual landforms. *By kind permission of GSNI.*

notches are preserved between Dunmore Head and the mouth of Carlingford Lough to the south where the corresponding marked readvance limit is termed the Cranfield Point moraine. The late-glacial emergent beach facies truncate earlier deposits and were deposited only in the ice-free area between the Dunmore Head and Cranfield Point moraines (Fig. 9.1). Therefore both ice limits are considered to be contemporaneous.

The Carlingford Bay Ice Lobe at Cranfield Point

A coastal section about 0.5 km east of Cranfield Point shows coarse-grained gravel interbedded with sand and diamict overlying marine mud which is dated to 14.7–15.6 ^{14}C kyr BP (Fig. 9.13) (McCabe and Clark, 1998). The crudely tratified gravel which is part of the Cranfield Point Morainic complex formed at the margin of the Carlingford Bay ice lobe during a major ice sheet readvance ~14 ^{14}C kyr BP. The moraine occurs as nested arcuate ridges which swing northwards from Cranfield Point to Mourne Park and Tullyframe (Fig. 9.1). Two major subparallel ridges in this moraine sector at Ballyardel and Drummanmore (30 m high, 300 m in width and 3–4 km long) determine the present course of the White river. From Tullyframe the line of moraine

extends westwards as terraces and ridges along the lower slopes of Formal Mountain and Knockshee Mountain at around 120 m OD. In this zone there is a very clear ice limit recorded by kettled ridges resting on rock with a transition northward over a few metres onto drift free bedrock. At Tullyframe the moraine consists of 10 m of very poorly sorted cobble and boulder gravel overlain by thick (~1–2 m) beds of massive and stratified sandy diamict, 20 m thick. Where meltwater erosion removed distal sediment between the moraine and the slopes of Knockchree Mountain, pull-apart structures in the core of the ridge record the withdrawal of lateral ice and sediment support (Fig. 7.19c). The moraine is about 12 km in length and records the north-lateral ice margin of the ice lobe. The pronounced eastward-facing bulge in the moraine morphology at Ballyardel records ice pressure due east onto the margin of the Mourne Plain. A series of stepped, steep ice-proximal slopes face west towards the Carlingford lowlands around Maghery and Benagh which formed the subglacial toe depression immediately behind the ice sheet terminus.

On the south side of Carlingford Bay moraines deposited by the same ice lobe are seen on the low ground at Ballagan Point (Fig. 9.1). An the northern margin of the Carlingford Mountains ice moving along the axis of

Fig. 9.13 Marine mud (14.7–15.6 ^{14}C kyr BP) overlain by glacigenic diamict, Cranfield Point, south County Down. The diamict was deposited when ice readvanced from Carlingford Lough across a marine embayment to form the terminal moraine known as the Cranfield Point moraine (see Fig. 9.1). *By kind permission of GSNI.*

the lough pushed south through the col at Windy Gap to form a set of nested moraine ridges in the upper Big River catchment at the Mass Rock (Fig. 9.3). Most of the moraine consists of large joint blocks derived from the local basic intrusions. At this site the ice reached 230 m OD, 12 km up-ice from the frontal terminus of the ice lobe at the mouth of Carlingford Lough. An exposure date from a large erratic perched on top of the moraine at the Mass Rock suggests deglaciation occurred ~17 cal ka BP (Bowen et al., 2002). Streamlined rock ridges and drumlins to the west of the moraine around War-renpoint and Newry record erosion, debris transfer and ice flow southeastwards to the ice lobe margin.

Northwards the Carlingford Trough narrows into the N–S tunnel valley at Poyntz Pass which is floored by torpedo-shaped drumlins cored with stratified deposits sealed by a diamict carapace (Fig. 9.1) (Dardis and McCabe, 1983). These N–S drumlins cross-cut earlier NE–SW ribbed moraine ridges and are aligned along the floor of the tunnel valley (Fig. 5.26). The tunnel valley and drumlin orientation record the regional ice flow lines which drained the ice mass centred in the Lough Neagh basin to the north and ended at the Cranfield moraine in the south, a distance of 70 km. The Poyntz Pass ice stream which moved through Carlingford Lough Narrows expanded laterally to form a pronounced ice lobe covering 60 km² at the mouth of Carlingford Lough (Fig. 9.14). The main feeder channel was only 1–2 km in width at the narrowest point of the lough, suggesting large ice volumes passed through the trough in order to maintain a tidewater ice front about 7 km across. The

widespread presence of steep, glacially-modified slopes flanking the trough testify to intense glacial linear erosion along a pre-existing geological weakness.

Although the limiting moraine at Cranfield Point has a ridged morphology, the feature grades imperceptibly eastwards into a flat plain some 200–300 m from the ridge crest. In the gravel pit at Sandpiper, pebble gravel is interbedded with horizontally laminated and rippled sand which contain isolated cobbles and boulders. At the eastern end of the pit a large boulder occurs in the pebbly sands about 0.5 km from the ice margin to the west at Cranfield Point (Figs. 9.15a, b). Above the pebbly sand is a tabular unit of stratified diamict consisting of sandy beds, 2–10 cm thick and laterally continuous for 20–50 m, flat-lying or sometimes wavy, and with graded and diffuse bed contacts and dispersed pebbles. Beds are occasionally cut by shallow (<1 m) scours infilled by vaguely-laminated pebbly-sand. The stratified diamict, which consists of pebbly-sand with a low, but variable, mud content, is characterised by two distinct clast fabrics. Most of the flat, pebble-sized clasts lie parallel to the bedding planes and dip gently to the southeast. Larger pebble to cobble-sized clasts occur sporadically throughout the deposit, deform underlying beds and may be vertically aligned. The stratified diamict grades up into 2–3 m of massive diamict with numerous lenses of sand and gravel up to a few metres wide. This tabular unit forms the surface of the outwash spread and is cut by small channels varying in cross section from sym-metrical, saucer-shaped depressions (4x1 m) to asym-metric forms (3x2 m) with overhanging sides. Channel

Fig. 9.14 Looking southeast (down-ice) along Carlingford Lough which is flanked by the southern edge of the western Mournes (left) and the Carlingford Mountains on the south side of the lough (right). *By kind permission of GSNI.*

Fig. 9.15a Shallow water stratified sediments in the Sandpiper pit, south County Down, which grade east for 0.5 km as a flat spread of glaciomarine deposits in front of the Cranfield Point moraine. The large lonestone is a dropstone released from floating ice.
By kind permission of GSNI.

Fig. 9.15b Detail of laminated, sandy diamict containing isolated cobbles and pebbles (dropstones), Sandpiper pit, south County Down. Some cobbles have their long axes in a vertical position whereas small pebbles are generally aligned parallel to bedding planes. These amalgamated beds were deposited rapidly from high-density underflows containing sand and silt admixtures with little time for sorting or grading (pen is 15 cm long).
By kind permission of GSNI.

margins are strongly erosional with a basal fill of massive pebbly gravel or cobble lags. Infills are mainly admixtures of laminated and wavy to contorted sand with lenses of granules and pebbly gravel. Thin lenses of massive, sandy diamict are occasionally present. The margins of some channels are poorly defined and channels may only be recognised by inner cores of better sorted granules and sand and outer margins of contorted sediment.

The large lonestones within the basal sand and gravel are interpreted as dropstones because the traction currents that deposited thin pebbly-gravel beds had not sufficient matrix to freight cobbles. There are no topographic highs locally which could generate competent flows capable of freighting or transporting boulders. The tabular geometry of the sandy diamict, the marked lateral continuity of individual beds, almost constant bed thickness and textural homogeneity suggests

deposition from dense underflows generated near the ice margin. These flows were characterised by similar magnitudes and flow competences. Flow strength was sufficient only to reorientate the smaller clasts parallel to bedding planes. Outsized clasts could not have been carried by the thin flows responsible for the stratified diamict, and are thus inferred to be ice-rafted. The general absence of silt and mud is attributed to active current winnowing and sediment bypass at this site. It is tempting to suggest that the cyclicity within the stratified diamict is similar to cyclopsams deposited by rainout from a buoyant jet (Mackiewicz *et al.*, 1984; Hunter *et al.*, 1996). However, an origin from hyperconcentrated sheet flows, which consisted of admixtures of sand and pebbles deposited rapidly, explains why graded beds are absent and pebble dispersal is present. Expanding flows may be strong enough to align small pebbles parallel to bedding planes prior to rapid areal freezing of laminae

Fig. 9.16a, b, c Facies and facies geometry at Derryoge, south County Down. **A)** General view of the diamict and mud facies in the cliff exposure truncated by raised shoreface gravels.
By kind permission of GSNI.

B) Detail of marine mud deposited in large channel cut into glaciomarine diamict. Note that the channel margin is marked by a cobble lag in the lower right part of the photograph.
By kind permission of GSNI.

C) Detail from raised late-glacial beach sand and gravel at the top of the cliff section. *By kind permission of GSNI.*

(McCabe and O'Cofaigh, 1994). The small channel fills which cut the flat-lying beds of diamict may be associated with cut and fill as channels prograded across the diamict surface. Channel overhangs and diapiric structures along channel margins suggest that the substrate was still soft and that some channels sank into the diamict, forming pseudonodules. In many cases disorganised pods like these, entombed within a finer-grained matrix, could be mistaken for iceberg dump deposits.

A continuous and well-defined late–glacial marine-cut notch (at 18–19 m OD) occurs for 3 km between Derryoge and the mouth of Carlingford Lough (Fig. 9.1). The fossil notch does not occur within Carlingford Lough itself but either fades out against the major line of moraine ridges or the subaqueous outwash deposited at the margins of the ice lobe which readvanced southeastwards to Cranfield Point (Fig. 9.1). The fossil cliff occurs inland from Derryoge where the associated beach gravel and the underlying surface of marine planation are exposed in section (Fig. 9.16). It is significant that the moraine ridges immediately to the east of the Cranfield Point Moraine formed during the earlier Clogher Head Readvance are also truncated by late-glacial marine erosion and replaced by a broad marine terrace. The raised terrace is well developed around Dunnaval seaward of the fossil beach notch. The location of this notch immediately outside the limits of the Cranfield Point Moraine shows that both were contemporaneous. This field evidence, comprising the close association between high RSLs and ice sheet limits, is recorded throughout eastern Ireland and supports the hypothesis that there is a close relationship between sea level history, ice-marginal sediments and ice sheet dynamics in the ISB.

The Cranfield Point Moraine marks the northern margin of the same ice lobe which deposited the Ballagan Point Moraine on the tip of the Carlingford Peninsula (Fig. 9.1). The raised beaches around Ballagan Point show an identical relationship to those outside the ice limit/moraine at Cranfield Point. This field relationship also suggests that both sets of moraines are ice-proximal and glaciomarine in origin because the ice acted as an effective barrier and prevented the late-glacial transgression from entering Carlingford Lough at this time. Therefore isostatic readjustment must have occurred rapidly following moraine and beach formation. The fossil beach seen outside (ie. to the east of) the Cranfield Point Moraine at Dunnaval postdates 14.7

[14]C kyr BP because the moraine overlies dated marine mud (Fig. 9.1). Correlation based on the ages of marine mud at the Cranfield Point, Linns, Rathcor and Killard Point moraines suggests that the maximum of this glacial readvance occurred ~14 ^{14}C kyr BP.

The Dundalk Bay Ice Lobe

The moraines and glacigenic deposits on the northern and southern margins of Dundalk Bay record deposition at the margins of a large (100 km^2) ice lobe centred in the bay (Fig. 9.17). These moraines form part of the much larger ice-marginal outwash system which borders the drumlin swarms of north central Ireland (Fig. 9.2). The regional pattern of drumlin orientations inland from Dundalk shows that drumlinisation was accompanied by substantial subglacial debris transfer along ice flow lines which reached tidewater on the margins of the Dundalk Bay ice lobe (Figs. 1.1, 5.1). Both ice lobe maxima in Dundalk Bay and Carlingford Lough were contemporaneous because the raised marine terraces and late-glacial notches immediately inland from Cooley Point record the same late-glacial sea level between the limiting moraines at Ballug Point and Ballagan Point (Fig. 9.17). There are thirteen major exposures around the bay used to identify the magnitudes and range of processes which contributed to moraine formation and facilitate ice lobe reconstruction around the bay margins (McCabe et al., 1987). Two main depositional settings have been identified. On northeastern bay margins moraines are composed mainly of ice contact deltaic gravel and diamict deposited in a narrow arm of the sea between the ice margin and the Carlingford Peninsula. On the southern margins of the ice lobe, open directly to marine influences from the Irish Sea Basin, moraines are dominated by diamictic muds and morainal banks.

On the northeastern margin of the bay at least four sub-parallel, nested ridges occur on low ground (~30 m OD) immediately south of the Carlingford massif (Fig. 9.2). The complex is about 8 km long by 1 km broad between Rockmarshall house and Ballug Point. Individual ridges trend WNW–ESE, are slightly sinuous, up to 25 m high and can be traced for 2–4 km before crossing the present coastline. Single, round crested ridges are the general rule though pitted, hummocky and flat to undulating gravelly spreads frequently form an integral part of individual ridge trends. Ridges at Annaloughlin, Rampark and Giles quay consist mainly of planar cross-bedded pebble and cobble gravel foresets dipping east-

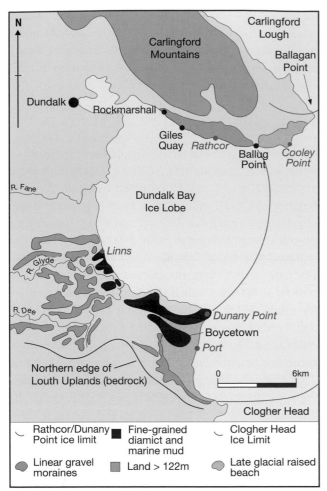

Fig. 9.17 Ice limits on the margins of Dundalk Bay. Ridge moraine patterns from the northern and southern margins of the bay are used to reconstruct the positions of the margins of the ice lobe which readvanced during the Killard Point Stadial. Mud dated to 14.2 ^{14}C kyr BP at Linns and Rathcor was tectonised by this ice advance across Dundalk Bay from north central Ireland. Older mud dated to >15 ^{14}C kyr BP occurs outside this ice sheet limit at Cooley Point and Port and was tectonised by an earlier ice sheet readvance to Clogher Head. Note that the main late-glacial raised beaches only occur immediately outside the inferred ice sheet limits between Dunany Point and Rathcor. *By kind permission of GSNI.*

wards at 10–15°. The foresets rest on flat beds of massive and laminated mud or rhythmically-bedded sequences of sand, diamict and massive pebble gravel. The ridge at Mountbagnall is similar but the foresets show a gradation distally (northeastwards) into massive, convoluted sands containing occasional cobble clusters and isolated pebbles. Overlying amalgamated beds of massive cobble gravel contain beds of massive diamict exhibiting a crude stratification. At the base of the Riverstown ridge massive boulder gravel is overlain by a thick (~8 m) unit of stratified diamict characterised by

discontinuous planar bedding planes, dispersed large clasts, a granule-mud matrix, gravelly pods and lenses, discontinuous sandy stringers and megaclasts up to 2 m long. Facies sequences show no evidence for glacitectonic thickening and clearly show that a range of processes occurred during ridge building in an ice-contact setting. A model is proposed involving point deposition and sediment thickening around the exits of glacial effluxes spaced along the WNW–ESE ice margin. These expanded laterally and coalesced into fairly continuous, ice-contact ridge moraines. Basal lithofacies can therefore vary from poorly-organised, amalgamated beds of gravel deposited near effluxes to resedimented packages of diamict, mud and gravel related to unstable slopes as the sediment pile grew. Transportion and deposition of the chaotic gravels must have involved maintenance of a heterogeneous sediment mix and suppression of grading or imbrication processes prior to frictional freezing. Most ridges are cored by gravelly foresets of a Gilbert-type delta which prograde northeast across other beds and record ponded water. Diamicts interbedded with well-sorted gravel foresets can be explained by debris flows from adjacent areas of unstable sediment piles. It is significant that parts of some ridges are dominated by stratified diamict interbedded with sand and gravel as at Riverstown. These ridge segments probably evolved some distance away from major sediment effluxes and represent mainly debris flow and subsidiary bottom-current activity into basinal sectors which developed between the sediment piles around glacial effluxes. Facies variability between diamict- and gravel-cored ridges therefore can occur along a single ridge and tends to reinforce the depositional geometry of the proposed model. Uppermost facies overlying foresets range from diamict to poorly bedded gravel and may locally record water shallowing and incipient topset formation. This essentially fan delta sequence containing thick, massive diamicts is consistent with deposition in a restricted, shallow-water body flanking the Carlingford Peninsula. However, because the eastern end of the system faced and was open to the Irish Sea, the water body along the ice margin was linked directly to high relative sea level at this time.

On the margins of Rathcor Bay the internal geometry of two prominent linear moraine ridges is exposed at the coast where their continuation southeastwards across the bay is marked by piles of cobbles and boulders on the foreshore (Fig. 9.18). They are lateral equivalents of the ridges immediately to the west, though their internal geometries and sedimentology are different. Morphologically, the outer moraine (Rathcor moraine) is a continuation of the ridge at Riverstown, and the inner moraine (Castlecarragh moraine) can be traced into the Mountbagnall ridge to the west. These moraines are up to 3 km long, are continuous and bordered by enclosed depressions. The sediments exposed suggest that these ridge elements represent successive positions of the ice margin as it oscillated along this part of the bay (Fig. 9.18). Oscillation one is recorded by deposition of a massive, overcompacted mud-rich diamict containing abundant striated clasts, shear and smudge structures at the eastern end of the Rathcor section. The overlying sediments forming the Rathcor moraine consist of a coarsening-upwards succession of parallel laminated sand, rippled sand, trough bedded sand and granules overlain by cross-bedded pebbly gravel. Pockets of ripple drift lamination, thin diamict beds, outsized clasts (<1 m long), discontinuous mud drapes and channel fills of massive sand also occur. The pebbly gravel is characterised by numerous cut and fill structures with good preservation of scoop-shaped erosional channel margins. Infills of crudely stratified pebbly sand contain concen-

trations of small cobbles though clasts are generally well dispersed. These beds are overlain by up to 2 m of rhythmically-bedded silt and mud which thicken eastwards towards Ballug Point (Fig. 9.18). Near the latter site, on the eastern side of Rathcor Bay in the upper cliff face, the rhythmites pass into sheared and slightly deformed mud containing a well-preserved marine microfauna. A sample of *Elphidium clavatum* provided an AMS ^{14}C age of 14250±130 years BP. Three-dimensional exposures a few metres inland (northeast) from the main cliff face show undeformed muds conformably overlain by sand and gravel. Clearly, the ice sheet advance to the moraine ridge at Rathcor/Ballug Point slightly postdates deposition of the marine mud (Fig. 9.19).

The lowermost part of the Rathcor section records shallow-water sediments with characteristics including channelled sands, massive sands and thick weakly-bedded sand beds similar to subaqueous outwash (Rust, 1977; Rust and Romanelli, 1975). The presence of outsized clasts, lenses of diamict and irregularly-shaped diamict clots suggests that ice rafting was present as the sequence evolved. The presence of pebbly sand is interesting because pebbles are dispersed throughout a

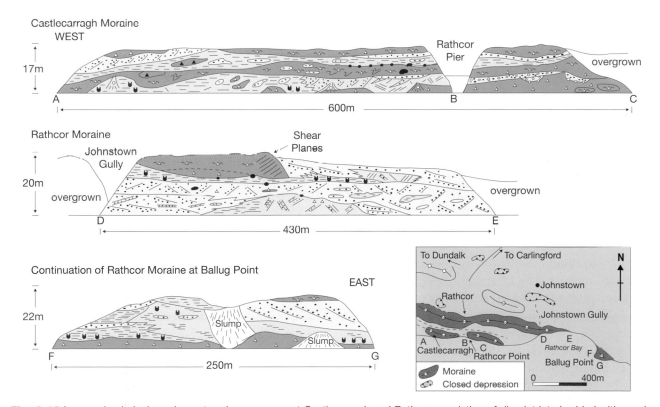

Fig. 9.18 Ice-proximal glaciomarine outwash sequences at Castlecarragh and Rathcor consisting of diamict interbedded with sand and gravel. Insets show the trends of associated moraine ridges. Note that marine erosion has removed part of the Rathcor moraine in Rathcor Bay but its easterly continuation can be recognised in the upper cliff face at Ballug Point. *By kind permission of GSNI.*

Fig. 9.19 Section (20 m high) in the Rathcor push moraine, Rathcor Bay. Note that sheared beds of diamict form the ridge crest and overlie fine-grained subaqueous outwash. *By kind permission of GSNI.*

sandy matrix. It seems to have been deposited *en masse* from turbulent suspensions which froze before grading could occur. The presence of cut and fill geometries also suggest that flow conditions were unsteady and possibly related to a low trajectory jet. Near points of jet detachment, boundary shear and turbulent mixing might have driven currents across the sediment apron, resulting in the variable cut and fill patterns. At the eastern end of the exposure rhythmically bedded fossiliferous marine mud interbedded with the sandy outwash confirms deposition in a marine setting.

At the top of the Rathcor section 5–6 m of diamict truncates the underlying stratified sediments (Fig. 9.19). Its base is massive but grades upwards into sheared diamict which thickens into a marked ridge crest that continues inland towards the westnorthwest from the cliff face at the centre of Rathcor Bay (Fig. 9.18). The moraine ridge has been eroded out from the centre of the bay, but to the southeast, the line of the moraine is picked up again as diamict forming a morainic high at the crest of Ballug Point. Along the western part of the section, nested shear planes rise steeply to the northeast to form the ridge crest (Fig. 9.19). Westwards, along the section, the shears flatten and lie parallel to bedding surfaces. Sediment thickening and shearing are therefore attributed to ice pressure from the south-

west. This scenario where successive, ice-marginal moraines are subparallel and record small oscillations as ice skimmed across its own subaqueous outwash is a common feature along the ice lobe margin. Such small-scale events possibly recording ice sheet readvance of a few hundred metres are characteristic of deposition at the margins of a large ice lobe when ice sheet flowlines were at a maximum extent. The pristine preservation of the nested set of ridges is explained by ice-marginal contraction in shallow water, whereas some meltwater dissection and erosion would occur during terrestrial recession.

The ridge at Castlecarragh records the second ice sheet oscillation which contains sediment more typical of basal and ice-proximal sedimentation. Three major sheets of diamict are interbedded with trough cross-stratified sands and gravels which contain beds of rhythmically-bedded silts and clays (Fig. 9.18). The lower two diamict beds are matrix-supported, muddy, contain dispersed glacially-bevelled clasts, overcompacted and occur as tabular units with erosive contacts. These characteristics suggest they are basal tills developed when the ice scavenged muds from the floor of the bay and moved across its own outwash (Fig. 1.3b). Because the uppermost diamict is stratified, contains sandy beds and gravelly lags, shows minor fold structures and drapes the underlying sediments, it probably formed by debris flows and with minor current winnowing. The significant differences in internal structure between the two ridges are related to proximity to the ice margin and the number of basal tills that are involved in ridge construction. The outer ridge at Rathcor contains only one till overlain by outwash, whereas the inner ridge at Castlecarragh contains two tills interbedded with outwash. It is likely that the lowest tills at both sites are lateral equivalents.

On the southern margin of Dundalk Bay, 7 km north of the Clogher Head ice limit, large ridges at Boycetown and Dunany Point mark the southern margin of the Dundalk Bay ice lobe which was open to influences from the main Irish Sea basin (Fig. 9.17). These and other smaller ridges at Linns consist mainly of muddy sediments which contrast with the stratified sequences

formed along the northern ice lobe margin (Fig. 9.20, 22a, b). The Dunany ridge is one of the largest and most continuous moraines in Ireland deposited at tidewater. It trends almost W–E and is continuous for 7 km inland from Dunany Point, ranges from 43 to 20 m in height, is 2–3 km in breadth, contains a rounded crest with a few saddles and is bounded on the north by steep, stepped slopes of ice-contact origin. Clearly, ridge development was accompanied by ice pressure from the north because of the pronounced southward bulge in the moraine. Inland from Mullin's Cross the large ridge on the north side of the River Dee between Stabannan and Ardee continues the line of the Dunany Point ice limit (Fig. 9.2). From Ardee discontinuous ridges and outwash record the ice limit to Kells and Mullingar, a distance of 70 km. The distal slopes of the moraine grade south-wards into an outwash spread which reaches the ice-

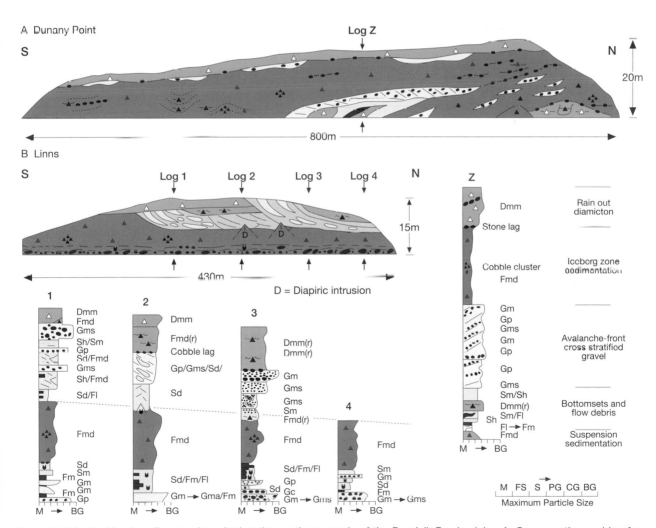

Fig. 9.20 Morainal bank sediments deposited at the southern margin of the Dundalk Bay ice lobe. A. Cross section and log from Dunany Point. B. Cross section and log from Linns. *By kind permission of GSNI.*

Fig. 9.21 Transverse section in the ice-marginal morainal bank at Dunany Point. This shows the deformed and disorganised gravelly beds overlain by a discontinuous stone lag which formed during ice front advance across subaqueous outwash. Glacially-bevelled lags throughout the sequence record repeated local oscillations of the ice front across the growing sediment apron. *By kind permission of GSNI.*

proximal slopes of the Clogher Head moraine. Around Port the outwash is notched by late-glacial marine erosion (Fig. 9.5).

The strongly asymmetric cross section in coastal exposures below the Mad Chair of Dunany are dominated by stacked beds of diamictic mud (Fig. 9.20). These beds generally dip southwards, are emphasised by discontinuous stone lags and discontinuous sandy and gravelly beds, massive and matrix-supported, and are characterised by extreme variations in clast content over a few metres (Fig. 9.21). Clasts are generally dispersed and are up to cobble grade. Three distinct sedimentary features characterise the muddy diamict. First, there are discontinuous clast lines up to 16 m in length and one clast thick where the upper clast surfaces are abraded, flattened and striated. Clast lines are stacked vertically at intervals (1–2 m) throughout the exposure. Where sediment beneath the clast line is gravelly it is generally disorganised and deformed. Second, there are circular to irregularly shaped pods of clasts entombed in diamictic mud beds. They seem structureless and generally are composed of cobbles and coarse pebbles. Third, lenses consisting of matrix-supported pebble gravel and clast-to-matrix supported cobble gravel are entombed within diamictic mud beds. In some cases isolated pods of pebbly gravel are present showing deformation of primary bedding structures. A major, undeformed wedge of stratified sediment thickens southwards for 150 m along the exposure and is entombed within beds of diamictic mud. It coarsens upwards from laminated silts to massive and laminated sands and into cross-stratified pebble and cobble gravel showing avalanche front geometry (Fig. 9.20).

The Dunany Point sequence and its internal geometry is similar to glaciomarine morainal bank deposits described by Powell (1984) and Domack (1983) and no units show any characteristics diagnostic of basal tills other than texture. The diamictic mud is a distinctive, ice-proximal lithofacies which forms by rapid ice-proximal deposition (rain out) of fines from meltwater stream plumes with widespread ice rafting of coarser debris. The essentially cohesionless and rapidly deposited sediments would be subject to almost continuous debris flowage and resedimentation which could account for some of the small lenses of gravelly debris, because winnowing can accompany flowage in a subaqueous setting. The presence of a coarsening-upwards wedge of stratified sediment supports an ice contact origin because direct meltwater input is required, possibly in the form of a small stream-head delta. Other stratified pods and lenses of gravel within the diamict also indicate that small streams and channels prograded across and sank within fluffy or unconsolidated diamictic mud beds. The discontinuous clast lines are more common in the succession than originally recorded (McCabe et al., 1987). These stacked boulder pavements are one clast thick and resemble the shallow water, glaciomarine pavements described from Middleton Island, close to the southern edge of the Alaskan continental shelf by Eyles (1988). It is likely, given the ice-proximal origin of the main sediments, that the boulder lags and concentrations within the succession are pavements formed in shallow water and were flattened, striated, pressed and aligned by successive minor advances of a partially buoyant ice margin or by sea ice floes. Ice pressure of this type would account for deformation of gravels below the boulder pavements. Essentially the ice margin readvanced over its own subaqueous outwash, possibly as a result of the growth of the ice-contact sediment pile itself, which would have tended to buttress the ice margin and reduce ice loss. The Dunany Point ridge clearly records most of the processes that have been identified from modern tidewater ice sheet margins including minor ice advance over the growing sediment pile and perhaps more continuous pressure from the centre of the ice lobe resulting in the pronounced southward bulge in ridge morphology.

A subdued ridge (~15 m high) at Linns, records ice-marginal dynamics after ice withdrew a short distance from the large moraine at Dunany Point (Fig. 9.17). The eastern part of this ridge has been truncated by marine action and to the west it is continued as large gravelly ridges at Drummeenagh, Wottonstown and Mansfieldstown. The ridge structure is different to that at Dunany Point because it records an ice readvance over fossiliferous marine muds (Fig. 9.20). A mono-specific sample of *E. clavatum* from the mud is dated to 14.2 [14]Ckyr BP, constraining the age of the read-vance. This site provides a critical index point for dating readvances in Dundalk Bay because the *in situ* muds record a non-glacial marine interval which occurred immediately after ice withdrawal from Dunany Point. The decay of ice from Dunany Point across this site is recorded by subaqueous outwash at the base of the exposure. Here, massive boulder gravels to crudely-stratified, cobble/pebble gravels show textural char-acteristics typical of high-density, gravelly debris flows including freighting of outsided clasts (~0.5 m) (Hein, 1982). These flows are interbedded with contorted mud beds indicating contemporaneity. Overlying draped laminated silts and sands containing partial Bouma sequences record deposition from density underflows. Well-developed dish and pillar structures, convolute bedding, and diapiric intrusions within the silts and sands are abruptly truncated by a thin (~10 cm) veneer of gravel and 5–6 m of mud along the entire section. The mud contains the greatest concentration of micro-fauna yet found in Ireland which is superbly preserved with glassy tests and the last apertures often in place. The muds are mainly massive though bundles of lami-nated beds occur at intervals in the unit which contain small dispersed pebbles. A sharp décollement separates the mud from 3–5 m of sand, gravel and diamictic mud which has been extensively sheared (Figs. 9.20, 9.22a, b, c). The diamictic mud occurs as thin (<0.7 m), discrete

lenses elongated along major shear zones. The nested set of subparallel shear structures rise southwards and is attributed to a readvance of the ice sheet margin over marine muds. Two cone-shaped mud diapirs between 3 and 4 m high have been injected into the sheared zone from below and may represent loading of the mud during ice readvance. The abrupt truncation of the liq-uefaction structures in the silts and sands lower in the section may also be a response to earlier ice loading and ice advance south over parts of the coarse-grained sub-aqueous outwash prior to deposition of the main mud beds.

Cooley Point

The Cooley Point exposure on the southeastern tip of the Carlinford Peninsula provides important infor-mation on relative sea levels before and after the ice sheet readvance to Rathcor/Ballug Point (Fig. 9.23) (McCabe and Haynes, 1996). Theoretically the site is also interesting because the sea level data contained within the stratigraphy can be related to the local ice sheet activity and ice limits. The section contains four main facies which were deposited outside the limit of the ice lobe, which advanced as far east as Ballug Point around 14 [14]Ckyr BP (Figs. 9.24, 9.25). Massive and lami-nated muds (1.5 m thick) are exposed at beach level for ~200 m (Fig. 9.26). The variably laminated mud has been deformed and contains contorted sandy laminae immediately below the overlying boulder facies (Fig. 9.26). The mud contains a marine microfauna domi-nated by *E. clavatum* (84%) and *R. globulifera* (15%). About 40 percent of the foraminiferas are well preserved with intact apertural faces. The majority of the rest are lustrous and transparent with broken apertural faces. Ninety-one per cent of the ostracods have both valves preserved and all the instars are present. Monospecific samples of *E. clavatum* yielded AMS ^{14}C dates of between 15.0–15.4 [14]Ckyr BP. The mud is part of a regional mud drape intermittently exposed along the northern shore of Dundalk Bay. At Giles Quay it is overlain by gravelly efflux deposits which were deposited during the last ice advance to the Rathcor/ Ballug Point moraine (Fig. 9.23). The largely clast-free mud therefore records pelagic sedimentation during a low energy, ice-free phase prior to the ice readvance marked by the Rathcor/Ballug moraine, with no extensive ice rafting. The Arctic micro-fauna is similar to that recorded from the mud around

Fig. 9.22a The structure of the Linns ridge, Dundalk Bay, County Louth. Transverse section across the ice-pushed ridge is indicated by shear planes rising southward. Section is 10 m high. *By kind permission of GSNI.*

Fig. 9.22b Detail of sheared package of gravel, sand diamict and marine mud, Linns. *By kind permission of GSNI.*

Fig. 9.22c Sheared mud and sand beds overlying horizontal, undeformed beds, Linns push moraine.
By kind permission of GSNI.

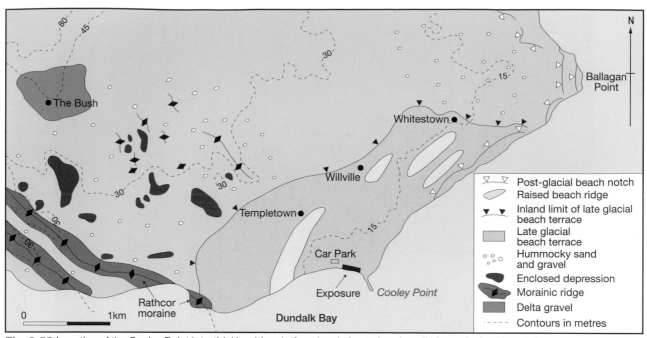

Fig. 9.23 Location of the Cooley Point intertidal boulder platform in relation to ice sheet limits and raised beach features (from McCabe and Dunlop, 2006). *By kind permission of GSNI.*

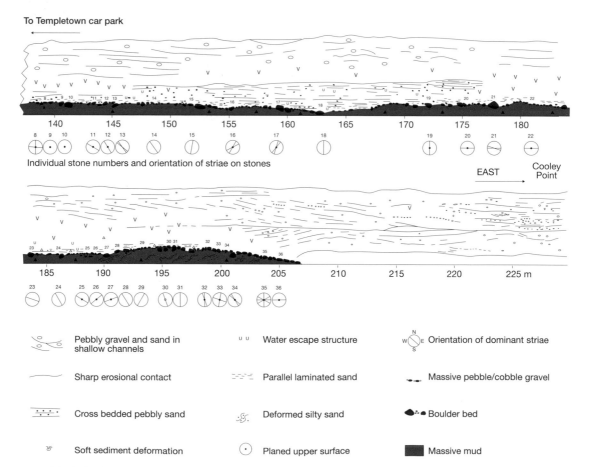

Fig. 9.24 Spatial relationship of facies exposed at Cooley Point and striae observations on upper surfaces of clasts in the intertidal boulder bed (after McCabe and Haynes, 1996). Note that the boulder bed is usually one clast thick. *By kind permission of GSNI.*

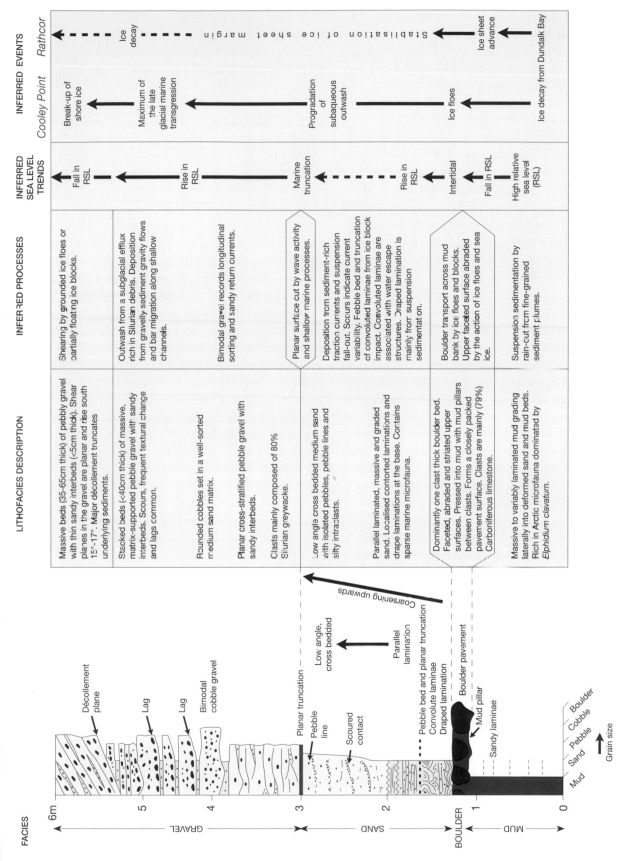

Fig. 9.25 Measured log from Cooley Point summarising the main sedimentary structures in the mud, sand, gravel and boulder facies, the contact relationships and inferred processes, sea level trends and events. *By kint permission of GSNI.*

Dundalk Bay though the mud at Cooley Point has older radiometric ages than the mud at Linns and Rathcor. Its age is similar to the older mud at Port which predates the Clogher Head readvance. This observation suggests that the mud at Cooley Point was overridden during the Clogher Head ice advance though no stratigraphic evidence of this episode remains at this site.

The boulder facies is mainly (70%) a bed of single clasts with local concentrations of near-touching to wedged clasts which form a pavement pressed into the top of the marine mud (Fig. 9.26b). Individual boulders are separated by mud 'wedges'. Cobble-sized clasts tend to occur in clusters of up to twelve cobbles measuring 1.3 m x by 0.6 m across. Surface undulations in the boulder pavement are up to 10 m across by 1 m high. Carboniferous limestone is the most common clast lithology (79%), most (91%) clasts are flat-lying (<10°) and individual clasts generally possess more than one bevelled surface. Upper surfaces of most (73%) clasts are strongly bevelled, contain variable sets of striae and form a prominent pavement surface. Dominant striae on the polished top surface of clasts do not show a unidirectional trend and do not show any simple relationship to the clast long axes. At the western end of the exposure the boulder pavement is replaced or continued at the same stratigraphical level by a sharp planar surface which is sometimes marked by disorganised cobble lenses. The boulder bed is draped by up to 10 cm of delicately laminated silty clay which passes up into parallel-laminated sand (<2 m) and pebbly sand (~2 m) (Fig. 9.25). Laminae are well-sorted, variable in thickness (0.25–1.0 cm), massive or normally graded, laterally continuous (<15 m) and contain mud rip-up clasts. Isolated ripple trains also occur. In places the horizontal laminations are broken by intraformational dish and pillar structures associated with water escape. The cross-bedded pebbly sand facies is characterised by shallow scours, pebble lags, dispersed pebbles, mud rip-up clasts and frequent textural changes. The sand facies is truncated by a sharp planar erosion surface overlain by pebbly gravel. This gravel which infills shallow scours which range from 10 to 30 m across by a few metres in depth. Stratification is marked by coarse pebble lags, cobble lags, frequent and abrupt textural changes and diffusely laminated sandy beds. Most beds are massive, poorly sorted and matrix supported. Occasional beds of openwork cobble gravel are present together with bimodal cobble gravel (cobbles set in a sandy matrix). These lithofacies are dominated (80%) by Silurian shales and greywackes which contrast with the limestones in the boulder facies below.

The origin and development of the boulder pavement at Cooley Point can be evaluated from four lines of evidence (Fig. 9.27):

a. Ice limits/stratigraphic position. The pavement occurs 1.8 km outside the Rathcor moraine, is not associated with any glacial diamict, is underlain by open water marine mud, and is overlain by draped lamination. The sequence context is therefore not subglacial but is suggestive of a shallow water, deglacierised area.

b. Boulder source. Boulders cannot be derived either by winnowing clast-free mud or from iceberg deposits, as residual lags would survive on the mud surface. High percentages of glacially-facetted and bullet-shaped clasts require a subglacial source. Iceberg rafting could have occurred as the ice advanced towards the ice limit at Ballug Point but this would have produced a more variable clast population than cobbles and boulders and would need sea levels higher than the mud bank. A more likely scenario is that boulders were transported to the ice front during ice advance associated with formation of the Ballug moraine. This ice flow would have crossed areas of Carboniferous limestones underlying the inner bay. Boulder transport may result from the break-up of winter ice followed by movement of mobile ice floes or refloated grounded ice blocks across the mud bank (Fig. 9.28). Rosen (1979) and McCann et al., (1981) have described how downward freezing of intertidal ice encases boulders as the ice cover is alternately raised and grounded on subarctic tidal flats. Observations from contemporary, subarctic, boulder-strewn tidal flats emphasise not only the role of wind- and current-driven ice flows in boulder transport but their rearrangement across the flats (Lauriol and Gray, 1980; Dionne, 1981; Hansom, 1983, 1986).

c. Sea level fall. An inferred fall in RSL of ~30 m must have occurred after the early deglacial glaciomarine highstand but prior to mud deposition and pavement formation. An abrupt change in biofacies from a well-preserved *in situ* open-water Arctic assemblage in the mud below the boulder bed, to an assemblage of poorly-preserved and derived specimens in the overlying mud drape laminae indicates a change in water depth. Stratigraphically this shallowing trend is mirrored by the depositional change from mud to sand. Tidal range is generally accepted to be an important variable in boulder transport and reorganisation (Rosen, 1979; McCann *et al.*, 1981). Tanner (1939) observed that

Fig. 9.26a Facies from the Cooley Point exposure, County Louth. Intertidal boulder bed one clast thick pressed into marine mud. Note the ice-bevelled upper surface of the boulders. Spade is 0.9 m long. *By kind permission of GSNI.*

Fig. 9.26b Facies from the Cooley Point exposure, County Louth. Mosaic-like packing of ice-bevelled boulders on top of mud and details of undulations in the pavement surface. Spade is 0.9 m long. *By kind permission of GSNI.*

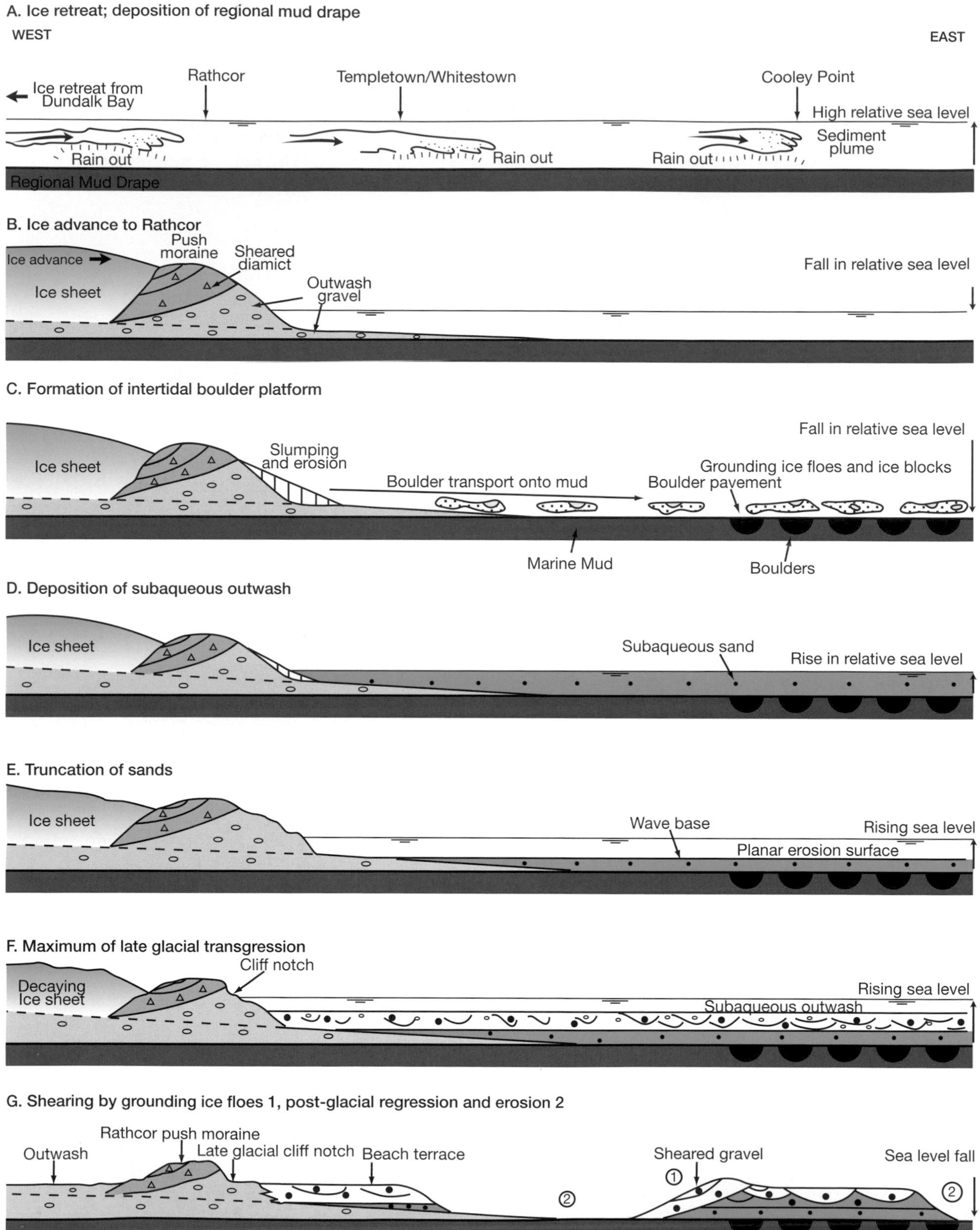

A. Ice retreat; deposition of regional mud drape

WEST   EAST

← Ice retreat from Dundalk Bay

Rathcor

Templetown/Whitestown

Cooley Point

High relative sea level

Sediment plume

Rain out   Rain out   Rain out

Regional Mud Drape

B. Ice advance to Rathcor

Push moraine
Sheared diamict

Ice advance →

Ice sheet

Outwash gravel

Fall in relative sea level

C. Formation of intertidal boulder platform

Ice sheet

Slumping and erosion

Boulder transport onto mud

Grounding ice floes and ice blocks

Boulder pavement

Marine Mud

Boulders

Fall in relative sea level

D. Deposition of subaqueous outwash

Ice sheet

Subaqueous sand

Rise in relative sea level

E. Truncation of sands

Ice sheet

Wave base

Planar erosion surface

Rising sea level

F. Maximum of late glacial transgression

Cliff notch

Decaying Ice sheet

Subaqueous outwash

Rising sea level

G. Shearing by grounding ice floes 1, post-glacial regression and erosion 2

Rathcor push moraine

Outwash

Late glacial cliff notch

Beach terrace

Sheared gravel

Sea level fall

①

②

②

Fig. 9.27 Cartoon representing the origin of the main stratigraphic elements at Cooley Point and the late-glacial development of the north shore of Dundalk Bay (after McCabe and Dunlop, 2006). *By kind permission of GSNI.*

boulder pavements in the fjords of the Barents Sea and in Labrador were best developed with tidal ranges of 2–4 m. If the late-glacial tidal range in Dundalk Bay is similar to the present day (~4.5 m) it provides a crude comparison with the tidal ranges associated with contemporary boulder pavement development.

d. Pavement characteristics. Boulder position and the presence of deformed mud pillars between clasts show that the boulders were pressed into the mud (Fig. 9.26a). There is an absence of clast-ploughing (Clark and Hansel, 1989) which might have suggested ice sliding over a subglacial deformable bed. Modern analogues show that the movement of multiple ice flows across boulder-strewn flats during successive seasons can result in highly variable boulder concentrations (Rosen, 1979). Pavement undulations are also similar in cross-sectional scale to the polygonal depressions recording stranded and rotating ice blocks in the sub-Antarctic (Fig. 9.28) (Eyles, 1994). The mosaic-like packing of the boulders and the variation in the orientation of striae along the pavement are also similar to those described from sub-Arctic intertidal settings (Hansom, 1983, 1986). Clast reorganisation is generally attributed to a combination of intertidal processes including sorting, stranding, pushing, pressing, dragging, rocking, and mutual interference between adjacent clasts. Striac variability, cross-cutting scratches and the absence of well-defined sets of subparallel striae on the upper clast surfaces suggest that abrasion was effected by a see-saw action typical of that produced by the frequent action of freely-moving ice floes. It is difficult to identify the position of the contemporary shoreline, but it was probably aligned from southwest to northeast, subparallel to the late-glacial beach ridges. If this is correct then about 70 percent of the dominant striae on the upper clast surfaces are aligned at a high angle to that trend. A similar relationship was noted by Hansom (1983) due to the onshore movement of floating ice or the rotating movement of tidally stranded ice blocks.

The main factors which contributed to intertidal boulder pavement development at Cooley Point include a fall in RSL from the glaciomarine highstand, a sub-

Fig. 9.28 Present day intertidal boulder pavement in the process of formation South Georgia, Antarctica which is a modern analogue for the interpretation of fossil pavements found within glacigenic sequences. Note the remanie blocks of shore ice. *By kind permission of GSNI.*

glacial boulder source at Rathcor, a mesotidal setting facilitating free ice movement in nearshore zones and an ice-dominated coastal environment with seasonal dynamic break-up, resulting in ice floes and refloated ice blocks, which allowed for boulder transport and vertical ice pressure and ice abrasion by freely moving ice floes or blocks, which reorganised the boulders into a pavement.

The sandy drapes immediately above the boulder bed and the overlying massive and graded horizontal beds are rhythmically-bedded and resemble turbidites and may indicate a rise in RSL. Disrupted laminae with dish and pillar structures are similar to water escape structures. The upward transition into pebbly sand within scour features marks an increase in current velocity. The sharp contact between the sand and overlying pebbly gravel facies is similar to wave-cut, transgressional surfaces and lags which occur below late-glacial beachface gravel in eastern Ireland. However, the gravel shows a variability in texture and sedimentary structures that is reminiscent of marine outwash within the shallow scours (Rust, 1977). Planar, cross-bedded sand and gravel are thought to be the product of bar migration along shallow channels, while matrix-rich gravel formed by sediment gravity flow (Lowe, 1979; Nemec and Steel, 1984). The juxtaposition of massive gravel and better-sorted, graded gravel facies can be rationalised by downslope evolution in flow support mechanisms (Postma, 1984). Bimodal gravel may record longitudinal sorting of cobbles to a brink point together with suspension sedimentation of sand from return currents. The presence of thick, amalgamated, poorly sorted beds together with prominent scour surfaces also reflect wide energy fluctuations and pulsating discharges. This variability in a few metres of sediment is common in marine outwash sequences recorded from incipient grounding line fans (Powell, 1990). Clasts in the outwash are mainly (79%) derived from Silurian bedrock, whereas the boulder bed is composed mostly of Carboniferous limestone. The most probable source of the sediment would be a subglacial efflux located immediately to the south of Cooley Point on the frontal margin of the Dundalk Bay ice lobe (Fig. 9.23). This palaeoflow agrees with the inferred south to north trend in channel axes. The late-glacial beach terrace, which occurs inland from Cooley Point, terminates at the Rathcor/Ballug Point ridge, and is backed to the north by a large beach terrace and shingle ridge at Templetown (Fig. 9.23). The latter postdates the boulder bed and shows that RSL rose to around 20 m OD after progradation of the outwash.

The evidence from Cooley Point provides sequence stratigraphical elements which have widespread applicability to deglacierised basins elsewhere. First, sea level records are contained within the overall sediment sequence, and individual sites rarely record the full spectrum of facies formed during emergence or submergence. Second, surfaces of erosion or their correlatives within adjacent glacigenic sequences are often overlooked. They provide information on changes in RSLs. Third, glaciomarine/marine mud around the ISB represents flooding zones as the ice sheet margins contracted. In a basinwide context they are important because the microfaunas are dominated by the cold water foraminifera *E. clavatum*. This has direct modern analogues with recently deglacierised Arctic tidewater environments. Common to both situations is the undoubted presence of major meltwater events during the deglacial cycle. Fourth, the boulder pavement at Cooley Point has a low preservation potential but provides evidence that shore-ice was present during the deglacial cycle. A variety of pavements and clast lines have been described from deposits around the ISB but their significance remains to be determined. Fifth, deformed sediments are often used to reconstruct ice sheet readvances in the ISB, or to infer processes of subglacial deformation. However, within a large basin that was subject to periods of catastrophic disintegration of ice masses, ice break-up and meltwater events, ice floes may have played an important role in sediment deformation and compaction of glaciomarine or other deposits in shallow water.

Rough Island, Strangford Lough

Strangford Lough trends NW–SE across the strike of bedrock which consists of Ordovician and Silurian greywackes and mudstone and is the southwesterly extension of the topographically elevated Southern Uplands of Scotland. Strangford Lough and the Ards Peninsula is dominated by bedforms which show variable thicknesses of glacial diamict masking rock cores. In the east, flow-parallel landforms with thick (~10 m) diamict dominate and often occur as erosional forms across the crests of large ENE–WSW ribbed moraines (Fig. 5.2). Westwards the thickness of the glacial diamict carapace decreases inland and in many exposures rock occurs near the surface (<1`m). The trend of the long axis of the lough parallels the local drumlin orientation, which suggests it may have been glacially overdeepened. The drumlin orientations record the last phase of drumlinisation and ice readvance southeastwards to the Killard Point limit (Fig. 9.7). Because the marine muds deposited at Rough Island occur within the limits of the Killard Point Readvance they provide limiting ages for deglaciation following the Killard Point Readvance.

In this part of County Down Stephens (1958, 1963) suggested that partial dissolution of the last ice sheet was accompanied by marine transgression which flooded parts of the isostatically-depressed land surface. Morrison and Stephens (1965) described up to 5 m of undated red marine mud beneath late-glacial solifluction deposits and post-glacial raised beach deposits at Roddan's Port on the east coast of the Ards Peninsula. More recent work shows that drapes of red mud occur on many of the drumlins flanking Strangford Lough, and can be traced upslope as a feather edge to about 16 m OD. Above this level the drumlin flanks have been notched (at 20 m OD) by marine erosion (Stephens and McCabe, 1977). Wave-cut notches are best preserved in Strangford Lough itself rather than on the east coast around Roddan's Port because of more intensive postglacial erosion around the exposed Irish Sea coast (Stephens, 1963). At the northern end of Strangford Lough well-defined late- and post-glacial marine terraces occur south and west of Newtownards (Fig. 9.29). The flat upper terrace surface lies around 20 m OD and is best seen around Milltown. At Longlands a larger post-glacial terrace cuts the earlier late-glacial terrace. Island Hill and Rough Island are erosional remnants of the upper terrace which must have been a higher and much more extensive feature at the north end of the lough during the late-glacial transgression. A similar degree of erosion occurs throughout the lough where individual drumlins are in all stages of destruction by present marine action, with some drumlins having being reduced to a pile of large clasts, known locally as pladdies. Clearly, post-glacial erosion of drumlins has provided sediment for redistribution and resedimentation within the post-glacial and present-day sediment cells of the lough.

Rough Island is an erosional remnant of the larger marine terrace which developed around the north end of the lough immediately after deglaciation of this site (Fig. 9.29). Three facies are exposed in the marine cut cliff on the southwestern side of the small island. Immediately above sea level a compact, unweathered diamict (<1.5 m exposed) is mainly derived from local Triassic sandstones but contains erratics of Cretaceous

chalk and flint, Tertiary basalt, red granite and Ailsa Craig microgranite, all of northern provenance. The clasts, ranging from pebbles to boulders, are set in a massive, red sandy-silt matrix. About 40 percent of the largest clasts show evidence for ice bevelling, shaping and striation. The sedimentary characteristics of similar diamicts found across this region suggests a subglacial origin. The unweathered top of the diamict is overlain by up to 3 m of rhythmically-bedded sand and mud, which show an upward decrease in mud and increase in sand content. In the lowest 0.5 m of the rhythmites the mud beds are massive (up to 20 cm thick) and contain an *in situ* assemblage of marine microfauna dominated by *E. clavatum* (95%) and *R. globulifera* (<4%). AMS ^{14}C dates of 12.7 and 13.1 ^{14}C kyr BP has been obtained from this level in the rhythmite succession (Figs. 4.13, 9.29). Higher in this succession microfaunal tests are weathered with at least half broken and discoloured, suggesting that resedimentation became dominant over *in situ* production. The upward increase in laminated sand beds may reflect enhanced marine current activity during a fall in relative sea level. Although the sand is generally thinly bedded with mud laminae, there are at least two massive sand beds (10–20 cm thick) whose precise mode of formation is unclear. The top of the rhythmite succession is sharply truncated and overlain by poorly-sorted pebble and cobble gravel with marine shell fragments and archaeological remains representing part of the post-glacial raised beach complex. Because the sequence coarsens upwards from massive mud to interlaminated sand and mud and finally to beachface gravel, it records an overall emergent facies sequence with an hiatus below the shoreface gravels possibly due to wave truncation.

The radiometric dates from this site are the only dates in the northern ISB to constrain final deglaciation and the timing of the late-glacial marine transgression. Similar deposits in an identical stratigraphic position are known in the Solway Firth (Wells, 1997) and may indicate a similar sequence of deglacial events. The regional distribution of the red marine clay exposed at Rough Island shows that glacioisostatic depression was greater that the eustatic sea level fall in the northern Irish Sea Basin at this time. A number of studies in northwest England (Merrit and Auton, 2000) and Wales (Austin and McCarroll, 1992) do not support the model, invoking high relative sea levels during deglaciation. However, there is evidence for emergent glaciomarine and late-glacial raised beach facies from around the

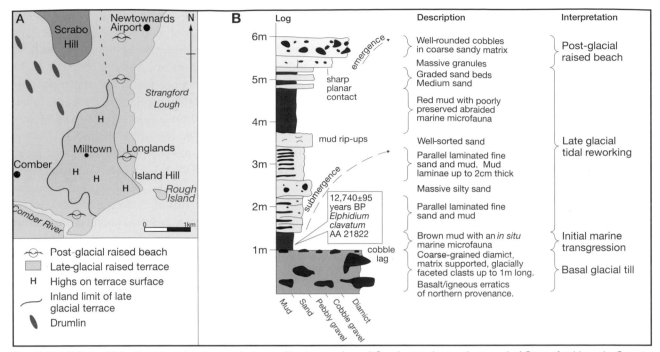

Fig. 9.29 A) Raised late glacial marine terrace between Newtownards and Comber at the northern end of Strangford Lough, County Down. B) The stratigraphic log from Rough Island represents a detached erosional remnant of the late-glacial raised marine terrace. *By kind permission of GSNI.*

coast of northeastern Ireland. If the minimum models of isostatic depression for western England and Wales are correct, then a major isostatic hinge-line was present along the axis of the Irish Sea during deglaciation and this may be a result of thick ice across the extensive lowlands of Ireland. It is difficult to envisage how the large ice dispersal centres in western Scotland and northern England did not result in similar isostatic patterns to that found in the western basin (Andrews et al., 1973). It is also possible that non-recognition of former sea level evidence in the eastern ISB is due to the lack of detailed information on the nearshore sediments which contain the most complete records of sea level fluctuations on glaciated shelves (Hunter et al., 1996).

Chronology of Ice Sheet Readvances

The AMS ^{14}C ages from marine muds at Linns (14,160+70 ^{14}C yr BP) and Rathcor Bay (14,250+130 ^{14}C yr BP) (Fig. 9.6) constrain the age of the younger ice limit in Dundalk Bay to be <14.2 [14]Cka BP. The age of this readvance is in good agreement with dates from muds interbedded in terminal outwash associated with an ice margin at Killard Point (13,785+115 ^{14}C yr BP, 13,995+105 ^{14}C yr BP) (Fig. 9.1) (McCabe and Clark, 1998), suggesting that this Dundalk Bay readvance occurred during the Killard

Point Stadial (Fig. 9.30). Two AMS ^{14}C ages from marine muds at Rough Island, north of Killard Point, indicate that deglaciation of the Killard Point readvance had begun before ~13.0 [14]Cka BP (Fig. 9.29). Four AMS ^{14}C ages from monospecific samples of *E. clavatum* taken from different levels in the vertical mud dyke in the till at Port indicate open marine conditions in Dundalk Bay ~15.3 [14]Cka BP (Fig. 9.6). Three AMS ^{14}C ages from marine muds on the north side of Dundalk Bay at Cooley Point (Fig. 9.6) provide additional evidence for open marine conditions at this time (McCabe and Haynes, 1996). Incorporation of these muds into the till at Port thus occurred in association with an ice readvance to the Clogher Head ice limit <15.0 [14]Cka BP (Fig. 9.30). Because the ^{14}C ages on marine muds at Linns and Rathcor Bay constrain a minimum age of deglaciation from this ice limit to be >14.2 [14]Cka BP, the Clogher Head readvance occurred between 15.0 and 14.2 [14]Cka BP, and is thus ~1000 years older than the Killard Point readvance (Fig. 9.30).

Stadials and Interstadials

Originally McCabe et al., (1998) defined the Cooley Point Interstadial as a nonglacial interval that occurred between 16.7 and 14.7 [14]Cka BP, and the Killard Point

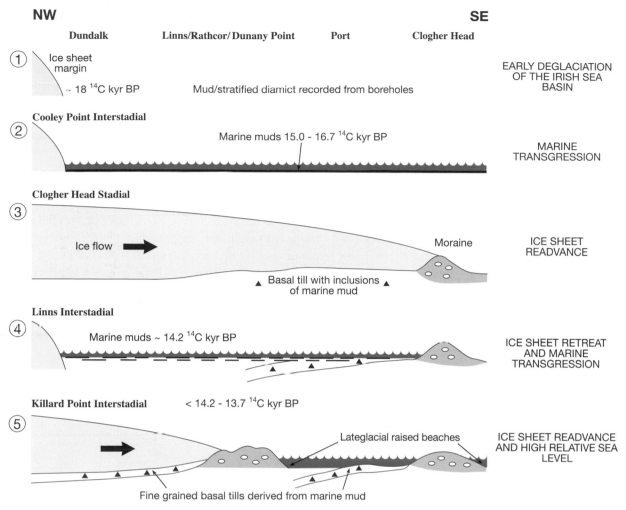

NW   **SE**

Dundalk   Linns/Rathcor/Dunany Point   Port   Clogher Head

(1) Ice sheet margin

~ 18 ^{14}C kyr BP   Mud/stratified diamict recorded from boreholes

EARLY DEGLACIATION OF THE IRISH SEA BASIN

Cooley Point Interstadial

(2) Marine muds 15.0 - 16.7 ^{14}C kyr BP

MARINE TRANSGRESSION

Clogher Head Stadial

(3) Ice flow →   Moraine

Basal till with inclusions of marine mud

ICE SHEET READVANCE

Linns Interstadial

(4) Marine muds ~ 14.2 ^{14}C kyr BP

ICE SHEET RETREAT AND MARINE TRANSGRESSION

Killard Point Interstadial   < 14.2 - 13.7 ^{14}C kyr BP

(5) Lateglacial raised beaches

ICE SHEET READVANCE AND HIGH RELATIVE SEA LEVEL

Fine grained basal tills derived from marine mud

Fig. 9.30 Schematic depiction of stages in the last deglaciation of Dundalk Bay. (1) Early deglaciation of the Irish Sea Basin (2) Deposition of marine muds in association with high relative sea level during the Cooley Point Interstadial (3) Readvance of the ice margin during the Clogher Head Stadial (4) Retreat of the ice margin and deposition of marine muds in association with high relative sea level during the Linns Interstadial (5) Readvance of the ice margin during the Killard Point Stadial (after McCabe et al., 2007b).

Stadial as a glacial readvance that began after 14.7 ^{14}C ka BP. On the basis of the ^{14}C ages that constrain the age of the Clogher Head readvance, the Clogher Head Stadial is defined as occurring ≤15.0 and ≤14.2 ^{14}C ka BP (Fig. 9.30). The Cooley Point Interstadial was terminated by this readvance, rather than by the Killard Point readvance, and is thus redefined as occurring sometime ≤16.7 ^{14}C ka BP and ≤15.0 ^{14}C ka BP The Linns Interstadial is redefined as the nonglacial interval that separates the Clogher Head and Killard Point Stadials on the basis of the ^{14}C ages on marine muds that were deposited in Dundalk Bay ~14.2 ^{14}C ka BP (Fig. 9.6). These ^{14}C ages for the Linns Interstadial also narrow the start of the Killard Point Stadial as occurring after 14.2 ^{14}C ka BP, while the ^{14}C ages from terminal outwash at Killard Point date the time when the ice margin was at its maximum extent (~13.8 ^{14}C ka BP).

Finally, the ^{14}C ages from Rough Island provide limiting ages for the end of the Killard Point Stadial (~13.0 14ka BP) (Fig. 9.29).

The ^{14}C ages on marine sediments that constrain the timing of the stadials and interstadials, as well as older ^{14}C ages from Kilkeel north of Dundalk Bay (Fig. 6.28) (Clark et al., 2004), also demonstrate that the northern part of the Irish Sea Basin was isostatically depressed below sea level throughout the last deglaciation until at least 12.7 ^{14}C ka BP (youngest age from Rough Island). Sustained isostatic depression occurred when eustatic sea level was >100 m below present, thus suggesting isostatic depression of at least 110 m (>100 m sea level lowering plus the height at which raised marine sediments occur today). This amount of isostatic depression requires that a significant ice mass remained over

Ireland during this time, with its margin experiencing two fluctuations that were likely on the order of tens of kilometres in length.

Records of ice-rafted debris (IRD) in marine cores off the coast of Ireland show a peak in IRD derived from the BIIS immediately prior to H1 (Fig. 9.31) (Zaragosi et al., 2001; Peck et al., 2006), or at the same time as the Clogher Head Stadial. Because an IRD signal alone cannot distinguish among the many potential mechanisms that can produce such a signal (McCabe and Clark, 1998), this correlation is significant in identifying the IRD peak as representing a readvance of the BIIS margin into the Irish Sea, delivering debris-laden icebergs. Rather than associating the increase in IRD with an ice sheet instability (Zaragosi et al., 2001; Peck et al., 2006), however, we attribute the IRD signal to an increase in the flux of icebergs associated with a steady-state readvance of the BIIS margin during the Clogher Head Stadial in response

to a reduction in the Atlantic meridional overturning circulation (AMOC) (McManus et al., 2004) and attendant cooling of the North Atlantic region associated with the start of the Oldest Dryas cold interval.

Retreat of the BIIS margin during the Linns Interstadial apparently occurred at the same time as Heinrich 1 (Fig. 9.31). Two possible scenarios might explain this association. First, the AMOC nearly collapsed during H1 (McManus et al., 2004), with an attendant expansion of sea ice that may have weakened the hydrological cycle, causing a more negative mass balance over the BIIS in an otherwise cold climate. Alternatively, the retreat of the BIIS during the Linns Interstadial may represent a dynamic response of the ice sheet independent of H1 that was caused by isostatic loading with an attendant increase in calving-induced retreat of its marine margin. Subsequent readvance of the BIIS into the northern Irish Sea Basin during the Killard Point

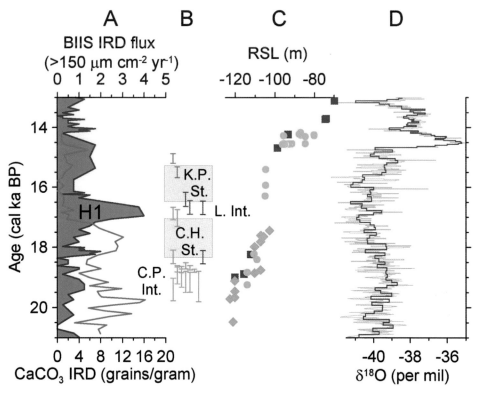

Fig. 9.31 A) Records of ice-rafted debris (IRD). In blue is the record of detrital carbonate IRD from core VM23-81 showing the IRD peak associated with Heinrich 1 (H1) (Bond et al., 1997), whereas in grey is the record from core MD01–2461 showing IRD derived from the BIIS (Peck et al., 2006). B) Calibrated radiocarbon dates (Stuiver et al., 2005) that constrain ages of stadials and interstadials in the Irish Sea Basin, with ages from Kilkeel (blue) (Clark et al., 2004), Dundalk Bay (green) (McCabe et al., 2005, 2007), Corvish (purple) (McCabe and Clark, 2003), Killard Point (red) (McCabe and Clark, 1998) and Rough Island (orange) (McCabe and Clark, 1998; McCabe et al., 2007b). Note that although a 400-yr reservoir age is assumed, the reservoir age may have varied during the last deglaciation in association with large changes in the meridianal overturning circulation (Bard et al., 1994; Waelbroeck et al., 2001; Bondevik et al., 2006; Peck et al., 2006). C) Far field records of relative sea level from Barbados (squares) (Bard et al., 1990), Sunda Shelf (Hanebuth et al., 2000), and Bonaparte gulf (Yokoyama et al., 2000). D) The Greenland GISP δ18O record (Grootes et al., 1993; Stuiver and Grootes, 2000).

Stadial may have occurred immediately after H1 in response to a warming climate suggested by the GISP2 δ^{18}O record (Fig. 9.31), resulting in a stronger hydrological cycle and more positive mass balance. On the other hand, if earlier retreat into the Linns Interstadial was a dynamic response to isostatic loading, subsequent readvance during the Killard Point Stadial may have similarly occurred when the ice margin stabilised on land and began to readvance again in response to the continued cold climate of the Oldest Dryas. In either case, there is no corresponding IRD signal such as the one associated with the Clogher Head Stadial (Fig. 9.31), even though the ice margin during the Killard Point Stadial also advanced into the Irish Sea Basin. This lack of a signal may reflect a flux of debris-poor icebergs or a different transport path than taken by those released during the Clogher Head Stadial. Regardless of the cause, this relationship (ice readvance, no IRD signal) points to the complex processes that must contribute to the formation of an IRD signal, emphasising caution in interpreting such a signal with respect to ice sheet behaviour and dynamics. In summary, the ^{14}C ages from raised marine sediments in the northern Irish Sea Basin requires that a substantial ice sheet remained on Ireland throughout much of the last deglaciation, with isostatic depression of at least 110 m. During this time, there were two readvances of the BIIS, which define the Clogher Head and Killard Point Stadials (Figs. 9.31, 9.32). These ice-margin fluctuations may have occurred in response to climate change associated with changes in the AMOC and attendant changes in sea ice and the hydrological cycle, or they may reflect dynamical responses of a marine-based ice margin to isostatic loading and unloading. Although these ice sheet readvances covered much smaller areas than maximum ice sheet limits they are important because they occurred after a major early deglaciation of the entire ice sheet when the Irish Sea was essentially a marine seaway with ice masses restricted to its margins. Following this catastrophic disintegration the reorganisation of the remaining ice masses records a prolonged ice sheet response to North Atlantic climate signals, possibly over 2–3000 years. The readvances bracketing the Heinrich 1 event are the only dated readvances known to occur between ice sheet maximum limits and the Nahanagan Stadial which began around 11 ^{14}C kyr BP. The latter cooling signal resulted in corrie glaciation only and followed the second major deglaciation which began before the onset of the Bølling/Allerød.

Fig. 9.32 Ice moulded and striated bedrock immediately below glaciomarine outwash, Killard Point, County Down (see Fig. 9.7). Note that the glaciomarine sequence is truncated by a raised marine cliff notch and beach terrace in the background.

10

Late-glacial Sea Levels and Ice Sheet History

Introduction

Many authors including Munthe (1897) attempted to relate high level morphological features such as deltas, terraces and beaches to former sea levels associated with the decaying margins of the last ice sheet in the British Isles. Although the origins and development of the theory of glacioisostatic and eustatic adjustments were based on the presence of raised marine landforms, there has been a drift away from ground truthing the results derived from both theory and models. Perhaps one of the most instructive papers on field relationships between raised marine forms, ice limits and sea levels was the essay on Pleistocene Shorelines by Stephens and Synge (1966b). A central theme still relevant to the field geologist was the complex interaction between the position of ice sheet margins, isostatic adjustment and global eustatic sea level trends. Interactions between these factors, together with postglacial marine erosion, contributed to the low preservation potential of many raised marine features, and the field investigator in most areas is confronted by fragmentary evidence along narrow coastal strips. Fragmentary data means that it is difficult to trace a particular landform over kilometres of coastline and in many cases morphological data is correlated without dating control. Nevertheless, in the British Isles as a whole there are suites of late-glacial beach systems and other raised marine indicators which are highest in the north where isostatic depression was thought to be greatest because of thicker ice loads (Stephens and Synge, 1965).

However, reconstructions of sea level history are much more than documenting relative heights of raised or buried sea level indicators, because the deglacial period, unlike later periods, was one of very rapid environmental change on submillennial timescales. Therefore we are dealing with a time period characterised by retreating ice margins, rapid land uplift, spikes of eustatic rise or meltwater events, ice readvances and shoreline reorganisation. Sea level history from different parts of the coastline will be unique because different ice sheet sectors will be influenced by discrete sets of local factors. The latter may vary from duration of ice cover to rates of relative change in local sea level. In addition because the British Isles consists of many islands it will be dominated by the development of marine seaways early on in the deglacial cycle which will tend to constrain and isolate major ice sheet masses to adjacent land areas. The N–S Irish Sea trough opened during early deglaciation as a large suture right in the geographic heart of the BIIS and undoubtedly influenced ice sheet activity and configuration and therefore sea levels (Fig. 6.2) (Eyles and McCabe, 1989a).

Ice sheet and isostatic depression

Traditional notions that there are 'pre-glacial' coastal slope elements and deposits preserved around the coastline has always attracted particular interest (Stephens, 1958). In many coastal sections a wave-cut rock platform is identified as the oldest Pleistocene feature even though there are no independent dates on such features. Even if raised platform fragments could be dated and the erosive agencies clearly identified there would still be major uncertainties concerning the age of the overlying drift sequences. In addition, combinations of structural controls on platform development and local wave climates can result in platform development over considerable height ranges (McKenna, 2002). Even more uncertainty has been introduced into the picture because most workers around the Irish Sea Basin and other parts of the coast assume that the stratigraphically often complex successions overlying rock platforms represent real Pleistocene stratigraphy, recording successive local and larger scale glacial or glacigenic events which have a basinwide signature. A result has been the proliferation of local stratigraphy and correlation between sites around the basin leading to models without geochronological control. Many critical sea level indicators such as beach gravel, marine mud and widespread subaqueous successions occur within or comprise most of the deposits resting on rock platforms around the coastline. The questions that arise from this observation therefore concern the role that sea level played during deglaciation and what the glacigenic successions around the coast actually do record.

In a much broader perspective observation shows that the most complex sedimentary successions in the island are located in coastal lowlands.These are the type sections that are subdivided and then used in a bas-inwide context. Exposures inland are generally stratigraphically simple and contain only one lithofacies, generally some form of diamict. The much thicker glacigenic sequences recognised from near-coastal locations are in the form of wedges or spreads, generally with a high degree of sedimentary variability, that must be related to different styles of marine influence because of their location alone (Fig. 10.1). These glacigenic sequences can occur as isolated spreads in areas such as western and northern coastal sectors where the surrounding topography is scoured free of drift (Fig. 10.2). They can occur on the margins of the Irish Sea Basin where drift deposits are extremely thick and continuous. Is there a common factor linking most of the complex glacigenic sequences on coastal locations? The most likely theme is that most were formed immediately when ice sheet margins contracted from the continental shelf and grounded on rock, re-equilibrated on topographic highs, became locked within local basins or subsequently readvanced.

This model is generally applicable because during the last glaciation the largest and thickest ice masses were located across the Irish lowlands (not highlands) with ice flow onto continental shelves. McCabe et al., (1986) suggested that ice thicknesses approaching 1000 m characterised the largest lowland dispersal centres. Although the largest centres of ice sheet dispersal were in the north and west, ice was still at least 740 m thick on the flanks of the Wicklow Mountains during deglaciation (Ballantyne et al., 2006) and even thicker in the west (McCabe et al., 2007a). This distribution of ice mass therefore resulted in deep isostatic depression across the entire island with the ice sitting on the depressed, saucer-like basin known as the Irish Lowlands. A partial analogy may be the Greenland ice sheet of today, sitting

in a depressed basin. When deglaciation commenced, extended ice flow lines on the shelf were sensitive to sea level changes and responded immediately. Ice margins retreated onto land, but because areas peripheral to the ice were still isostatically depressed below global eustatic levels, many ice margins ended at tidewater. Configurations and depths of peripheral troughs in association with positions of ice sheet margins may in part determine precise locations of glacigenic deposition. Many of the more complete glacigenic sequences found on the coastline contain either emergent or submergent facies sequences whereas others are mainly subaqueous in origin. Hiatuses within successions inevitably occur and stress the rapidity of environmental change driven by isostatic/eustatic interactions at this time. This variability both in facies successions and landform characteristics is consistent with the general model of high relative sea levels during deglaciation around the coastline of Ireland (Fig. 10.3).

It is also highly significant that these glacigenic sequences are essentially coarse-grained, ice-contact sediments and as such are generally unfossiliferous, containing reworked fossils. Critically the deglacial stratigraphy also contains *in situ* marine beds containing opportunistic biocoenoses which are radiocarbon dated and provide a framework for the general model of sea level history. This framework extends from ~30 to 14 cal ka BP and is based on field data from coastal exposures around the island. Traditionally, raised beach and other indicators of sea level history were thought to occur only in northeastern Ireland. New data from widely spaced sections and moraine positions on the continental shelf therefore require a more holistic approach to sea level history based on dated, realistic ice sheet data, rather than on theoretical models (McCabe, 1997). It is also noteworthy that successive glacial terminations are generally abrupt events, each perhaps spanning a few thousand years at most. For example, the sedimentary record from the Irish Sea Basin records rapid evacuation of the basin accompanied by rapid uplift (Eyles and McCabe, 1989a, b; Clark et al., 2004). Rapid environmental changes of this nature will result in the preservation of relatively restricted areas containing sedimentary evidence of events which in reality affected much larger areas across the basin. When this evidence is combined into an event stratigraphy recording the evacuation of ice from the basin as a tidewater ice margin it provides an important interpretative tool for environmental reconstruction and sea level history.

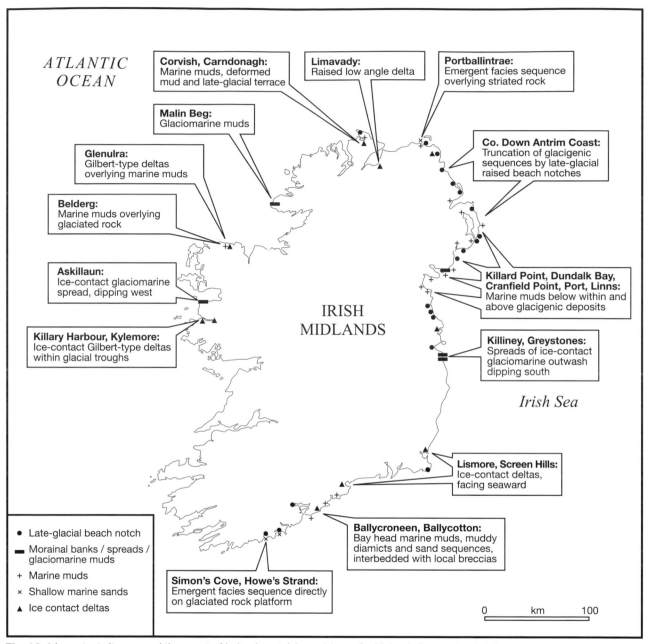

Fig. 10.1 Important sites around the coast of Ireland containing evidence for high relative sea level during the last glacial termination.

The following text labels appear on the map:

Corvish, Carndonagh: Marine muds, deformed mud and late-glacial terrace

Limavady: Raised low angle delta

Portballintrae: Emergent facies sequence overlying striated rock

Malin Beg: Glaciomarine muds

Co. Down Antrim Coast: Truncation of glacigenic sequences by late-glacial raised beach notches

Glenulra: Gilbert-type deltas overlying marine muds

Belderg: Marine muds overlying glaciated rock

Askillaun: Ice-contact glaciomarine spread, dipping west

Killard Point, Dundalk Bay, Cranfield Point, Port, Linns: Marine muds below within and above glacigenic deposits

Killary Harbour, Kylemore: Ice-contact Gilbert-type deltas within glacial troughs

Killiney, Greystones: Spreads of ice-contact glaciomarine outwash dipping south

Lismore, Screen Hills: Ice-contact deltas, facing seaward

Ballycroneen, Ballycotton: Bay head marine muds, muddy diamicts and sand sequences, interbedded with local breccias

Simon's Cove, Howe's Strand: Emergent facies sequence directly on glaciated rock platform

ATLANTIC OCEAN

IRISH MIDLANDS

Irish Sea

Legend:
- ● Late-glacial beach notch
- ▬ Morainal banks / spreads / glaciomarine muds
- + Marine muds
- × Shallow marine sands
- ▲ Ice contact deltas

0 km 100

Types and nature of field evidence

Fossil shorelines around a small island group like the British Isles should provide a wide range of raised land-forms and sediments recording former sea levels during the last glaciation and deglaciation. Even though the BIIS was sensitive to climate signals it could not have contributed significantly to global sea level change (<2 m water equivalent) so that raised features may be expected to record both near and far-field effects. For

these records to be preserved and document sea level trends from different ice sheet sectors, local amounts of isostatic depression must exceed global sea level fall. In addition postglacial marine erosion has in some cases undercut and removed higher late-glacial shoreline fragments. In countless exposures from northern Britain the marine erosion surfaces which separate glacigenic diamicts from overlying postglacial beaches simply record major gaps in the record. The fragmentary nature of the evidence is also influenced by the topography

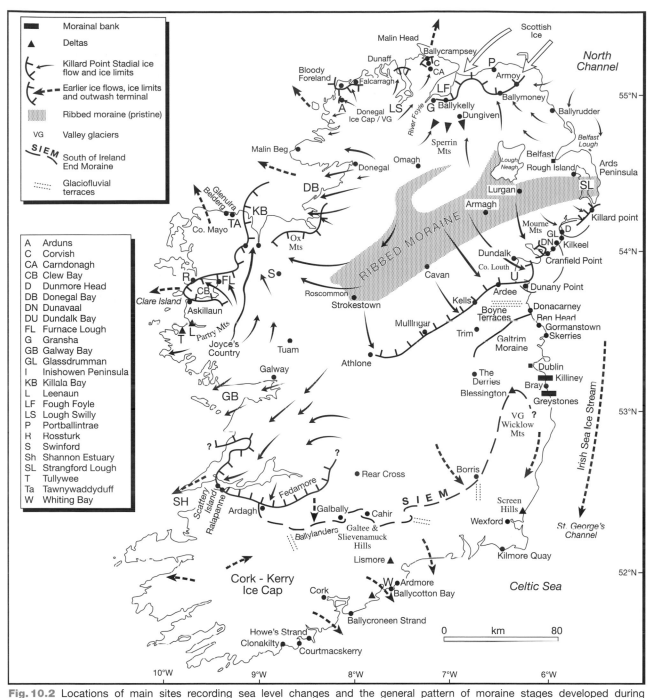

Fig. 10.2 Locations of main sites recording sea level changes and the general pattern of moraine stages developed during deglaciation.

of the coastline during a transgression. For example, an increase in sea level by 30 m would create broad shallow marine embayments over low-lying coasts such as County Down, Clew Bay, Donegal Bay and Dundalk Bay but would have very little impact on steep, cliffed or rocky coastal tracts (Fig. 10.3). It could therefore be argued that because of the fragmentary nature of the field data it is difficult to correlate beach remnants over large coastal tracts. Most correlations of raised shoreline systems at this time from northern Britain using shoreline relation diagrams have been made on incomplete data sets, especially regarding the timing and extent of the ice sheet system which strongly conditions isostatic compensation.

Fig. 10.3 Subglacial bedform landscape from Clew Bay flooded by the postglacial sea level rise.

It seems more logical to examine the general sea level trends and their close relations to the position of ice sheet margins where sufficient field evidence is available. Because prominent ice sheet limits have now been age-constrained by radiocarbon dates, this methodology can now be applied in order to examine relationships between sea level history and ice history. The fact that ice sheet margins can remain or readvance onto a particular coastal tract during deglaciation will determine the extent of a particular transgression. If a particular ice margin then retreats, allowing the transgression to proceed, then raised beach forms will be metachronous and will not form a synchronous system as deglaciation proceeds. This relationship stresses not only the temporary nature of raised marine systems but the importance of ice sheet limits as deglaciation proceeds. The temporary nature of the field data (formed quickly) means that a particular marine high-stand may not be preserved either across bedrock or in areas of steep relief. However, near an ice margin, where glacial effluxes deliver large quantities of sediment, the preservation potential for notching of drift banks and sediment reorganisation into nearshore coastal systems is enhanced. Perhaps because of the emphemeral nature of late-glacial sea levels the best evidence will be preserved proximal to ice sheet limits. Preservation will also be enhanced at these localities because the depocentres are the precise areas where isostatic uplift will be accelerating as ice mass decreases. In Ireland the

patterns of isostatic shorelines clearly record deglaciation of coastal segments and their altitudes are related to former thicknesses of ice over the lowlands.

Contemporary shorelines are zones with an altitudinal range where a variety of sediments and morphological forms occur between high and low water marks. Different coastal materials and possibly structure will also give rise to a variety of morphological forms which will also be influenced by degree of exposure, local relief, dominant wave climates, extreme events, tidal variations and beach dynamics. Thus, past shorelines need to be accurately described including cliffs, platforms and beach material. In northern and eastern Ireland there are a wide range of raised landforms and sediments which record former high relative sea levels. The main forms include raised marine terraces, beach ridges, swash gullies, cliff notches, storm ridges, ice-contact deltas and washing limits. Some of these features can form the marine limit in a particular area, but in general are used to identify minimum sea levels. In north central Ireland these late-glacial forms occur between present sea level and ~30 m above sea level.

In the past individual raised features proved difficult to date with any precision because beach fragments,, even though they may occur at similar heights, are likely to be metachronous along a coastal strip due to differential crustal depression away from the area of greatest loading and timing of isostatic uplift. However, some of the main late-glacial raised shorelines can be dated with some precision because they are contemporaneous with ice sheet readvances. Where these raised strandlines end at major moraines and do not occur inside the line of the moraine then both are contemporaneous. The late-glacial marine-cut notch and platform at Dunnaval, County Down, is exactly the same age as the marine-cut platform at Killard Point some 50 km to the northeast because both end at moraines recording the ice margin at 14 ^{14}C kyr BP (Fig. 9.1). Furthermore, both are cut across subaqueous outwash formed at tidewater ice margins and therefore record emergence at that site. This field relationship also shows that the actual timing of beach formation occurred shortly after subaqueous deposition and that isostatic uplift had begun to occur as the ice sheet margin still stood at its terminal position. This observation means that ice wastage was already in progress at the ice sheet maximum during the Killard Point Stadial, possibly related to ice sheet thinning, wastage further inland, and downdraw. Clearly, field data of this type show that the ice sheet system was dynamic and that links can be established between the sediments deposited at the ice margin and the subglacial sediment flux as bedforms are eroded/modified/overprinted. Within any one area major differences in the heights of raised marine indicators may be used to infer rates of changes in regional processes during deglaciation. These include changes in ice loading during deglaciation, timing of isostatic uplift, high relative sea levels and ice sheet history. The presence of marine muds within glacigenic successions from around the island permits phases of marine transgression to be identified, though not the precise height of associated sea level.

Significant progress has been made in the identification of the subaqueous processes associated with deposition of morainal banks and spreads along the coast of eastern Ireland. The associated landforms can then provide minimum heights for former sea levels at that point because either the morainal bank was ended at tidewater or time was insufficient for development into a shallow-water, Gilbert-type delta. Morainal banks contain a wide variety of facies but are characterised by channelled geometries, rapid facies changes from interbedded mud, gravel and diamict, variable bed contacts and facies continua (Eyles and McCabe, 1989b). Proximal facies may include clast-supported gravel, locally boulder rich, that is crudely bedded with clast a-axes dipping up-current (Rust, 1977).These generally disorganised ice contact deposits may translate distally into subaqueous outwash which is often character-

ised by tabular beds, planar massive sand beds, indistinct lamination, cross-bedded beds of pebbly gravel, vaguely-stratified and thinly bedded diamicts, and matrix-supported pebbly gravel. Perhaps diagnostic criteria for subaqueous outwash is the presence of all or most of these facies within the same succession. In the majority of cases subaqueous sequences contain outsized clasts deposited as dropstones confirming an iceberg source (Figs. 9.15a, b). In an ideal facies sequence the subaqueous outwash should be truncated as wave base is reached and shoreface gravel and sand are deposited (Fig. 9.16). This relationship is well developed along the coast of south county Down, west of Kilkeel. The upper beachface deposits consist of laterally-continuous, thin beds of well-sorted, sometimes alternating beds of sand and pebbly gravel. Stratification is emphasised by discontinuous pebble lags with clasts flattened along bedding planes. The gravel clasts are both shape and size sorted with well-rounded edges. In exceptional cases fairly complete emergent facies sequences containing wave-influenced rhythmites are preserved as at Portballintrae, County Antrim (Fig. 10.4). However, in most cases facies sequences are only partially preserved but can identify the general sea level trend.

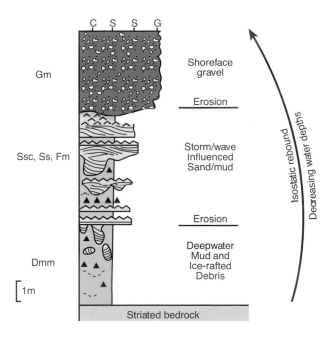

Fig. 10.4 Emergent facies sequence from Portballintrae, north County Antrim. The sequence is important because it is almost complete, containing the three main facies expected (ice-contact or glaciomarine, shallow marine wave-influenced, upper beach face) when land emerges from the late-glacial sea.

Sea level records from both individual sedimentary successions and along specific coastal sectors associated with a particular ice sheet history often have quite different degrees of resolution. In western sectors of the ice sheet on the southern margins of Donegal Bay former sea levels are well-constrained by radiocarbon dating because fossiliferous muds provide critical index points in the stratigraphic succession between 28 and 19 cal ka BP (McCabe et al., 2005; 2007a). The large moraines around Clew Bay at Furnace Lough and Askillaun, Donegal Bay along the northern flanks of the Ox Mountains and at The Ballycrampsey Meeting House are all part of the ice sheet readvance system that brackets the Heinrich 1 event between 15 and 18 cal ka BP (Fig. 10.5). Similar ice sheet readvance maxima in east central and northern Ireland are age-constrained to a few hundred years. Along the western Irish Sea basin south of Drogheda the sea level data are from morainal banks and deltas which formed at tidewater during early deglaciation (22–19 cal ka BP). Along the south coast of the island ice moved from the lowlands into the Celtic Sea (Fig. 10.6). Therefore the raised beach gravel and related deposits which overlie the glaciated surface of the rock platform postdate the earliest phase of deglaciation when ice withdrew north from the continental shelf. The main sedimentary and topographic characteristics of this distinctive group of marine indicators will be assessed starting with the south coastal sequences and moving anticlockwise around the coastline (Fig. 10.1). Sea level history is therefore not considered in precise chronological order because each ice sheet sector contains different sea level records related to variable preservation potentials, distinct deglacial histories and types of raised landforms.

The south coast of Ireland

Wright and Muff's (1904) descriptions of a marine rock platform buried by combinations of beach gravel, sand, gravel, diamict, mud and head along the south coast have led to a succession of theories on the origins and ages of the platform and overlying deposits (Figs. 10.7, 10.8) (Synge, 1981; O'Cofaigh and Evans, 2001). In most cases and without the help of any reliable dates or faunal remains various reports conclude that the sediments range in age from the last interglacial (gravel/beach) to muds deformed by the last ice sheet. However, it is possible to take a more rational view of the stratigraphy because the last ice sheet moved from the interior generally southeastwards possibly 40 km into the Celtic Sea (McCabe and O'Cofaigh, 1996). The intense erosional marks on the shore platform show that the platform was scoured of sediment after ice moved onto the continental shelf (Fig. 10.9a). This ice flow event is also documented by similar patterns of ice flow recorded by ice directional indicators in the coastal zone between Cork and west Wexford (Figs. 10.6, 10.7). All of the deposits overlying the rock platform must therefore postdate this ice flow. Coeval north to south ice flow lines identified from satellite imagery suggest that the regional ice flow originated from the north Irish lowlands and occurred during the last glaciation. This timing is also consistent with early deglaciation dates ~22 cal ka BP from County Waterford (Bowen et al., 2002) and later deglaciation of thick ice from the borders of the Wicklow Mountains (Ballantyne et al., 2006). Glacial scouring across a range of bedrock forms are a common feature on coastal margins around the island because ice generally moved onto the continental shelf (Fig. 10.6). Resulting facies

Fig. 10.5 Erosional remnant of a glaciomarine spread, Askillaun exposure, south side of Clew Bay, western Ireland. At a maximum this sediment spread is 40 m in height and up to 500 m long. The internal geometry is dominated by beds of massive diamict (debris flows) which slope westwards (to the right) towards the open Atlantic ocean. Stratified gravel, sand, silt and clay beds contained within shallow channels comprise about 15 percent of the deposit formed from glacial effluxes. Note dispersed lonestones and cobble lags throughout the sequence. Figure for scale is 1.6 m high.

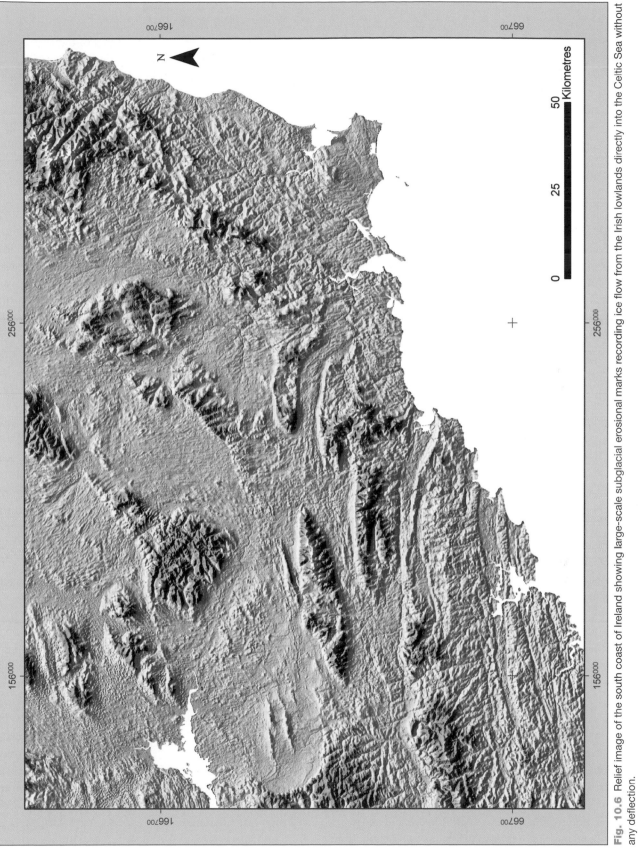

Fig. 10.6 Relief image of the south coast of Ireland showing large-scale subglacial erosional marks recording ice flow from the Irish lowlands directly into the Celtic Sea without any deflection.

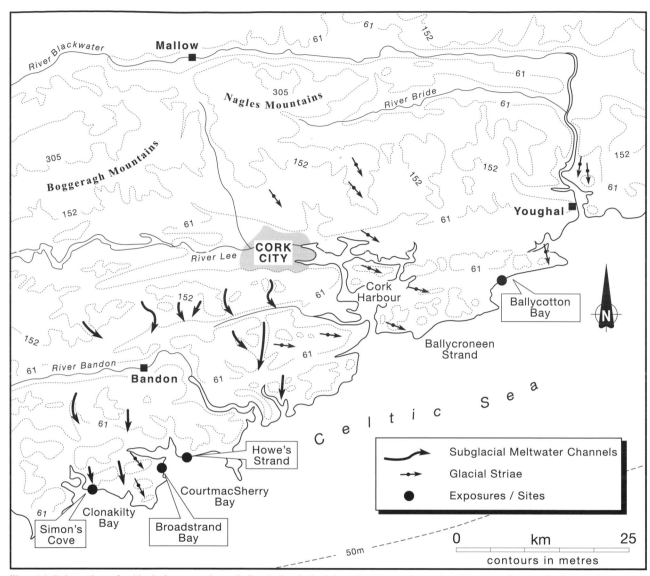

Fig. 10.7 Location of critical sites, southward directed subglacial meltwater marks and striae, east County Cork, south coast of Ireland. Note that the directional features observed in the field are consistent with the larger-scale erosional features recorded in Figure 10.6 (after McCabe and O'Cofaigh, 1996). *By kind permission of SEPM (Society for Sedimentary Geology).*

sequences overlying ice-eroded surfaces therefore can be interpreted in terms of deglaciation events unless independent dating is available.

The extensive marine rock platform (~4 m above sea level) along the south coast rises landward into a buried cliff (Fig. 10.8) (Wright and Muff, 1904; Farrington, 1966). In Clonakilty and Courtmacsherry Bays the platform surface is cut by closely spaced (1–5 m), north-to-south aligned, subparallel furrows up to 20 m long by 0.5–2.0 m deep (Fig. 10.9a). Long profiles are undulatory punctuated by potholes and some are incised across topographic highs on the platform. All furrows cut across the W–E strike of thin-bedded, compact shales and are not

related to or deflected by either joint patterns or other structural weaknesses. Farrington (1966) and Mitchell (1972) suggested that the platform was trimmed by the sea immediately before the last ice advance. The morphology of the furrows including up-and-down long profiles, pitted long profiles, close spacing and incision across highs shows they were cut subglacially beneath an ice sheet (cf. Kor et al., 1991). The furrows could only have been eroded by unidirectional southward water flow with the presence of a bounding medium (ice) to maintain separate subparallel meltwater flows. The stratigraphic position of the furrows suggests they relate to significant subglacial meltwater erosion when the

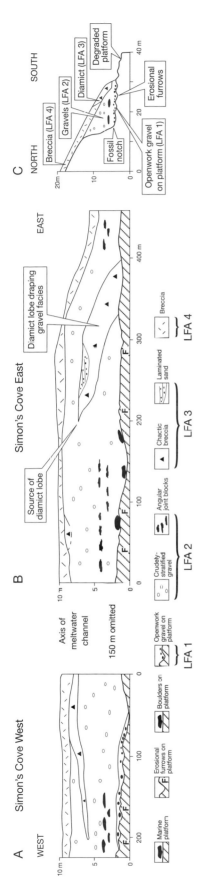

Fig. 10.8 Spatial relationships of the lithofacies associations, Simon's Cove, County Cork. Exposures A and B are separated by a north–south meltwater channel. General relationships of the depositional facies with the buried cliff, local slopes and platform are shown in C (after McCabe and O'Cofaigh, 1996). *By kind permission of SEPM (Society for Sedimentary Geology).*

outer seal of meltwater reservoirs within the ice sheet were broken. Because the overlying deposits contain subaqueous beds interpreted as marine it is possible that rising sea levels triggered ice wastage and vigorous subglacial meltwater flow.

The most complete and informative stratigraphic sequences overlying the rock platform occur at Simon's Cove and Howe's Strand (Figs. 10.10, 10.11). Stacked, subhorizontal beds of largely openwork, densely packed gravel are present on the glacially eroded platform and in furrows both of which are striated (Fig. 10.9d). Diversity in bed structure, textural changes, shape and size sorting and cobble lags provide crude stratification. The a-axes of large (~3m) erratic boulders are shore parallel and some are granites derived from the east (Fig. 10.9c). These facies are part of a raised beach present at many localities along the south coast (Wright and Muff, 1904) though no models take into account the fact that the beach complex directly overlies an unweathered glaciated platform which still preserves delicate striations beneath the stratified drift sequence. Therefore a major glacial event occurred between initial platform development by shoreline processes and deposition of the raised beach. Although some (O'Cofaigh and Evans, 2001) have arbitrarily assigned the beach to the last interglacial there are at least five arguments which show that the beach was emplaced in an Arctic environment immediately after the last ice sheet withdrew a short distance inland. First, the beach was deposited on delicately ice-moulded surfaces without any intervening time for either shoreline erosion laterally along the strike or surficial chemical weathering. This observation indicates short-lived rapid emplacement of pebbles and cobbles on the platform surface during rising sea level. Second, no marine faunas have ever been recorded from the beach despite careful searches for over one hundred years. It could be argued that the faunas have been weathered out, but beach material is unweathered and there are no signs of reprecipitation of carbonate. Third, the beach and related sediments are interbedded with breccias (head) derived from local slopes (Fig. 10.9c). Angular clasts organised as beds would not be typical of interglacial beach sequences where there would be time (years or less) for sorting and rounding to occur. Theoretically maximum production and preservation of breccias would occur immediately after an area is vacated by ice when periglacial mechanical and slope processes can operate effectively into subjacent depositional sinks. Fourth, the presence of large boulders

Fig. 10.9 Facies from the Simon's Cove and Howe's Strand areas, south coast of Ireland.
A) Subglacial N–S aligned erosional furrows cut into the surface of the marine platform. Note that these are closely spaced and are at right angles to the east–west strike of the rock. The erosional furrows are overlain by outcrops of scour- and drape- hummocky cross-stratified medium sand, recording shallow marine conditions. Poorly-bedded slope deposits occur at the top of the section.

B) Detail of swaly and hummocky cross-stratified sand with scours overlying the subglacially furrowed platform.

C) Unfossiliferous raised beach sand and gravel containing large erratics and vertically stacked, discontinuous boulder pavements which are mainly locally derived angular joint blocks. Most joint blocks are preferentially aligned along bedding planes suggesting reorientation either by waves or shore ice. Relative sea level fall is recorded by the upper beds of angular slope breccias. Note figure for scale.

D) Detail of clast packing from the raised beach gravel, Simon's Cove. Note the size sorting, shape sorting and the presence of some angular detritus derived from local cliffs.

E) Chaotic diamict from the front of a debris flow lobe, Simon's Cove east. Note the highly variable arrangement of angular clasts, rounded clasts, faceted clasts and marked textural changes. Clasts with long axes also exhibit a type of chaotic fabric around the margins of the flow.

resting on the platform and single beds of boulders one clast thick interbedded with the beach gravel indicates contemporaneity. Some of these clasts are exotic and from the east and must have been delivered to the site after subglacial erosion of the platform because southward moving ice and subsequent meltwater flow scoured the platform clean of debris. Clearly, the presence of only large boulders within well-sorted pebbly gravel means that the boulders are not a lag feature but were transported and emplaced by a mechanism distinct from the usual shoreline processes. The most likely agents that transported the boulders to the immediate area are icebergs and shore ice. The boulders are not derived directly from pre-existing glacigenic sediment because they are subangular to angular, are not ice-bevelled and there are no remanie glacigenic deposits

on other parts of the platform. Finally, the position of boulders entombed within finer beach gravel and with their a-axes aligned subparallel to the former coastline is consistent with the action of shore ice pressure similar to that observed from contemporary Arctic coasts (Fig. 10.10). Fifth, the beach is overlain by shallow subaqueous facies that also contain boulder beds one clast thick, and breccia beds. The presence of these facies indicates that the local boulder-sized, angular clasts were still being moved downslope and that once they reached sea level some were aligned by the pressures of winter shore ice. Therefore the sequence context of the underlying beach gravel is consistent with formation in an Arctic beachface setting subject to shore ice formation and breakup.

Sequences of laminated sand up to 4 m thick at

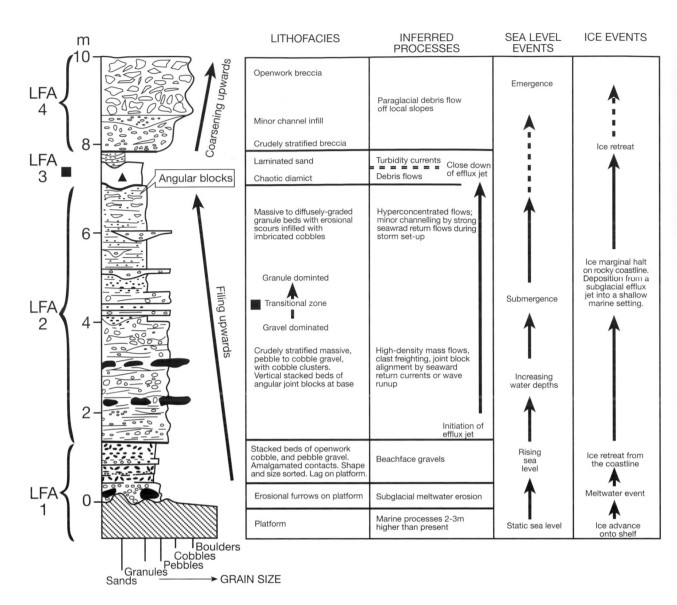

Fig. 10.10 Measured log from Simon's Cove summarising the contact relationships, the main sedimentary structures in LFAs 1, 2, 3 and 4, inferred processes and inferred events (after McCabe and O'Cofaigh, 1996). *By kind permission of SEPM (Society for Sedimentary Geology).*

Howe's Strand provide critical evidence for a phase of submergence following beach formation on the platform (Fig. 10.11). The base of this facies consists of erosional hummocky cross-stratification (HCS), with truncated laminae (<0.4 m thick) with apparent spacings of 1–2 m and heights of 10–15 cm (Fig. 10.9b). Higher in the sequence other types of HCS (scour/drape, accretionary) and nested sets of swaly cross-stratification (SCS) are more common together with lines of isolated pebbles. At the base of one hummock an irregular lens of angular clasts set in a sandy matrix infills a shallow depression and shows evidence of reworking of subjacent laminae along its margins. At Broadstrand Bay both form-concordant and discordant HCS types occur in the same stratigraphic position to that at Howe's Strand. The geometry and internal structure of the laminated HCS and SCS are similar to examples described by Dott and Bourgeois (1982). HCS is thought to be the product of both oscillatory and combined flows, large suspended loads and rapidly changing hydraulic stability fields below fair-weather wave base (Brenchley, 1985). Continuous reworking and scouring together with the absence of structures typical of fair-weather pelagic sedimentation is confirmed by the absence of structures such as ripples, mud caps, massive sand beds and by textural homogeneity of laminae. Episodic input

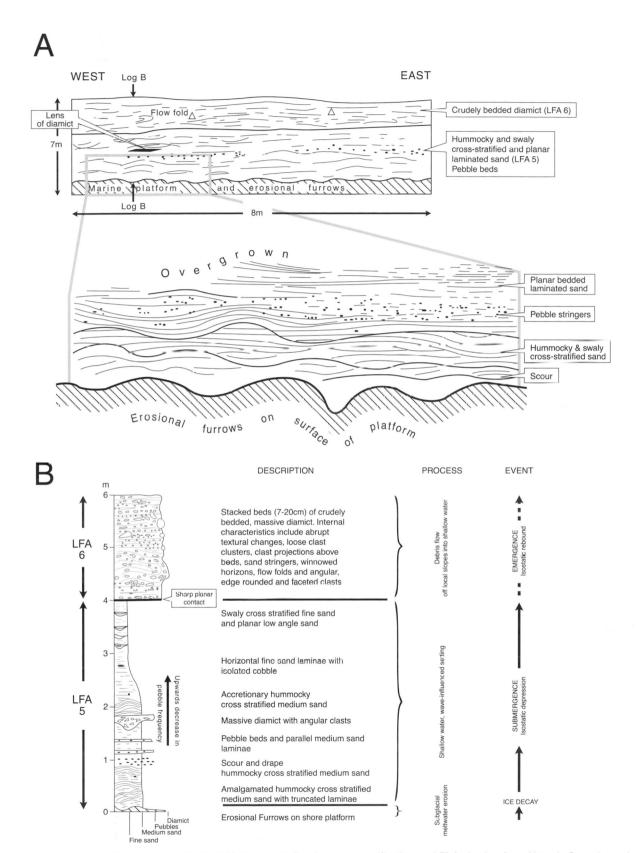

Fig. 10.11 General stratigraphy and detail of A) hummocky/swaly cross-stratification and B) facies log from Howe's Strand, south coast of Ireland (after McCabe and O'Cofaigh, 1996). *By kind permission of SEPM (Society for Sedimentary Geology).*

of coarser-grained debris by mass flow from adjacent slopes is marked by thin lenses of angular clasts and pebble lines.

Submergence during deglaciation of the Cork coast is also recorded by the 5–7 m thick sequences of crudely stratified pebble/cobble gravel beds which fine upwards into diffusely-stratified beds of graded granules at Simon's Cove (Fig 10.10). Pebble beds are up to 40 cm thick with amalgamated contacts marked by abrupt textural changes, discontinuous cobble lines and shallow scours. Well-sorted granule beds are lensate to tabular, 10–30 cm thick and laterally extensive. The location of the gravels indicates they were derived from a subglacial efflux located at the apex of the cove (Fig. 10.8). The facies sequence (pebbles/cobbles → granules) is inferred to record subaqueous sedimentation from a turbulent jet (McCabe and O'Cofaigh, 1996). Stacked beds with crude stratification, amalgamated contacts, abrupt textural changes, fabric variability and little edge rounding probably reflect suspension sedimentation as the jet lost its competence (Powell, 1990). Lateral continuity and textural similarity of granule beds can be related to pulsed flows associated with hyperconcentrated, expanding flows, strong enough to align small pebbles along bedding planes (Gorrell and Shaw, 1991). The gravel facies contains angular joint blocks with a-axes up to 2 m long arranged as discontinuous, subhorizontal lines vertically stacked ~1 m apart. The a-axes are aligned parallel to the present beachface which is assumed to be similar to the glacial one. The blocks are likely derived from free fall or erosion from the adjacent fossil cliffline as meltwater eroded the channel at the head of the cove. Their strong a-axes alignment suggest that their final alignment is related to either shore-normal wave runup or backwash during storms. This origin is to some extent supported by seaward transport of cobbles with northward imbrication of ab-axes set within discrete scours in the granule beds. All structures within these facies are consistent with shallow marine sedimentation with the ice margin grounded a short distance to the north. The presence of a large fan of diamict with an apex to the northwest is consistent with the build up of heterogeneous debris at the ice grounding line and resedimentation southwards across the subaqueous outwash (10.9e). The fan consists of amalgamated lobes of sheared, contorted and chaotically arranged, angular to subangular and faceted clasts set in a highly variable admixture of matrix-supported glacigenic debris and contorted sandy to pebbly units.

Breccias up to 2 m thick overlie the ice-contact marine sequence and can be traced into bedrock sources near the fossil cliff (Fig. 10.8). Clearly the breccia is locally derived, paraglacial in origin and denotes final emergence (Fig. 10.9c). Similar, crudely-stratified breccias are widely recognised from exposures along the south coast where they are interbedded with resedimented glacigenic sediments close to the fossil cliffline (Fig. 10.9c) (Farrington, 1966).

Many of the bays east of Cork City are plugged with fine-grained, muddy sequences between 5 and 20 m thick, up to 1.5 km across the bays and no more than 0.5 km inland. These facies contain fragments of marine shells and foraminifera but do not occur on intervening interfluves between the bays. They also are interbedded with a range of other coarse-grained diamicts, laminated deposits, gravel and sand. At Ballycotton Bay the mud drape consists of frequent changes in lithofacies separated by sharp to gradational contacts and shallow scours. Mud beds are flat-lying with numerous wispy, sandy stringers and isolated pebbles (Fig. 10.12a). Lithofacies variations include mud-streaked sand, sand-streaked mud, and wavy bedded sand. Frequent lithofacies changes of this type are strongly suggestive of wave agitation associated with storms (Reineck and Singh, 1972). Eyles and McCabe (1989a) interpreted the sequences as glaciomarine and related to bay head deposition, near grounded ice margins which had retreated onland from the Celtic Sea. The overall facies sequences and sediment continua are consistent with deposition from turbid plumes, debris flow, density flows, soft sediment deformation and wave-reworking. Deformation of these sediments including attenuation, folding and thrusting at metre scales was directly related to onshore ice advance from the Irish Sea ice stream (O'Cofaigh and Evans, 2001). However, there is no direct evidence that the last Irish Sea ice flowed across the coast of east County Cork. Furthermore, if this were the case then there should be striae recording not only this hyopothetical ice flow of Irish Sea ice to the northwest, but also deflections in regional striae patterns related to ice flow from inland sources. Significantly, all the known striae and other directional indicators point to a uniform, undeflected regional ice flow southwards across the coast of southern Ireland into the Celtic Sea (Figs. 10.6, 10.7). Erratics can be delivered to the area by floating ice and all of the sediment can be sourced locally by reworking along grounding lines associated with inland ice (Fig. 10.12). Radiocarbon dates from

Fig. 10.12a Laminated and massive beds of marine mud from Ballycotton Bay, south coast of Ireland.

Fig. 10.12b Stratified secquence of debris flows off local slopes which are interbedded with muds, Ardmore Bay, south coast of Ireland.

derived marine shells are consistent with glacial deformation of fine-grained sediment after 26–20 ^{14}C kyr BP (O'Cofaigh and Evans, 2007) as inland ice grounded in the bays of southern Ireland.

Two critical stratigraphic markers support a model based on advance across the coast followed by ice retreat onto land. First, all the directional indicators along this coastal segment show that the last ice sheet passed offshore into the Celtic Sea and there is no recorded evidence for an onshore advance of Irish Sea ice. Striae at Kilmore Quay in southeastern County Wexford record this southward movement right on the western marginal limit of Irish Sea ice. As this ice retreated local, coarse-grained debris was released at the ice margin forming coarse-grained diamicts which

always show evidence for resedimentation within or above the muddy sequences. Sequence context considerations therefore show that sediments are intimately related and that the deformation features which are present beneath the coarse-grained diamict are a direct result of local ice-marginal oscillations. Second, beds of head or angular debris derived from local slopes on bay margins dip towards the axes of the bays and are interbedded with the fine-grained sediments plugging the embayments (Fig. 10.12b). Facies relationships of this type indicate that both the subaqueous fine-grained sediments and the coarser heads (breccias) which are bedded parallel to local slopes are penecontemporaneous and proglacial. The preferred depositional model therefore includes grounded ice margins within major bays, ice flow from the Irish Midlands, ice-marginal reworking of older sediments, minor ice-marginal movements resulting in sediment overriding/shearing rather than homogenisation, and delivery of significant quantities of local debris towards ice margins.

It is argued that the fine-grained facies assemblages from Ballycroneen Strand and Ardmore Bay are lateral equivalents of the shallow-water marine deposits around Simon's Cove and Howe's Strand. They were probably deposited in deeper water recording a depressed isostatic surface that sloped eastwards. This pattern of isostatic deformation is consistent with the potential amounts of isostatic depression associated with the relative small Cork/Kerry ice cap to the west and the much larger and thicker ice masses emanating from the Irish Midlands to the east (Fig. 10.6). In the valleys around and to the east of Cork City there are ice-contact, Gilbert-type deltas containing gravelly foresets that dip consistently to the south and southeast (Synge, 1981). Deltas record temporary halts in ice retreat westwards from the coast towards centres of ice dispersal in Cork and Kerry. The location and sedimentary geometry of these features is also consistent with high relative sea levels during deglaciation of this coastal zone. In terms of facies trends, the deltas record a slightly later, deeper and more extensive phase of submergence than that documented from the facies sequences along the coast. Features of this type formed quickly and their preservation is linked to rapid isostatic uplift. The importance of the facies sequences along the south coast is that they record submergence for a short time immediately following deglaciation of the coast. As such they represent part of an event stratigraphy rather than type sites of 'Irish Sea tills' (Mitchell, 1960, 1972).

Early deglaciation of the east coast

The raised marine landforms along the east coast provide evidence for high relative sea levels during the early deglaciation of the southern Irish Sea Basin around 22 cal ka BP. Landforms recording minimum late-glacial sea levels occur sporadically between the Screen Hills delta moraine in the south to County Meath in the north (Fig. 6.10). North of the River Boyne most marine indicators formed during early deglaciation have been cut out by the lower raised shoreline system associated with readvance moraines in County Louth. The latter formed much later in the deglacial cycle around 18–16.5 cal ka BP.

The large (<70 m thick, 20 km long, 150 km² in area) ice-contact delta complex known as the Screen Hills moraine formed on the western margin of St. George's Channel in a deglacial suture between ice in the Irish Sea and ice advancing from the Irish Midlands (Fig. 10.13). The interlobate complex prograded over marine muds with gravelly foresets facing full-frontal into the open Celtic Sea suggesting deposition in water up to 30 m OD. Because this high water level is very close to and faces directly into the adjacent, open deglaciated Celtic Sea (ie. grounded ice margin was situated in St. George's Channel to the northeast of the delta) it is unrealistic to invoke ice-dammed glaciolacustrine bodies without reconstructing improbable ice sheet geometries/margins (cf. Thomas and Summers, 1983). The position of the Screen Hills ice margin is also consistent with the regional pattern of deglaciation along the south coast when ice contracted onshore from the continental shelf (Fig. 10.13c). A relative sea level around 30 m OD in southeastern Wexford is also consistent with probable isostatic deflection patterns which resulted in a surface that sloped east from Cork towards the southern Irish Sea basin. The fine-grained muds and ice contact features deposited by inland ice on this surface probably record a similar sea level to that at the Screen Hills. Both sets of evidence consisting of deltas, muds and glaciomarine deposits directly face the open Celtic Sea for 150 km of coast and there are no moraines known offshore that could have dammed a lake in the vicinity of the present coastline.

To the north of the Screen Hills morainal banks and glaciomarine spreads tend to occur either on the lee side (ie. south) of bedrock ridges or within topographic lows. Contrary to some ice sheet reconstructions on the coast of county Wicklow or south County Dublin there are no lines of continuous, large moraines recording a pulsed northward decay of the ice sheet margin. The thick glacigenic sequence at Greystones which thins southward from the rock ridge of Bray Head is a good example of a morainal bank (Figs. 6.13, 6.14). The coarsening-upwards sequence records deposition by tidewater processes when the ice margin re-equilibrated on Bray Head. Ice-marginal pinning on a topographic high facilitating growth of a significant sediment pile. The sediment spread failed to evolve into a shallow-water Gilbert-type delta possibly because the ice wasted rapidly from the temporary halt. It therefore records a minimum sea level of around 25 m OD (McCabe and O'Cofaigh, 1995). Ten kilometres to the north ice-

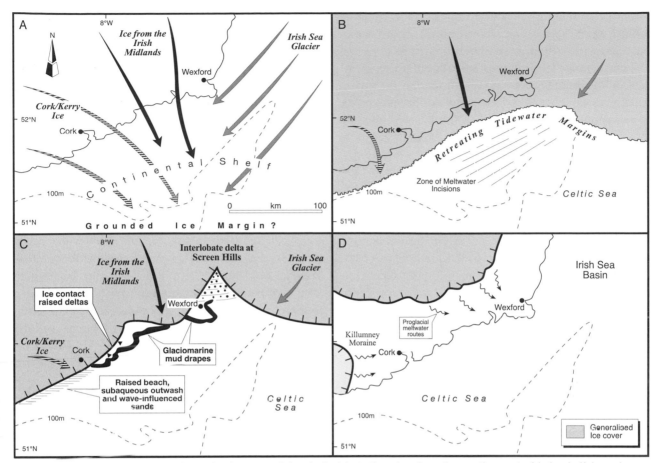

Fig. 10.13 Cartoons illustrating the main elements of the deglacial stratigraphy along the south coast of Ireland. A) Ice advance onto the inner shelf. B) Isostatic depression and rising relative sea level followed by rapid ice loss by iceberg calving and regional meltwater events including furrowing of the coastal marine platform. C) Re-equilibration or short halts during ice retreat. During this phase narrow coastal zones were submerged by rising sea level. Raised beach gravel, subaqueous outwash and wave-influenced sands were deposited in southeastern County Cork. Deeper water marine muds accumulated along the coast between Cork Harbour and County Wexford near ice sheet margins. Local advances of these ice margins have in places deformed the muds. Farther east in County Wexford the Screen Hills delta prograded over marine muds containing dropstones (Thomas and Summers, 1983). D) Terrestrial emergence and periglacial sedimentation off local slopes. The ice margin retreated westwards towards the Cork/Kerry dispersal centre and northwards into the southern Irish lowlands. The 'South of Ireland End Moraine' represents later outwash from the retreating ice margin (after McCabe and O'Cofaigh, 1996). *By kind permission of SEPM (Society for Sedimentary Geology).*

contact deposition at tidewater, backfilling the depression between Bray and Killiney also records a minimum sea level around 20 m OD (Eyles and McCabe, 1989b). This deposit is important because it links the onshore sequence with an offshore tunnel valley network, stressing the importance of the onshore tunnel valleys directing subglacial water and poorly-sorted sediment to point sources at the ice margin.

The picture of rising sea level as a result of glacio-isostatic downwarping of a large foreland basin around the ice sheet and the uncoupling of the ice sheet from its bed is consistent with the offlapping sedimentary wedges in the western Irish Sea basin. The small delta near Skerries in north County Dublin probably formed in a very shallow marine setting as the ice margin wasted northwards. To the north, E–W moraines at Delvin, Ben Head and Donacarney show few characteristics of glaciomarine conditions and are mainly terrestrial in origin even though distal outwash at Gormanstown is fine-grained and rhythmically bedded. Because the moraines are traceable inland for many kilometres they are significant landscape elements and may be associated with either a minor readvance or halt in ice-marginal retreat. It is unlikely that these moraines were controlled by climate because deglacial climate warming in the North Atlantic had commenced and was not reversed until the advent of Heinrich event 1. Within the context of rapid deglaciation it is suggested that where isostatic uplift locally exceeds eustatic sea level fall the ice margin becomes terrestrial forming linear push moraines instead of glaciomarine spreads.

The opening of the Irish Sea Basin as a marine seaway early in the last deglaciation is confirmed by the raised glacigenic sequences exposed in the northern Irish Sea Basin along the coast of south County Down, west of Kilkeel (Clark et al., 2004) (Fig. 6.28). Early deglaciation (>20 cal ka BP) of coastal zones in southeastern Ulster was accompanied by high relative sea level (~30 m OD) due to isostatic depression when global eustatic levels were low. The marine limit consists of a prominent notch at 30 m OD which can be traced continuously from Glassdrumman southwestwards to near Ballymartin, a distance of 4 kilometres. Global sea level was 130–140 m lower than at present during this time (Yokoyama et al., 2000), indicating that the marine limit records a net isostatic uplift of 160–170 m (Clark et al., 2004). As deglaciation proceeded the glaciomarine diamicts underlying the Mourne Plain emerged and were channelled by subaerial streams draining south from the Mourne Mountains. Because these channels are graded at least to present sea level they indicate that the land had emerged isostatically at least 30 m when they formed. Each of the six channels filled with marine mud formed during a major transgression which is dated to ~19,000 years BP. The presence of marine mud within subaerial channels requires a rapid rise in eustatic sea level to flood the channels on what was otherwise still an isostatically emergent coast (Fig. 6.25). This far-field signal is important in identifying the termination of the LGM associated with the partial collapse of one or more continental ice sheets, as measured by the first rise in global sea level (Clark et al., 2004). The data from County Down confirm that the rapid rise in sea level first inferred from Bonaparte Gulf (Yokoyama et al., 2000) may have been shorter than 500 years. Peaks in ice-rafted detritus throughout the North Atlantic and Nordic Seas at this time suggest that possible sources for the meltwater event were the Greenland, Iceland and Fennoscandian ice sheets (Bond et al., 1997; Elliot et al., 2001). Finally, it is argued that the 19,000 year event freshened the North Atlantic, caused rapid reduction in water mass ventilation, led to decreased Atlantic Meriadinal Overturning (AMO) and heralded the start of prolonged cooling during the Oldest Dryas. Clark et al. (2004) also suggest that because the AMO causes northward heat transport in the Atlantic Basin, a reduction in the AMO should induce a compensatory warming in the tropical Atlantic and Southern Ocean (the bipolar thermal seesaw).

Directional indicators and ice-contact sediment wedges deposited at tidewater record ice-marginal withdrawal north along the axis of the ISB. The presence of thick ice in the north and ice flow from Scotland into northeastern Ireland show that the North Channel was ice locked during early deglaciation. This field evidence demonstrates that the ISB was flooded from the south through St. George's Channel. However, because early deglaciation was extensive and affected the entire ice sheet (Bowen et al., 2002) the Irish Sea Basin eventually became an open marine seaway ~21 cal ka BP with major ice masses restricted to basin margins. Six stratigraphically consistent radiocarbon dates from a core (Cir 5/82) recording fine-grained marine sedimentation between the Isle of Man and the Ulster coast from 20 to 12.4 ^{14}C kyr BP (Kershaw, 1986) support the presence of an open seaway unaffected by glacigenic sedimentation. Later ice readvances reached the margins of the basin but did not advance far offshore.

The age-constrained field evidence from the north ISB has three implications for studies on the history of the last ice sheet in the British Isles. First, evidence for substantial isostatic loading well in excess of eustatic sea level lowering means that relative sea level was high and the ice sheet must have been strongly influenced by marine downdraw during deglaciation. A corollary is that subglacial landforms patterns on adjacent land margins are in part developed under the ice streams draining the ice sheet. Second, the current geophysical models that predict sea levels are not supported by a growing body of evidence that deglaciation of the last ice sheet was accompanied by high relative sea levels (eg. Thomas and Chiverrell, 2006). Furthermore, rock exposure cosmogenic dating from the Wicklow Mountains shows that thick ice (>750 m) occurred during deglaciation (Ballantyne et al., 2006) which means that isostatic depression was greater than eustatic sea level fall (eg. Yokoyama et al., 2001). Third, given the fact that the British Isles consists of a group of islands influenced by deep isostatic depression during deglaciation then it is reasonable to conclude that seaways developed around islands and major inlets once deglaciation commenced. This general model is consistent with major ice flow indicators around the islands and stresses the impact of marine downdraw on ice sheet dynamics and the roles played by ice stream drainage. In this context the very early deglaciation across the whole of the ice mass may have been triggered by rising sea levels in peripheral troughs around the ice sheet. Rapid tidewater evacuation of the Irish Sea Basin, which lay astride the geographic centre of the ice sheet, early in the deglaciation contributed significantly to the initial destabilisation of the ice sheet. Field evidence shows that the ice sheet lost about two thirds of its mass and seaways characterised by direct Atlantic influences separated the remnants of the ice sheet.

After this early deglacial event most major ice centres of ice sheet dispersal were isolated with relatively small outlets into peripheral marine embayments. A central theme to this model is that the maximum ice sheet model which covered most of Ireland, Scotland, Wales and northern England consisted of amalgamated ice streams from different centres of ice sheet dispersal. The resulting composite ice sheet was inherently unstable because major flow lines sited in topographic lows and marine embayments were essentially sutures along which the mechanisms of downdraw were perpetuated upglacier from retreating tidewater ice sheet margins. The available field evidence suggests that the extent of ice wastage was strongly controlled by high relative deglacial sea levels which depleted terrestrial parts of the ice sheet by marine downdraw. More detailed isostatic/eustatic interactions also document the dynamic interactions between local and far-field effects along certain parts of the coast.

Sea level curve for northeastern coastal sector

During the last glacial termination details of sea level records from glaciated coasts are fragmentary because they are destroyed by very rapid environmental change and readvances during deglaciation (Stea et al., 1998). Consequently there are very few longterm records (eg. 10,000 years) of height and age-constrained deglacial sea levels along mid latitude glaciated coasts. This constrains the ability of many geophysical and other models to describe and predict sea level tendencies during deglaciation (Lambeck, 1995,1996; Lambeck and Purcel, 2001; Shennan et al., 2006a, b). Along northeastern coasts evidence for former sea levels has a high preservation potential because the sensitivity of the BIIS is marked by very early deglaciation when about two thirds of the ice mass wasted catastrophically just after 22 cal ka BP, allowing marine waters to flood the isostatically depressed surface (Bowen et al. 2002). Marine transgressions are dated from marine microfaunas obtained from muds which formed distal to retreat-

ing ice sheet margins (Clark et al. 2004; McCabe et al. 2005) and associated shoreline deposits (Carter, 1993; Kelley et al., 2006). The marine deposits formed mainly in embayments distal to ice sheet margins close to the contemporary coast (Fig. 10.14). Data from northeastern Ireland (Fig. 10.15) provide robust evidence for relative sea level changes during the last glacial termination (early deglaciation to Bølling/Allerød) extending the length of sea level records back 7000 years to ~21 cal ka BP, a time period not documented in northern Britain (eg. Shennan et al., 2005, 2006a, b).

Ice-directional indicators and bedforms suggest that the inland ice domes which remained after early deglaciation were bordered by a peripheral zone which was influenced by similar isostatic and eustatic effects recording common sea level trends (Fig. 10.14). The sites used to construct the sea level curve occur in this area peripheral to the ice mass inland (McCabe et al. 2007c) (Fig. 10.16). Resulting isobase patterns are earlier than and different from the later generalised, postglacial patterns which slope southwards. Implicit in this methodology is that patterns of uplift may change during the course of a deglacial cycle, especially when the ice sheet re-establishes mass and configuration. In addition, traditional notions of monotonic ice retreat towards centres of ice dispersal are now replaced by a highly dynamic ice sheet characterised by large and rapid fluctuations on millennial timescales (McCabe & Clark, 1998). The deglacial marine stratigraphy contains evidence for at least five marine transgressions during the last termination. These are age-constrained by dating monospecific samples of *Elphidium clavatum* from *in situ* microfaunas contained in raised marine mud. These opportunistic biocoenoses have been recorded from modern Arctic–subarctic areas recently vacated by tidewater glaciers (Hald et al. 1994). Additional geomorphic evidence including mappable relationships between raised beaches and ice-marginal moraines (Synge, 1977; Stephens & McCabe, 1977) and raised sea level indicators (eg. intertidal boulder pavement) from isostatically emergent coasts is added to the other marine data used to construct the sea level curve (McCabe and Haynes, 1996).

The relative sea level record is described from four major time slots each containing different sea level trends (Fig. 10.16):

a) The earliest part of the sea level curve (Interval 21–19 cal ka BP; Sites 1-4) is obtained from raised diamict sequences infilling the Mourne Plain when glacio-

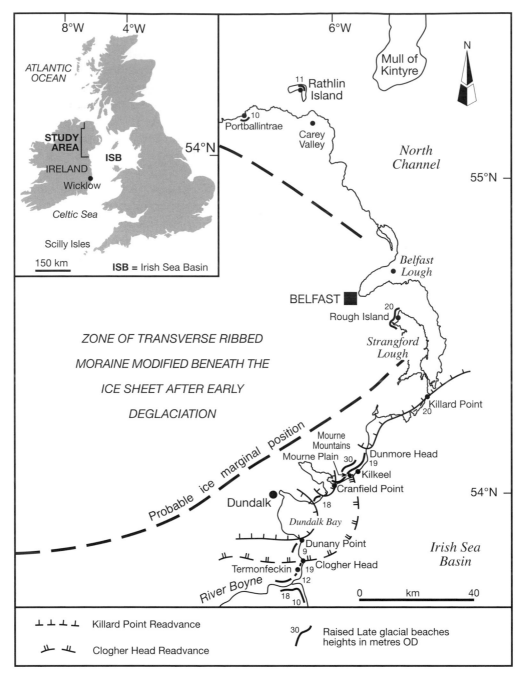

Fig. 10.14 Location of northeastern Ireland and sites used to reconstruct a sea level curve for northeastern Ireland (see Figure 10.16). The peripheral relationship of sea level reference sites to the position of the ice mass should be noted (after McCabe et al., 2007c).
By kind permission of The Geological Society.

marine conditions record final ice sheet withdrawal from the Irish Sea Basin (McCabe, 1986). Concurrent raised beach notches at 30 m OD formed along 4 km of the Kilkeel coast recently vacated by ice (Stephens and McCabe, 1977). A fall in relative sea level of around 40 m must have occurred around 20 cal ka BP when coastal emergence resulted in subaerial channels graded to below sea level (Clark et al. 2004). A subsequent abrupt

rise in global sea level filled the erosional channels with fossiliferous marine mud dated to 19,000 years BP (Fig. 6.28). The mud crops out at elevations up to 10 m OD which represents a minimum level for sea level at this time. A rapid fall in relative sea level of ~8 m over the subsequent 500 years following the global meltwater pulse is recorded by shallow marine muds on the shores of Dundalk Bay. The timing and small magnitude

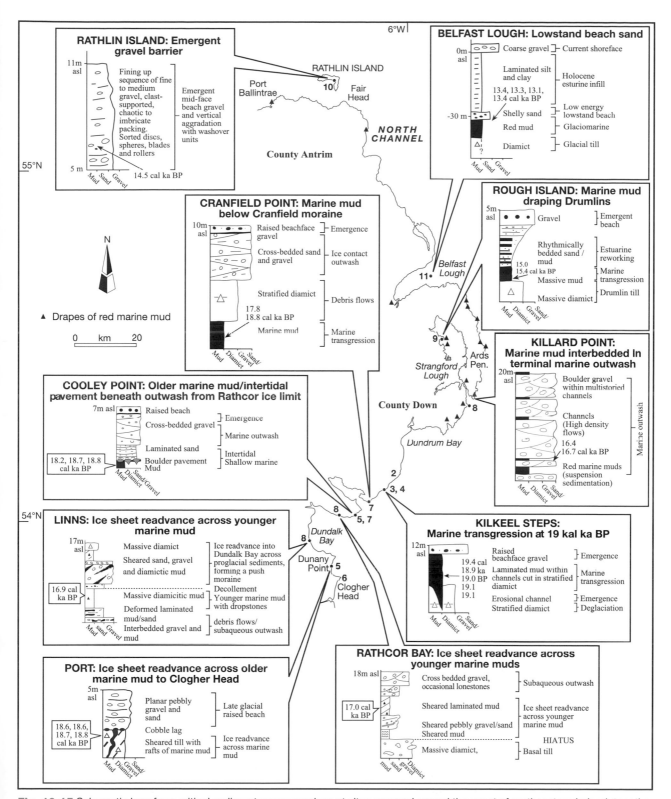

Fig. 10.15 Schematic logs from critical sedimentary successions at sites exposed around the coast of northeastern Ireland, together with the stratigraphic position of the dates used to identify and age constrain the sea level curve shown in figure 10.16. (after McCabe et al., 2007c). *By kind permission of The Geological Society.*

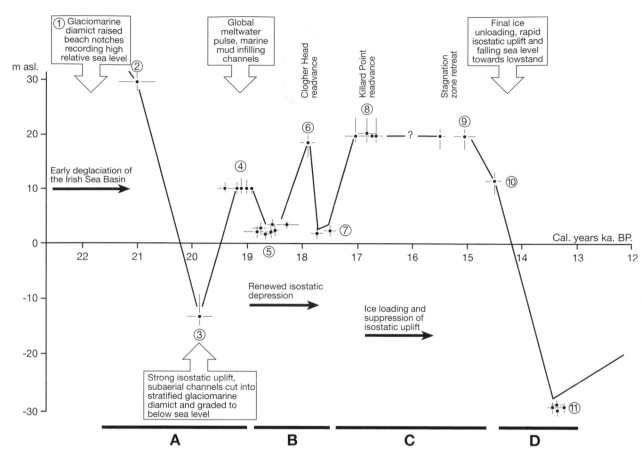

Fig. 10.16 Sea level curve for northeastern Ireland during the last glacial termination. Sea level index points are assessed and age-constrained as follows. 1) Early deglaciation (Bowen et al. 2002) and deposition of marine muds in the northern basin >22 cal.ka BP (Kershaw, 1986). 2) Raised beach notches cut into glaciomarine diamict formed during early deglaciation (McCabe, 1986). 3) Lowstand recorded by subaerial channels graded to below datum along the Kilkeel coast, Clark et al. 2004. 4) Meltwater pulse recorded by mud infilling the channels at Kilkeel (Clark et al. 2004). 5) Marine mud records sea level higher than present at Cooley Point and at Port (McCabe and Haynes, 1996; McCabe et al. 2007). 6) Ice sheet readvance to Clogher Head (McCabe et al. 2007b). 7) Marine mud at Cranfield Point and intertidal boulder pavement at Cooley Point records sea level above present (McCabe et al. 2005). 8) Ice sheet readvance to Killard Point limits. 9) Catastrophic ice retreat, drapes of red marine clay on subglacial bedforms and formation of late-glacial raised marine terrace, Rough Island, Strangford Lough (McCabe et al. 2005). 10) Falling sea level and formation of gravel barrier on Rathlin Island (Carter, 1993). 11) Marked lowstand with submerged beach/intertidal sands, Belfast Lough (Kelley et al. 2006).
By kind permission of The Geological Society.

of this RSL fall implies that continuing isostatic uplift was masked by the rising waters of the global meltwater event rather than any change in the rate of uplift.

b) The middle part of the curve (Interval 19–17.5 cal ka BP; Sites 5–7) is marked by sea levels a few metres above present recorded by marine mud around Dundalk Bay followed by a marked rise in sea level (>14 m) around 18.5 cal ka BP which occurred at about the same time as a major ice sheet readvance as far south as Clogher Head (Fig. 10.14). Raised beach terraces at Termon-feckin south of the ice sheet limit and delta terraces at the mouth of the river Boyne record sea levels up to 19 m OD at this time (Synge, 1977). It is argued that this increase in relative sea level represents a millennial

timescale build up of ice on the lowlands followed by a centennial timescale readvance and a renewed spike of isostatic depression. Raised marine muds at Cranfield Point show that sea level remained above datum after the spike because the new ice load suppressed accelera-tion in isostatic uplift at this time.

c) A second ice sheet readvance (Interval 17–14.5 cal ka BP; Sites 8–10) which occurred around 16.5 cal ka BP was less extensive than the earlier readvance to Clogher Head. It is marked by morainal banks between Killard Point and Dunany Point, a distance of 130 km (McCabe and Clark, 1998). The ice advanced across shallow water marine muds a few metres above ordnance datum. A cold water, intertidal boulder pavement at 1 m OD.

formed in Dundalk Bay immediately prior to the readvance and was buried by subaqueous outwash (McCabe and Haynes, 1996). Terminal outwash interbedded with marine mud and contemporaneous raised beach gravel from widely spaced sites along this ice limit consistently record minimum sea levels at 19–20 m OD at Killard Point, Dunmore Head and Cranfield Point. These levels persisted for about 1000 years and probably are related to prolonged ice sheet loading because the ice margin remained at its terminal limits for a similar length of time. Records of the late-glacial marine transgression within the Killard Point ice limit at Rough Island on the northern shores of Strangford Lough suggest that initial transgression occurred between 14.8 and 15.6 cal ka BP). Sea levels up to 19 m OD determined from the raised marine mud at Rough Island, extensive marine terraces and contemporaneous raised notches on the northern margin of Strangford Lough, are much younger (~1000 years) but are similar in height to those at the Killard Point ice maximum, 30 km to the south (Fig. 9.29). One possible explanation for this continued high relative sea level is that the ice disintegrated catastrophically during stagnation zone retreat from Killard Point and marine inundation of Strangford Lough was instantaneous with little time for isostatic compensation. Alternatively the continued high relative sea level may be partly due to increased global meltwater flux immediately prior to meltwater pulse (MWP) 1a if significant isostatic compensation had occurred. Therefore in this instance it is difficult to decide whether or not constrained local rebound or the contemporaneous global meltwater flux determined sea level at this time.

d) Rapid uplift of 50 m in about 1500 years (Interval 15.0–13.5 cal ka BP; Sites 11–12) occurred immediately after catastrophic ice wastage at the onset of the Bølling/Allerød warming. In broad terms, this marked lowstand followed suppression of an early isostatic uplift signature because of ice sheet re-establishment after early deglaciation and two major readvances during deglaciation. Falling sea level from the highstand is recorded by radiocarbon dated shells (14.4 cal ka BP) from a raised gravel barrier on Rathlin island (Carter, 1993). On the mainland the shallow marine emergent facies sequence containing wave-influenced rhythmites formed below about 10 m of water is probably the distal facies equivalent of the gravel barrier on Rathlin Island (McCabe et al., 1994). The lowstand is recorded by submerged intertidal deposits dated to 13.5 cal ka BP from Belfast Lough (Kelley et al., 2006).

The sea level curve depicts rapid initial uplift following massive deglaciation, suppression of isostatic uplift as ice sheets re-established/readvanced and rapid isostatic uplift when the ice wasted catastrophically at the onset of the Allerød/Bølling warming. The saw-tooth shape of the curve represents superimposition of far-field and other effects on a sensitive, deeply-depressed surface which had undergone early and extensive deglaciation (Fig. 10.16). Evidence from a variety of raised late-glacial marine indicators along the Antrim coast illustrated by shoreline diagrams has not been used in the construction of the sea level curve for northeastern Ireland (Prior, 1966; Stephens and Synge, 1966a). However, much of the data is within the height ranges for marine activity when Scottish ice finally withdrew from this coastal sector.

Unlike most other current models (Lambeck, 1996) this curve does not depend on floating late-glacial sea level index points or on presumed retrogressive projections from early postglacial sea level trends. This data for the first time in the British Isles provides age-constrained field evidence recording at least four successive time intervals characterised by discrete sets of local and far-field processes (Fig. 10.16). The key to interpretation of the main processes driving sea level change is the ^{14}C constraints on RSL changes that reveal the timing of isostatic and eustatic changes that contributed to the RSL record. Isostatic components are resolved by subtracting eustatic sea level estimates from the age-constrained field observations (Fig. 10.17).

The marine transgressions between 21 and 14 cal ka BP have important basinwide implications because they occurred when global eustatic levels were low (Yokoyama et al., 2000). Sea levels were above datum for 70 percent of this time span (7000 years) when full marine conditions have been age-constrained in the central and northern parts of the Irish Sea Basin. Cosmogenic rock exposure dates from the Wicklow Mountains suggest that the Irish Sea ice stream was 600–700 m thick at 18–19 cal ka BP (Ballintyne et al., 2006) and must also have resulted in deep (~200 m) isostatic depression in the southern basin. The amount of depression greatly exceeds global eustatic fall and is consistent with our deglacial evidence that thick ice existed across the Irish lowlands during the last glaciation. The evidence for substantial isostatic depression along the basin reinforces the hypothesis of Eyles and McCabe (1989a) that the glacial geology of the Irish Sea basin is an event stratigraphy recording the entry of marine waters into

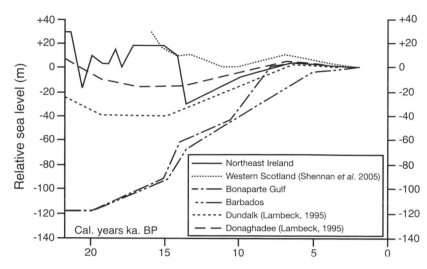

Fig. 10.17 Sea level curves from NE Ireland and Scotland. Eustatic curves and modelled curves are shown for comparison (after McCabe et al., 2007c). *By kind permission of The Geological Society.*

the southern basin causing dynamic thinning and the retreat of the Irish Sea glacier northward as a tidewater margin.

The predictions of palaeowater depths, shoreline positions, ice-marginal positions and ice sheet loading from geophysical models (Lambeck, 1995, 1996; Lambeck & Purcell, 2001) for the Irish Sea and adjacent areas are quite different from the ground truth contained in the late-glacial sea level curve (Fig. 10.16). The mathematical models are flawed because they are based on an amalgam of undated, generalised flowlines, assumed monotonic ice retreat, thin ice and static ice sheet models. The sea level curve is consistent with a thick ice sheet, significant fluctuations in the ice margins, readvances and changes in mass balance. Because of deep isostatic depression the local sea level elements were overprinted on two occasions by far-field effects. The sea level curve provides a yardstick for testing and comparing models on relative sea level changes during deglaciation that include both near- and far-field effects.

Ice sheet limits and sea levels in north County Donegal

Stephens and Synge's (1965) classic account on the raised beaches of north County Donegal first introduced the hypothesis that an end moraine mapped as the limit of fresh drift was closely related to the highest late-glacial shoreline. They observed that gravelly topography and thick tills occur at intervals across

north Donegal between Moville and Bloody Foreland (Fig. 10.2). These fragmented occurrences have been linked together and termed the north Donegal moraine because they border the drumlin fields to the south and occur in areas that are otherwise relatively drift free. To the south of Bloody Foreland block moraines mark the position of the ice which advanced offshore. This drift limit marks the southern edge of the highest marine shorelines (~25 m OD) or shingle bars well seen on Malin Head and Fanad Head. Because the higher shorelines end at moraines or drift limits and are not present within the inner sea loughs both are considered contemporaneous (Stephens and Synge, 1965). Three late-glacial shorelines were recognised, each thought to represent a distinct transgression on a depressed land surface. However, it is extremely difficult to date raised morphological forms such as washing limits, shingle bars and notches or to link forms from site to site along an indented coastline.

At the mouth of Lough Foyle a push moraine complex near Moville consists of current-bedded sand and gravel interbedded with red clay (Stephens and Synge, 1965). The thickest beds (~2 m thick) of clay occur towards the base of the exposed section and contain marine diatoms. Because these beds are contorted by ice pushing from the southwest they were deposited on an isostatically depressed surface prior to ice sheet readvance to the outer push moraine. High relative sea level persisted as delta formation progressed because marine muds are interbedded with outwash throughout the exposure. An inner push moraine with a crest at 32 m OD consists of diamict overlying deformed outwash. Red clays occur up to 27 m OD where they drape the outer moraine which has been flattened by marine erosion. Similar marine clays have been observed within the Culdaff depression 13 km to the northwest and below a shingle ridge at Doon, 1 km east of Malin Head (Fig. 10.18). Because the marine clays occur on both sides of the Inishowen Peninsula and record similar heights (27–25 m OD) for minimum sea level they were probably formed at approximately the same time. No fossils from these exposures have been radiocarbon dated. However, because they are interbedded with the push moraine at Moville, at least part of the marine clay is contemporaneous with maximum limits of the ice sheet lobes that readvanced into Lough Foyle and Lough Swilly (Fig. 10.2). Stephens and Synge (1965) considered that these morainic limits were part of the Drumlin Readvance ice sheet system in the north and west of the island. Raised beach notches cut into the distal parts of the Moville moraine occur within the inner lough as far south as Drumskellan and record a fall in relative sea level to ~18 m OD after ice had withdrawn southwards to the Gransha moraine.

The Inishowen Peninsula is the most northerly point in Ireland and is open to high energy marine systems

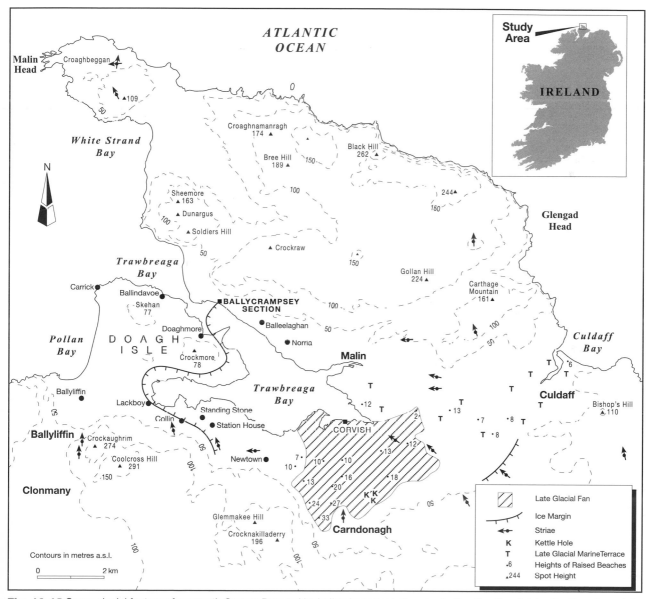

Fig. 10.18 Some glacial features from north County Donegal including important sites, ice limits, directional data, the Carndonagh fan and sites surveyed for sea level history. All heights and contours in metres above Irish datum. Inset shows study area in Ireland (after McCabe and Clark, 2003). *By kind permission of The Geological Society.*

resulting in widespread erosion of swash gullies along geological weaknesses and formation of shingle bars up to 26 m OD (Stephens and Synge, 1965). A variety of raised marine landforms occur from the marine limit right down to near present sea level but it is difficult to date these forms with any accuracy unless they can be linked with datable marine deposits or moraines. Areal patterns of glacial erosion show that ice from lowland centres of ice dispersal and from the Donegal ice cap coalesced and advanced northwards onto the continental shelf. A single cosmogenic date (Bowen et al., 2002) suggests that the peninsula was deglaciated around 25 cal ka BP and some of the raised beach features could have formed shortly afterwards. Within Trawbreaga Bay, basal tills on bay margins record the position of an ice lobe that readvanced into the inner bay after initial deglaciation (Fig. 10.18) (McCabe and Clark, 2003). Near this ice limit at Ballycrampsey a flat-topped ridge consists of glacitectonised thrust slabs of schist separated by pebbles and cobbles with an upper irregular surface overlain by pervasively deformed cobbles containing angular rock fragments and detached slabs of schist (Fig. 10.19). The imbricate pattern of thrusts and aligned cobbles (dip SE at 65°) was generated by an overriding shear to the NW imparted by an overlying ice sheet. Field relations and the occurrence of till only within the postulated ice sheet limits also suggests that the ridge, till and glaciotectonic deformation are part of

the same subglacial system below the margin of the ice lobe centred on the inner part of Trawbreaga Bay. It is unusual to find glaciotectonically sheared, well-rounded and sorted cobble gravel within thrust bedrock slabs. A possible analogy occurs on the low (2–5 m OD) raised Holocene beach platform immediately west of Malin village (Fig. 10.18). This upper shoreface setting consists of detached sea stacks separated by swash gullies, cobble beach flats and low beach ridges. Loosened slabs from the stacks together with the cobbles provide the appropriate facies for glaciotectonic compression during an ice sheet readvance across a deglaciated coastline. An important inference is that relative sea level was slightly higher that at present immediately before the ice sheet readvance, and also that the present day platform at Malin formed mainly after final deglaciation because stacks with detached slabs (~4 m high) are preserved on its surface.

At the head of Trawbreaga Bay, immediately north of Carndonagh a late-glacial outwash surface slopes northwards from ~30 to 7 m OD (Fig. 10.18). Ice contact features including kettle holes and boulder/cobble gravel within cross-cutting channels record ice-proximal sedimentation when the ice margin pinned on rock at Carndonagh village. Because this feature is one of the largest sediment spreads in north Donegal it probably represents a discrete and later ice limit than the Ballycrampsey readvance. This inference for two ice sheet events during

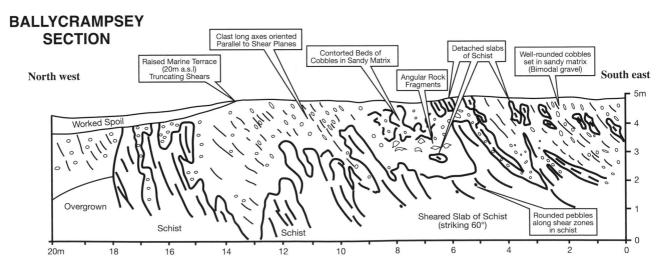

Fig. 10.19 Field sketch of section exposed at Ballycrampsey, County Donegal. Sketch shows slabs of schist and well-rounded pebbles compressed by ice pressure into a ridge (km scale) which may be traced across Trawbreaga Bay, causing narrowing of the bay. All of the shear planes, tectonic lineations and ab planes of the cobbles dip consistently southeastwards. Tectonic structures are abruptly truncated at the top of the section. Note the presence of rounded quartzite cobbles towards the base of the section and the presence of isolated slabs of schist entirely within the cobble facies at the top of the section (from McCabe and Clark, 2003).
By kind permission of The Geological Society.

deglaciation is supported by a section in marine muds at the distal or northern end of the Carndonagh spread at Corvish (Fig. 10.20). Four main facies are identified on the basis of vertical position in the section, textural changes, degree of interbedding and lamination, degree of sediment deformation, changes in microfaunal assemblages and AMS ^{14}C dating of marine microfaunas (Fig. 10.20). The oldest facies consists of finely-laminated mud containing a circum-Arctic microfauna. A monospecific sample of *Q. seminulum* was AMS dated to 17,140±110 years BP at -2.2 m above sea level and a sample of *E. clavatum* to 15,025±95 years BP at -1.2 m above sea level. The older date shows that early deglaciation occurred >20 cal ka BP and that the land surface was deeply isostatically depressed up to ~18 cal ka BP when eustatic sea levels were low (-120 m) (Yokoyama et al., 2000). Overlying mud facies contains clasts, lacks bed continuity upsequence, contains numerous slickensides and irregular sandy inclusions, contains centimetre scale shears and is pervasively sheared towards the top. Although the marine microfaunal species are similar to the undeformed mud below, three AMS ^{14}C dates are consistently older than the date (15,025±95 years BP) obtained from the top of the laminated mud. Because these dates are stratigraphically inverted with respect to those from the undisturbed laminated mud below, it is probable that the deformed muds record reworking of the lower laminated mud by an ice sheet readvance over the site. It is thought that relative sea level was similar to that at present because there is no evidence for any significant change in the microfaunal population. Shallow marine sedimentation probably continued during ice advance into the inner bay as far west as Ballycrampsey. The stratigraphic position of the mud beds show that the readvance to Ballycrampsey occurred between 15.2 and 14.0 ^{14}C kyr BP. The deformed gravels and schists at Ballycrampsey represent an upper beachface shoreline overrun by ice and suggest that relative sea level was similar to that at present immediately before the readvance. Renewed isostatic deflection at maximum readvance limits into Trawbreaga Bay is recorded by the swash gullies and washing limits (~30 m OD) outside of these ice limits (Fig. 10.21). Undeformed, rhythmically-bedded marine mud overlying the deformed mud beds records quiet water sedimentation and drapes the deformation structures caused by ice advance across the bay. A monospecific sample of *E. clavatum* dated to 14,045±100 years BP shows that the inner bay was now free of ice with little evidence of ice rafting. Because the muds

occur at the distal end of the ice contact spread and the fact that muds (2 m thick) have been recorded (Stephens and Synge, 1965) from other parts of the spread it is likely that the ice limit at Carndonagh marks a significant ice-marginal event associated with high relative sea level around 14 ^{14}C kyr BP. The pattern of late-glacial raised beaches outside the Ballycrampsey moraine are up to ~30 m OD whereas inside the moraine they reach 20 m OD and fade out against the outwash spread at Carndonagh. This pattern suggests that increased isostatic uplift had occurred during ice recession south from Ballycrampsey. Lower shorelines record farther isostatic uplift though these are generally less well developed that the main late glacial notch (Fig. 10.21).

There are field records for two significant ice sheet limits in Trawbreaga Bay (Ballycrampsey and Carndonagh) and in Lough Foyle (Moville and Gransha) which are associated with well-defined, late-glacial raised shorelines. The radiocarbon dates and stratigraphy from Trawbreaga Bay indicate that the earlier readvance to Ballycrampsey occurred sometime between 15.2 and 14.0 ^{14}C kyr BP and the later readvance to Carndonagh occurred around 14 ^{14}C kyr BP. This chronology is based on fossiliferous marine muds and is correlative with similar ice-marginal fluctuations recognised from the north Irish Sea Basin (McCabe et al., 2007b). Together these records show that eastern and northern sectors of the Irish Ice Sheet oscillated at the same time during the Oldest Dryas cold interval and bracket Heinrich Event 1. The records of high relative sea level and fluctuations of ~30 m during readvance events strongly show that the lithosphere was responsive to increased ice sheet loading on millennial timescales. It could also be argued that the similarity in marine limits over about 3–4000 years of the deglacial cycle means that relatively thick ice persisted in northwestern sectors and suppressed isostatic depression. Net amounts of isostatic depression during the ice sheet oscillations was in excess of 140 m and may have been greater at maximum ice sheet limits on the continental shelf.

Sea levels/ice sheet history prior to the 'LGM'

Glacial deposits and landforms from the west coast of Ireland have conventionally been interpreted as recording advance of the Irish ice sheet (IIS) during the Last Glacial Maximum (LGM = 19–23 calibrated ka (cal ka); Mix et al., 2001) onto the continental shelf followed by

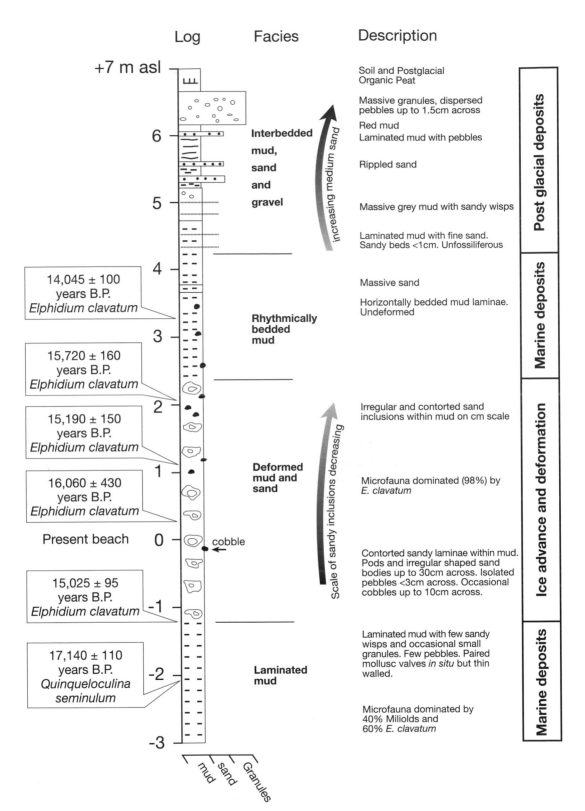

Log **Facies** **Description**

+7 m asl

Soil and Postglacial
Organic Peat

Massive granules, dispersed
pebbles up to 1.5cm across

6
Red mud
Interbedded Laminated mud with pebbles
mud,
sand Rippled sand
and
5 **gravel** Massive grey mud with sandy wisps

Laminated mud with fine sand.
Sandy beds <1cm. Unfossiliferous

4

14,045 ± 100 Massive sand
years B.P.
Elphidium clavatum Horizontally bedded mud laminae.
Undeformed
3 **Rhythmically**
bedded
15,720 ± 160 **mud**
years B.P.
Elphidium clavatum

2 Irregular and contorted sand
15,190 ± 150 inclusions within mud on cm scale
years B.P.
Elphidium clavatum

1 **Deformed**
16,060 ± 430 **mud and** Microfauna dominated (98%) by
years B.P. **sand** *E. clavatum*
Elphidium clavatum

Present beach 0 ← cobble
Contorted sandy laminae within mud.
Pods and irregular shaped sand
bodies up to 30cm across. Isolated
pebbles <3cm across. Occasional
15,025 ± 95 cobbles up to 10cm across.
years B.P.
Elphidium clavatum -1

Laminated mud with few sandy
wisps and occasional small
granules. Few pebbles. Paired
17,140 ± 110 mollusc valves *in situ* but thin
years B.P. -2 **Laminated** walled.
Quinqueloculina **mud**
seminulum
Microfauna dominated by
40% Miliolds and
60% *E. clavatum*
-3

increasing medium sand

Scale of sandy inclusions decreasing

mud sand Granules

Post glacial deposits

Marine deposits

Ice advance and deformation

Marine deposits

Fig. 10.20 Stratigraphic log from Corvish at the distal end of the late-glacial spread at Carndonagh. Flat lamination shows that the upper and lower parts of the succession are undisturbed by ice sheet advance. The ^{14}C dates from the middle part of the section are consistently older than the date obtained from the top of the underlying, undeformed mud. This pattern is consistent with ice deformation and shearing of muddy sediment across the site prior to deposition of undeformed rhythmically-bedded marine mud at the top of the succession (after McCabe and Clark, 2003). *By kind permission of The Geological Society.*

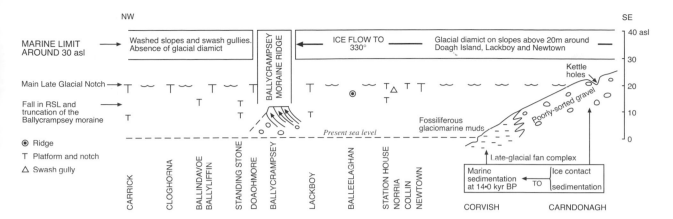

Fig. 10.21 Relationships between sea level changes, ice readvance to the Ballycrampsey limit and the late-glacial spread at Carndonagh. Poorly-drained soils within this ice limit are developed on basal diamict, which is absent to the NW (outside) of the ice limit at Ballycrampsey. High marine limits around 30 m above sea level are found outside the Ballycrampsey moraine, suggesting their removal from the head of the bay during final ice advance to Ballycrampsey. The main late-glacial strandline seems to be contemporaneous with part of the late-glacial spread at Carndonagh and is higher than the muds dated to 14.1 ^{14}C kyr BP. Heights surveyed using Trimble 4200 GPS. *By kind permission of The Geological Society.*

monotonic ice retreat (Synge, 1968; Mitchell et al., 1973). New chronological constraints, however, demonstrate that the IIS was a more dynamic system with several ice-margin fluctuations over the last 45 ka comparable to the LGM event (McCabe et al., 1986, 1998; McCabe and Clark, 1998, 2003; Bowen et al., 2002; McCabe et al., 2005). Ice flow indicators suggest multiple westerly ice advances from dispersal centers in central Ireland (Knight et al., 1999; Clark and Meehan, 2001), but none of these are dated. Although existing ^{14}C and [36]Cl ages suggest that the glacial and relative sea level history of the western Irish ice margin may be similar to that from the Irish Sea Basin, most of these ages relate to events since the LGM (McCabe and Clark, 2003; McCabe et al., 2005). Radiocarbon ages on macro- and microfossil shells from glacial and raised marine sediments preserved on the south side of Donegal Bay, western Ireland significantly improve and extend our understanding of the glacial and relative sea level history in western Ireland prior to the LGM (Fig. 10.22) (McCabe et al., 2007a).

Two sites exposed in the lower part of the Glenulra valley on the south side of Donegal Bay contain fossiliferous deposits initially interpreted as tills (Hinch, 1913) (Fig. 10.23). On the west side of the valley, at Glenulra Farm, four fossiliferous facies are exposed in a river cut (Fig. 10.24). The lowermost facies comprises locally derived, brecciated sandstone bedrock. Above this facies is 1.5 m of matrix-supported, overconsolidated massive diamict containing dispersed, glacially-bevelled and edge-rounded clasts. Derived shell fragments of *Arctic islandica* are present in both of these units. The diamict facies is overlain by up to 7 m of laminated mud containing an *in situ* marine microfauna. This facies is overlain by 4 m of low angle, poorly-sorted matrix-to-clast supported pebble and cobble gravel beds. The gravel, which contains derived marine shell fragments, forms a distinct terrace on both sides of the valley at about 80 m above sea level. A river cut on the east side of the valley near a sand and gravel quarry, 400 m downstream from Glenulra Farm, exposes five facies. Only one contains fossils of *A. islandica*. The lowermost facies is a compact diamict similar to that found upstream (Fig. 10.25). The facies overlying the diamict consists of amalgamated beds of massive, mainly clast-supported, disorganised boulder gravel, with clast long axes suggesting deposition from the south (Fig. 7.4). Derived shell fragments are present throughout the unit. This facies is overlain by massive, laminated and cross-bedded sand which contain three mud beds up to 1.5 m thick. A marked erosional contact separates the sand from the overlying pebble gravel foresets which dip consistently at 15° to the northwest. The foresets are truncated by crudely bedded, disorganised boulder gravel. The surface of this sequence forms a terrace at ~80 m OD, which is paired with the terrace across the valley at Glenulra Farm.

Three kilometers southeast of Glenulra, near Ballycastle, is a large gravel deposit at least 40 m thick, with enclosed depressions on its southern surface and steep ice contact slopes on its southern margin (Fig. 10.22). McCabe et al. (1986) described three facies in this deposit from an exposure at Brookhill that indicate deposition

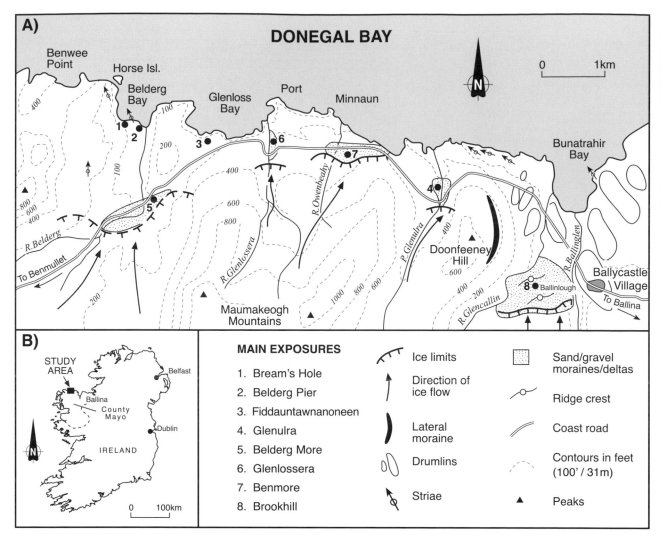

Fig. 10.22 Glacigenic features on the south side of Donegal Bay between Belderg and Ballycastle (from McCabe et al., 1986).
By kind permission of The Geological Society of America (GSA).

as a Gilbert-type delta. The basal facies consists of parallel laminated and rippled medium sand containing isolated cobbles and lenses of bimodal gravel consisting of rounded cobbles set in a sandy matrix (Fig. 10.26). The sand is overlain by pebbly foresets dipping north at 15–20°. The top of the gravelly foresets are between 74 and 78 m OD. Poorly-sorted topset beds of matrix-supported cobble and pebble gravel are contained within shallow scours.

Fifteen AMS ^{14}C radiocarbon dates from shells contained within the fossiliferous units exposed at Glenulra extend our understanding of glacial and relative sea level history from the west of Ireland by ~20,000 years beyond the LGM (Fig. 10.24). Only ages on foraminifera from the marine mud facies at Glenulra Farm are *in situ*, with the remainder being on individual fragments of reworked

shells. All ages on foraminifera from the marine mud are on hand-picked samples of pristine tests of *E. clavatum* which often had the last aperture intact, and they are in stratigraphic order. The ages on reworked shells range from 26,375±115 cal yr BP to 43,890±520 cal yr BP, but because they are reworked, these ages are not necessarily in stratigraphic order. The *in situ* ages on foraminifera at Glenulra Farm, however, provide a known-age datum against which ages on reworked shells from units above and below must stratigraphically conform. Accordingly, the fact that the ages on the uppermost unit are similar to or older than the ages of the underlying marine mud indicates that the event which caused deposition of this uppermost unit reworked pre-existing shells from some location upvalley after deposition of the marine mud. On the other hand, the ages on the lowermost unit must

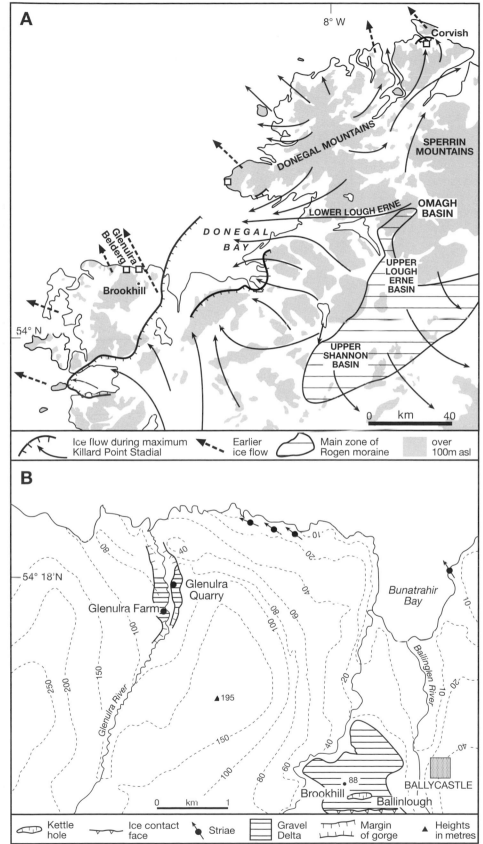

Fig. 10.23 Ice-sheet flow lines during the Killard Point Stadial from northwestern Ireland and sites discussed in text. Note the locations of exposures at Glenulra and Belderg which are as much as 30 km to the west of the Killard Point Stadial ice sheet limits (from McCabe et al., 2007a). *By kind permission of The Geological Society of America (GSA).*

A

8° W

Corvish

DONEGAL MOUNTAINS

SPERRIN MOUNTAINS

Glenulra Belderg

LOWER LOUGH ERNE

OMAGH BASIN

D O N E G A L
B A Y

UPPER LOUGH ERNE BASIN

Brookhill

54° N

UPPER SHANNON BASIN

0   km   40

Ice flow during maximum Killard Point Stadial	Earlier ice flow	Main zone of Rogen moraine	over 100m asl

B

54° 18'N

Glenulra Quarry

Glenulra Farm

Bunatrahir Bay

Glenulra River

▲195

Ballinglen River

▲88

Brookhill

BALLYCASTLE

Ballinlough

0   km   1

Kettle hole	Ice contact face	Striae	Gravel Delta	Margin of gorge / Heights in metres

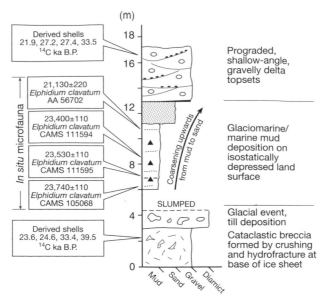

Fig. 10.24 Stratigraphic log from Glenulra Farm, north County Mayo (from McCabe et al., 2007a).
By kind permission of The Geological Society of America (GSA).

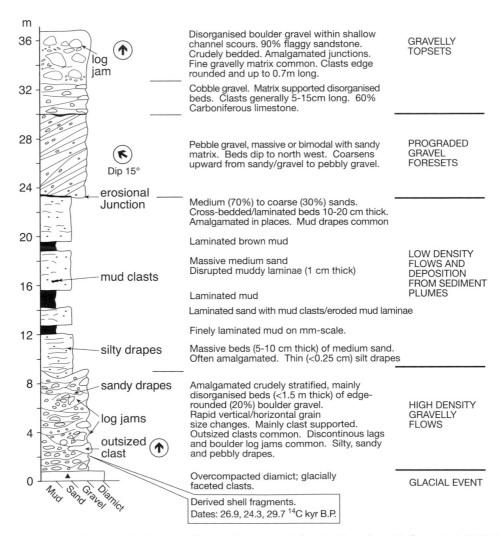

Fig. 10.25 Stratigraphic log from Glenulra Quarry, north County Mayo (from McCabe et al., 2007a).
By kind permission of The Geological Society of America (GSA).

Fig. 10.26 Bimodal gravel consisting of rounded cobbles set in a sandy matrix, Brookhill delta, north County Mayo.

be older than the marine mud simply to conform to the law of superposition. The four radiocarbon ages from the marine mud follow this rule.

Along the south coast of Donegal Bay at Glenulra, striae and streamlined bedrock record a northwesterly offshore ice advance onto the continental shelf (Fig. 10.18). The diamict preserved within the Glenulra valley also suggests an offshore advance. Because this evidence for glaciation is ~20 km west of the limit of the most recent major ice sheet readvance in western Ireland associated with the Killard Point Stadial (~15.5 cal ka BP) (Fig. 10.23) (McCabe et al., 1998, 2005), on stratigraphical grounds alone it represents an earlier glacial event. Because all ice-directional indicators demonstrate that ice flow was offshore, the shells from the diamict cannot be derived from Donegal Bay to the north but are probably derived from locations to the south of the Glenulra sites. These shells must therefore record an ice sheet situated on the western seaboard of Ireland with attendant isostatic loading sufficient to maintain high relative sea level for much of the ~20-kyr interval bracketed by our radiocarbon ages. This is consistent with [36]Cl dating and records of ice-rafted debris which suggest that several large fluctuations of the ice sheet occurred between ~40 and 25 ka (Bowen et al., 2002). It is not possible to distinguish from the data

whether relative sea level remained continuously above the Glenulra site from 26 to 45 cal ka BP or fluctuated to lower elevations in response to changes in eustatic level and ice loading. Nevertheless, when compared to the record of eustatic changes, the data indicates that the Donegal coast was isostatically depressed by 150 m to more than 180 m during much of this interval (Fig. 10.27). The youngest ^{14}C age in the lower diamict at Glenulra Farm indicates an ice advance across the site after 28,180±130 cal yr BP, whereas the oldest ^{14}C age from the overlying marine mud indicates ice retreat before 28,295±145 cal yr BP Accordingly, these ages constrain an extremely rapid ice-margin fluctuation. The age of this event indicates that it might correspond to a brief stadial event seen in the Greenland GISP2 ice core between interstadials 3 and 4 (Grootes et al., 1993; Stuiver and Grootes, 2000) (Fig. 10.27), although uncertainties in the ice-core chronology and in reservoir-age corrections preclude a firm correlation.

The four AMS ^{14}C ages obtained from the *in situ* marine microfauna indicate that after the ice margin retreated from the continental shelf, relative sea level remained high (>70 m above sea level) from 25,360±310 cal yr BP to 28,295±145 yr B.P (Fig. 10.27), suggesting the persistence of a thick ice mass over western Ireland throughout this time. The presence of thick, stratified

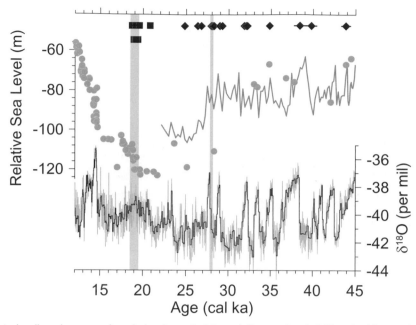

Fig. 10.27 A). Calibrated radiocarbon ages from Ireland constraining relative sea level at Glenulra (diamonds), Belderg (squares) and Kilkeel (triangles). B). Relative sea level data from far-field sites (Bard et al., 1993; Chappel et al., 1996; Cutler et al., 2003; Hanebuth et al., 2000; Siddall et al., 2003; Yokoyama et al., 2000). C). The [18]O record from Greenland Ice Sheet Project 2 (GISP20 ice core (Grootes et al., 1993; Stuiver and Grootes, 2000). *By kind permission of The Geological Society of America (GSA).*

and coarse-grained facies at Glenulra Quarry downvalley from the marine facies at Glenulra Farm suggests that the Glenulra Quarry facies were deposited in association with ice retreat from the coast. The lower three facies are subaqueous and record rapidly retreating ice from the immediate area, loss of accommodation space largely by sediment deposition, and progradation northwards of gravelly topsets. The range of sedimentary structures within the lower gravel indicates subaqueous deposition from high density flows along the valley (Lowe, 1982; Postma et al., 1988). The overlying horizontal sand and mud beds record low-density flows and quiet water sedimentation from sediment plumes. The third facies, consisting of northward-dipping foresets, records prograding coarser-grained sediment and decreasing water depths. The uppermost facies of boulder gravel was deposited within shallow channels deposited in shallow, high-energy channels. The foresets at this site record sea level at ~74 m above sea level. The ice-contact slopes and kettleholes along the southern margin of the Brookhill delta suggest that it was deposited in contact with an ice margin that had withdrawn only a few kilometers inland from Donegal Bay (McCabe et al., 1986). The gravelly foresets of the delta dip northwards and record a water plane infilling Bunatrahir Bay and the Ballinglen valley to ~78 m above sea level.

Bowen et al. (2002) first documented evidence for extensive pre-LGM ice cover over Ireland. In particular, [36]Cl ages indicate ice extent at ~35 [36]Cl ka that was similar to or perhaps greater than that associated with the LGM event. Records of ice-rafted debris provide additional temporal constraints that support the near-continuous presence of a marine ice margin off the coast of western Scotland since at least to 45 cal ka BP (Knutz et al., 2001). The ^{14}C ages confirm the presence of extensive ice for ~20 kyr prior to the LGM. Moreover, evidence for an ice advance over the Glenulra site at ~28 cal ka BP, as well as the evidence that Glenulra was never overridden again, indicates that at least for this sector of the ice sheet, the LGM margin was less extensive than the 28-kyr event. Although the western LGM margin was less extensive than the earlier advance, two lines of evidence suggest that it remained in Donegal Bay during the LGM. First, the presence of marine muds recording high relative sea level at Glenulra Farm requires ice loading until at least 25,650±330 cal yr BP Second, fossiliferous glacio-marine muds at the Belderg and Fiddauntawnanoneen sites ~5 km west of Glenulra have ^{14}C ages that range from 18,740±180 to 20,840±310 cal ka BP (Fig. 10.28) (McCabe et al., 2005). In this case, the marine sediments only constrain relative sea level to be at least 10 m above sea level, whereas the Glenulra record suggests that

BELDERG PIER, COUNTY MAYO, IRELAND

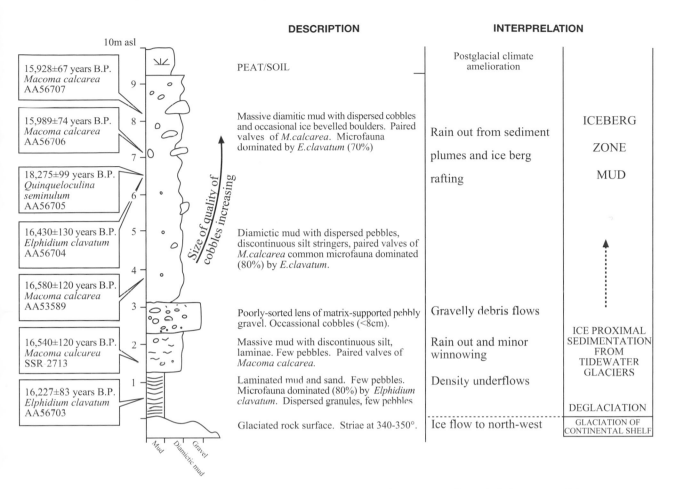

Fig. 10.28 Stratigraphic log and radiocarbon dates from Belderg Pier, north County Mayo (from McCabe et al., 2005).

it never exceeded the 80 m deltas there. Accordingly, because these muds were deposited during the LGM, when eustatic sea level was ~130 m lower than present (Yokoyama et al., 2000) (Fig. 10.27), isostatic depression was at least 140 m but not more than 210 m. The existing sea level constraints thus indicate significant isostatic depression of the south coast of Donegal Bay for much of the interval between ~40 cal ka BP until at least 19 cal ka BP, with little difference between pre-LGM and LGM ice loading.

The Belderg *Tellina* Clays

For over a century shelly drifts have been known from the Belderg area of north County Mayo on the south side of Donegal Bay west of Killala Bay. Early interpretations considered the Belderg '*Tellina* clays' were deposited from floating ice (Hinch, 1913)

though some (Synge, 1968; Colhoun, 1973; Davies and Stephens, 1978) considered the clays to be till deposited by an onshore ice movement of Scottish origin (e.g. Warren, 1985). These ideas are outdated because they are based largely on textural similarities between widely separated lithological beds or units, the mistaken concept that shelly drift is somehow related to Scottish ice, poor sedimentological parameters and the belief that individual beds can be correlated or compared over wide areas without geochronological control. Stratigraphic complexity within the deposits along the coastal segment between Ballycastle village and Benwee Point was recognised and explained by iceberg zone mud and ice contact deposition from tidewater glaciers around 17 ^{14}C kyr BP (McCabe et al., 1986). Later work has shown that the shelly drift at Belderg records a much later glacigenic event than the pre-LGM shelly deposits from Glenulra 5 km to the east

(McCabe et al., 2005). Therefore the hypothesis that the high relative sea levels around 80 m recorded by the raised marine deltas in this coastal sector at Glenulra and Brookhill are correlatives with the marine muds at Belderg cannot be sustained (Fig. 10.23). It is significant that the older fossiliferous muds at Glenulra are only preserved within a bedrock valley whereas the younger muds occur as a drape along the Belderg coast.

The schist bedrock around Belderg Pier was moulded and striated by northwesterly ice flow onto the continental shelf (Fig. 10.28). This surface is overlain directly by laminated mud and sand containing occasional pebbles, massive mud, lens-shaped bodies of cobble gravel and a massive tabular bed of diamictic mud which can be traced discontinuously for about 2 km along the coast. The basal part of the succession may record an ice-proximal environment because of the rapid facies changes. Rain-out, gravelly debris flows and density underflows are all typical processes operating when ice vacates a site, with textural variability reflecting continuously changing transport mechanisms (Powell, 1984). The thick and uniform aspect of the diamictic mud forming the upper part of the section is best interpreted as near-continuous rain-out from sediment plumes together with iceberg rafting and surpression of bottom current activity. There are at least twenty three species of microfauna which are dominated by *Elphidiun clavatum* and *Quinqueloculina seminulum*. These are typical members of the circum-Arctic, cold-water fauna especially in waters of lowered salinity adjacent to the snouts of glaciers (Loeblich and Tappan, 1953; Nagy, 1965). In addition all of the less common species such as *Buccella frigida* and Cibidides fletcheri also belong to the circum-Arctic cold-water fauna. The low diversity microfauna is very similar to those recorded by Nagy (1965) from shallow marine stations off Spitzbergen. Although the fauna contains both pristine and damaged specimens the occurrence of undamaged single valves of delicate ostracods suggests that little reworking has taken place.

McCabe et al. (1986) originally dated this site by conventional ^{14}C dating of paired valves of *Macoma calacrea* (16,540±120 ^{14}C yr BP). More recent AMS ^{14}C dates support the original date and record glaciomarine sedimentation between 16,580±120 and 15,928±70 ^{14}C yr BP (McCabe et al., 2005). However, one older date (18,275±100 ^{14}C yr BP) from *Quinqueloculina seminulum* differs from the other dates measured on both macro- and microfossils and may be derived from older sediments. The base of the glaciomarine sediments at Belderg Pier may be incomplete and do not record the onset of glaciomarine sedimentation. A date of 16,970±100 ^{14}C yr BP on *M. calcarea* from the eastern part of the mud drape at Fiddauntawnanoneen, one kilometre to the east, suggests earlier glaciomarine sedimentation. There is no doubt that the Belderg sediments postdate the last major ice sheet advance from northwestern Ireland onto the continental shelf and document glaciomarine sedimentation on an isostatically depressed land surface on the southern margin of Donegal Bay. Relative sea levels were in excess of 10 m but more probably in the region of 20–30 m in order to facilitate free movement of icebergs and the supply of cobble and pebble ice-rafted detritus to the site. This phase of high relative sea level occurred between 1000 and 2000 years after the initial deglaciation of the whole continental shelf when ice margins retreated landwards. It is therefore unlikely that glaciomarine conditions at this time were directly related to early deglaciation because of immediate and continued isostatic uplift as the ice thinned and loading was reduced. Because the ages of the Belderg sediments are similar to marine muds formed during a global eustatic rise ~19 cal kyr BP recorded from the Irish Sea Basin, it is likely that both are related to the same event. This correlation is supported by the fact that ice margins had contracted onto land following widespread early deglaciation and the apparent absence of any strong climatic forcing at this time which could have initiated farther glaciomarine sedimentation along the outer margins of Donegal Bay. The iceberg zone muds at Belderg may be related to destabilisation of the ice sheet margin which was probably grounded in the inner bay following the initial deglaciation. Although there are no ice contact sediments or moraines associated with this event, sedimentation from tidewater is clearly possible because of the proximity of the ice margin to the major centres of ice dispersal in the Omagh, Shannon and Erne basins (Fig. 10.23a).

The presence of age-constrained high relative sea levels and glaciomarine sedimentation in western Ireland raises a number of issues centred on the thickness of the ice sheet, its local history and whether or not the land surface was in equilibrium with ice sheet loading at any time since MOI 3. Isostatic effects of the last ice sheet in western Ireland cannot be overlooked, given that the large size (ca. 421,000 km[2] and perhaps 1000 m thick) of the ice sheet at 17 ^{14}C kyr BP. (McCabe et al., 1986). Isostatic depression cannot be entirely explained in terms of an ice cover over the present coast because much of the west had been already deglaciated by 17 [14]Ckyr BP. Possibly the high relative sea levels at this time can be explained by depression inherited from an earlier larger event which was partially maintained by ice on coasts near large centres of ice dispersal. Uplift may have been suppressed and the land surface was not in equilibrium with existing ice mass. Alternatively the general absence of a wide range of raised shoreline phenomena may be linked either to complex areal deformation patterns or to the presence of a thin ice cover on some western coasts following deglaciation. In a general sense the latter explanation is consistent with the idea that a global eustatic rise in sea level triggered deglaciation of remaining marine-based ice margins off the western Irish coast. Iceberg zone sedimentation similar to that recorded at Belderg would have been the most probable ice-marginal response. Rapid deglaciation of a continental shelf would normally be followed by ice-marginal re-equilibration within major bays and pinning points simply because of a decrease in the rate of ice wastage.

Ice sheet limits and sea level in Clew Bay

Northwest pointing horns on the composite subglacial bedforms at the head of Clew Bay record the last northwesterly ice flow across the bay and parallel the most recent set of glacial striae on the solid cores of some bedforms (Fig. 4.11) (McCabe et al., 1998). Bedform patterns inland from Westport show that the south to north ice flow lines of the final ice movement in western Ireland extended from Joyce's Country for 75 km to Crossmolina, clearly documenting the presence of a major ice dispersal centre in Connemara (Fig. 10.2). The moraine ridges on the northern side of the bay between Dooghbeg and Furnace Lough mark a terrestrial ice sheet limit and glacigenic sediments at Askillaun on the southern side record tidewater sedimentation (Figs. 10.5, 10.29). The scale (20–45 m thick) and lateral extent (continuous for 50 km) of these ridges are very similar to those on the southern margin of Donegal Bay and are similar in age. Satellite imagery shows that ice flows during the Killard Point Stadial were contiguous across lowland Ireland from Dundalk Bay to Clew Bay. Overprinting of earlier bedforms by the latest ice flow shows that the flow lines of this ice sheet system are coeval with the moraines bordering the ice sheet in Clew Bay. The tidewater sequence at Askillaun is therefore part of the Killard Point Stadial and represents sediment transfer

and output from this readvance system. The sediment pile at Askillaun must be associated with a regional sea level in excess of 45 m OD.

Originally Hanvey (1988) and Warren (1992) mapped drift ridges along the southern margin of Clew Bay as drumlins which implies that they were carved by ice flowing west onto the continental shelf. However, drift ridges that exist in the area around Old Head and Askillaun are not typical drumlins because they have been carved from a sediment spread and are bounded by scars of meltwater erosion, their internal structure is stratified and undisturbed, they are intimately related to adjacent spreads of stratified diamict recording subaqueous deposition, they do not form either a linear group or a swarm and they show no obvious relationship either to regional ice flow or to the composite bedforms at the head of the bay. The internal geometry of the large sections along the coast is very similar, recording similar depositional settings. At Askillaun, the exposure consists mainly (75%) of stacked sheets of diamict that dip consistently westwards at ~5°. Although diamict beds (1–2 m thick) are massive they contain numerous sandy wisps, laminated beds, outsized clasts (<1.5 m across) and in some cases coarse-tail grading patterns. Individual beds are traceable for tens of metres along section. The crude stratification of the diamicts is emphasised by lenses of pebble gravel a few metres in length and more discrete erosional channel margins up to 5 m across infilled with stratified gravel. At intervals in the succession laminated muds up to 0.5 m thick contain small pebbles and extend for 2–4 m. The most noticeable characteristic feature of the succession are the laterally discontinuous (4–20 m) boulder and cobble lines, one clast thick, that occur at vertical intervals of 3–5 m. The upper clast surfaces are striated, abraded and flattened and although the clast lines dip gently west some lines are wavy to irregular. Sediment transitions are common between gravel and diamict end members. The westward dips are primary and unrelated to glaciotectonics, suggesting progressive sediment release and buildup from an ice margin immediately to the east pinned on the rock ridges around Louisburgh.

Because sediments are up to 45 m thick it is difficult to generate and accommodate this thickness either subglacially or beneath a debris deficient ice shelf. The sediment geometry, rapid facies transitions, channel infills, textural variability, crude stratification and grading are all typical of ice contact, glaciomarine deposition. The presence of striated boulder lines strongly suggests

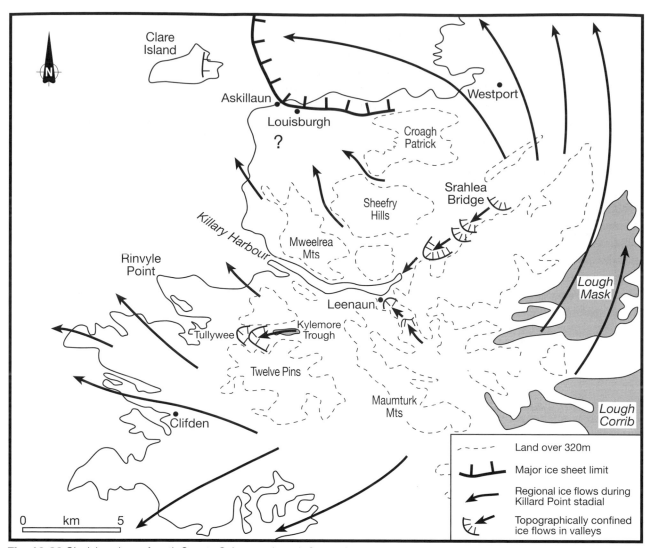

Fig. 10.29 Glacial geology of north County Galway and south County Mayo showing general ice flows and ice limits during the Killard Point Stadial (after Thomas and Chiverrell, 2006). *By permission of Elsevier.*

that the ice periodically readvanced locally over the growing sediment pile possibly on a seasonal basis. A good analogy consisting of diamicts, boulder pavements and stratified deposits had been described from the Yatataga Formation in Alaska by Eyles (1988). The final ice sheet margin lay to the east of this site and local ice flow indicators show that it is impossible to reconstruct an ice dam across the bay which might have created a water body and a depositional sink for these sediments. Therefore it is likely that the sequences are glaciomarine and associated with high relative sea level in excess of 45 m OD. On the basis of the bedform patterns and age of the limiting moraines it is inferred that this occurred when ice readvanced northwestward across Clew Bay during the Killard Point Stadial. Although marine shell

fragments occur in the diamicts, none have been radio-carbon dated. High relative sea levels in the west of Ireland must be expected during the period of initial deglaciation because of the existence of the large centre of ice dispersal in Connemara that was sufficiently powerful to advance north to Ballina (Fig. 1.4). However, because there are no obvious raised beach landforms, this scenario suggests that very rapid uplift followed rapid deglaciation of the western coastline.

Ice contact deltas in Killary Harbour and the Erriff Valley

Killary Harbour is a glacial trough about 7 km long opening directly (westward) into the Atlantic Ocean.

Eastwards the trough continues in a more subdued form along the Erriff valley (Fig. 1.9a). This overdeepened trough is bounded by compartmentalised mountain blocks up to 820 m OD and is fed by smaller valleys some of which end in corries. At Leenaun the small valley leading southeastwards from the southern margin of Killary Harbour is choked with sand and gravel associated with a halt in ice-marginal retreat (Fig. 10.30). A major feature in the valley consists of a terrace surface at 78 m IOD which ends in a cross-valley moraine fed by eskers along the valley axis (Fig. 10.31) (Thomas and Chiverrell, 2006). Sections in the terrace show that basal till passes upwards into 1.5 m of parallel laminated sand and clay containing lonestones termed basin floor sediment and 10 m of large-scale, planar gravelly foresets dipping 15–20° to the northwest (Thomas and

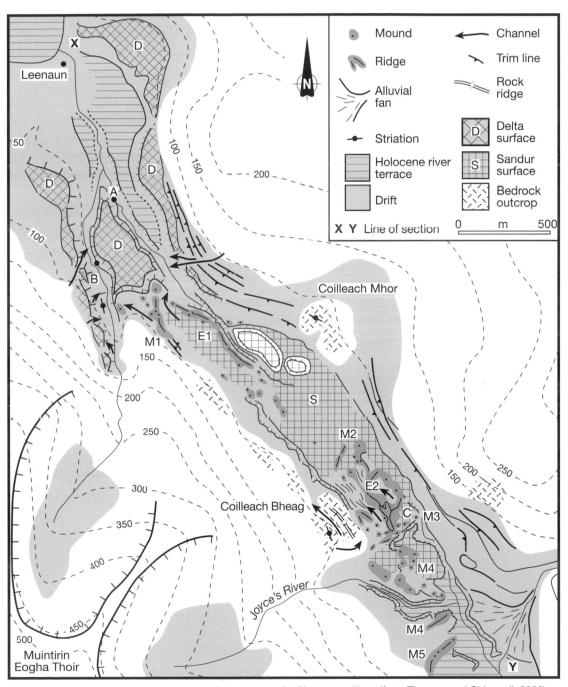

Fig. 10.30 Geomorphological map of the area south of Leenaun village (from Thomas and Chiverrell, 2006).
By permission of Elsevier.

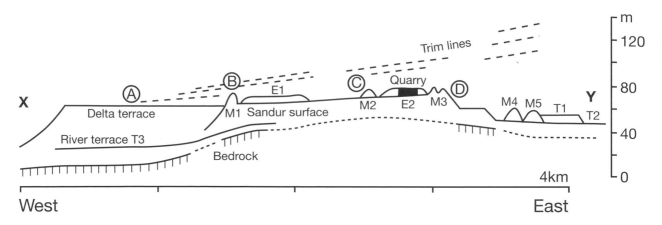

Fig. 10.31 The Leenaun fan-delta. Down-valley section along line X–Y shown in figure 10.30 shows the relationships between the delta terrace, trim lines, cross-valley moraines (M1–5), sandur surfaces (S1–2), eskers (E1–2) and Holocene river terraces (T1-3). (from Thomas and Chiverrell, 2006). *By permission of Elsevier.*

Fig. 10.32 The Leenaun delta terrace surface and cross-valley moraines viewed from the north shore of Killary Harbour. The height of the apex of the outer part of the delta terrace is at 78 m Irish OD (Thomas and Chiverrell, 2006).

Chiverrell, 2006). The terrace surface is underlain by thin (<1 m), flat-lying beds of pebble gravel. The basal till and local striation pattern is consistent with ice flow from the major centre of ice dispersal in Joyce's Country which is of Killard Point Stadial age. Overlying rhythmites record density flows and iceberg zone sedimentation once ice vacated the site and the main part of the foresets formed during progradation of a Gilbert-type delta fed by esker tunnels from the southeast (Fig. 10.32). Farther ice retreat southeastwards along Joyce's River saw abandonment and dissection of the delta when water levels fell.

Eastwards, the Erriff River occupies a narrow valley 15 km long which opens out into the Irish lowlands. Three kilometres upvalley, Colonel's Pool Quarry is cut into a cross-valley, ice-contact moraine marking a halt in ice retreat to the east (Fig. 10.29). Exposures show sediments within four troughs which are stacked and incised up to 30 m wide, 60 m long and 10 m thick (Thomas and Chiverrell, 2006). Facies infilling the troughs consist of graded pebble gravel, sheets of gravel with parallel laminated sand, ungraded pebbly gravel and isolated lenses and clusters of cobbles. Matrix-rich diamicts also occur. Gravelly slump folds occur at the base of the troughs.

Overall the sequences are consistent with subaqueous sedimentation built downvalley from a grounded ice sheet margin. The sediments aggradated in water with a minimum surface at least 35–40 m IOD. Two kilometres upstream at Glennacally Bridge a delta surface lies at 70 m IOD probably formed in the same water body (Thomas and Chiverrell, 2006). Farther upvalley opposite Srahlea Bridge three quarries in terraces contain similar sedimentary assemblages. These are dominated by poorly-sorted gravel mounds (tunnel fills) buried by graded gravels and sands (trough fills), foreset sands and gravels and thin topset beds. Thomas and Chiverrell (2006) demonstrate that the Srahlea exposures record complex subaqueous sedimentation at the ice sheet margin as it retreated eastwards into central Ireland. Esker sediments are buried by tunnel mouth sediments and foresets. Only in isolated cases did vertical accretion reach the shoreline height of 78 m IOD during evolution into a delta (Fig. 10.33).

Only at Leenaun and Srahlea does vertical sediment accumulation reach the water line to form ice-contact deltas. The pattern of water levels at heights of 78 m IOD for around 20 km in the Erriff valley upvalley from Killary Harbour cannot be easily explained by an

Fig. 10.33 Srahlea Quarry section, 20 m high. Two sets of delta foresets dipping south (1) and west (2) respectively, and overlain by sharply unconformable delta topsets. Photograph taken 30 m up-current of delta break-point at a height of 75 m IOD (after Thomas and Chiverrell, 2006).

ice dam because the last ice sheet advanced from the southeast. In this respect the Killary Harbour/Erriff trough system was to a large extent shielded by the SW–NE trend of the Maumturk/Partry Mountain range from extensive glacierisation related to the large dispersal centre in Joyce's country to the south. Bedform patterns show that the mountains deflected the main ice flows northeastwards towards Castlebar. Therefore, troughs were probably occupied by valley glaciers and the deltas may simply record glacial limits during the Killard Point Stadial. Thomas and Chiverrell (2006) consider several explanations but arrive at the conclusion that the water levels must be related to high relative sea levels during deglaciation. They place emphasis on the fact that the directional elements of the deltas slope consistently westwards, individual sites face full frontal into the open Atlantic Ocean and a common water level between 60 and 80 m IOD is more likely related to former sea level against a retreating ice margin rather than glaciolacustrine levels which would be variable in a valley setting.

Ice limits and the Tullywee delta

Flat-topped gravelly drift terraces occur near the western margin of the Kylemore trough at Tullywee Bridge (Warren, 1988; Coxon and Browne, 1991; McCabe and Dardis, 1992; Thomas and Chiverrell, 2006). The first two interpret the available sections as Gilbert-type deltas and the latter two as subaqueous fans. Thomas and Chiverrell (2006) show that the ice-marginal terraces were fed from subglacial tunnels now preserved as eskers along the floor of the Kylemore trough (Fig. 10.29). Basal sediments in the succession consist of crudely-stratified cobble and boulder gravel arranged in a sequence of anticlines. Although Thomas and Chiverrell (2006) considered these to be a result of drape folding and faulting it is possible that they are mainly a result of megadune migration along tunnels towards meltwater exits. Overlying facies are variable but include large trough-bedded sets of cobble gravel, normally graded beds of pebble gravel and laminated sand, interbedded lenses of chaotically-bedded cobble, parallel laminated sand and silt with outsized clasts and boulder gravel, diamict and muddy sediment. Probably the sediments accumulated close to the exit of a major glacial efflux which vented into standing water. Facies variability is a result of high energy gravelly flows which translated laterally into less dense flows. Stacking of debris from repeated mass flows may be a result of either slope failure as sediment piles grew at the grounding line or from concentrated sediment discharges from the subglacial tunnels. Although some facies resemble foresets of a Gilbert-type delta the overall facies assemblage is more typical of a grounding line spread that failed to evolve into a Gilbert-type delta. This inference suggests that the sediment pile grew rapidly and there was insufficient time available to develop vertically into a Gilbert-type delta during ice sheet grounding. Thomas and Chiverrell (2006) suggest that the minimum water-surface height of the fronting water body was at least 60–65 m IOD though lower terraces at 35 and 45 m IOD may record lowering of the water surface.

Thomas and Chiverrell (2006) clearly show that the subaqueous fans extend west across the Dawros river basin towards the adjacent open Atlantic ocean. They argue that a glaciolacustrine origin for the Tullywee water level, at a minimum of 65 m IOD, is extremely unlikely as this site faces full frontal into the open Atlantic Ocean. Only a tortuous and highly improbable ice margin involving an eastward ice movement from the continental shelf would be able to impound water in the vicinity of Tullywee. These authors confirm the original suggestion of Coxon and Browne (1991) and McCabe and Dardis (1992) that the deposits record high relative sea levels during ice wastage. The very presence of thick drift sequences positioned at the western exits of glacial troughs such as Kylemore undoubtedly record important ice sheet limits. Moreover the drifts occur in areas which are otherwise deeply-eroded, ice scoured and generally drift free. Regional patterns of ice directed landforms strongly suggest that ice at this time was largely confined to valleys as outlet-type glaciers from the main ice sheet on the adjacent lowlands and mountains. The gross morphology of the compartmentalised mountain blocks with major through valleys and cols facilitated the westward expanse of outlet glaciers towards the coastline. If this is the case then the deltas and fans represent a readvance of the ice margin rather than a simple withdrawal from an earlier LGM type limit, 30 or 60 km out on the continental shelf. Satellite imagery shows that the regional ice sheet flow lines leading to this ice limit are consistent with the development of a major centre of ice dispersal in Joyce's Country during the Killard Point Stadial (McCabe et al., 1998).

Ice sheet history and sea levels

Bowen (1991) concluded that well over a century after their formulation, some of the ideas of early workers on the origins of glacigenic drifts still have considerable relevance for fundamentals of glaciation in the British Isles. He concluded that many 'glacial deposits' should be re-evaluated in a glaciomarine paradigm. More specifically, topics such as the role of sea level in the build-up, extent and decay of ice sheets in the British Isles including the degree of isostatic depression, were fundamentals in any interpretation of glacial systems near Atlantic margins. Bedform patterns and reconstructed ice sheet limits clearly show that Ireland was ice-covered during maximum glaciation and estimates of ice thickness exceed 700 m and approach 1000 m (McCabe et al., 1986; Ballantyne et al., 2006, 2007). The presence of depocentres containing sediments recording ice contact, glaciomarine sequences and high relative sea levels are found along southern, eastern, northeastern, northwestern, and western coasts. Their apparent absence along the northern coast may reflect the longer duration of ice cover and the late readvance of Scottish ice across the coastline of County Antrim. The general model depicts ice withdrawal from the continental shelf because of deep isostatic depression and the fact that ice sheet flowlines were at their most sensitive and longest at this time. As ice margins re-equilibrated on land margins, complex and rapidly changing facies successions formed in peripheral troughs and topographic lows. Most of these deltas and subaqueous fan systems formed over short periods and invariably document sediment point sources near ice limits. This model is improved when a chronology based on eight well-dated sites is added to ice sheet events and sea level evidence, providing a more holistic picture of changes in ice sheet configuration, mobility and history (Fig. 10.1).

The fact that Amino Acid, radiometric and relative dating frameworks are now available to constrain relative sea levels, means that isostatic responses and some eustatic interactions can be traced over the course of the last glacial cycle between 30 and 14 cal ka BP. At least six major isostatic provenances are recognised from the sedimentary records and these are closely linked with ice sheet mobility and history:

1. The Genulra site confirms the presence of extensive ice for ~20 kyr prior to the LGM, an ice advance over the site at ~28 cal ka BP, and similar ice loading until ~25 cal ka BP. Isostatic depression of up to 180 m is dependent on the build up of large ice sheets in the west early in the last glaciation. The transport of Galway granite erratics (Warren, 1991) and Tyrone igneous erratics eastwards is also consistent with the early build-up of extensive ice masses across extensive areas of western Ireland because these erratics are commonly found reworked in tills associated with later ice moving westwards. The large esker system across the Irish Midlands between the Derries and Athlone recording westerly ice sheet decay is also strong evidence that large centres of ice dispersal in the west remained active during early deglaciation.

2. Instantaneous lowland glacierisation and major shifts in the centres of ice dispersal eastwards occurred immediately after initial ice sheet development in the west. This phase must have occurred before the main period of early deglaciation ~21–20 cal ka BP and approximates to what is traditionally known as the LGM (LGM = 19–23 cal ka BP; Mix et al., 2001). In the west ice lobes remained within the larger bays but did not extend far offshore, probably because of sustained isostatic depression and accelerated ice loss along marine embayments. The Glenulra site confirms that the pre-LGM ice was more extensive than the later LGM limits. Bedform patterns show that ice from lowland centres of ice dispersal filled the Irish Sea Basin in excess of 800 m (Ballantyne et al., 2006) and extended for a few kilometres off the south coast. Clearly the main focus of isostatic loading increased eastwards as glaciation proceeded and was accompanied by a sensitive lithosphere response.

3. Early deglaciation and loss of over two thirds of the ice sheet occurred before 20 cal ka BP, when ice masses contracted onto land margins from the continental shelf and Irish Sea Basin. The location of temporary ice-marginal halts during ice wastage along the Irish Sea Basin records sediment wedges offlapping from south to north on sub-millennial timescales (McCabe, 1996). Deep isostatic depression with estimates in excess of 170 m persisted in central parts of the basin because global eustatic sea levels were possibly at ≥140 m at this time. Timing of early deglaciation is constrained by radiocarbon dates in northwestern, eastern and western ice sheet sectors.

4. Deglaciation of the northern Irish Sea Basin was followed by isostatic uplift allowing meltwaters to cut subaerial channels graded to below sea level. Because a global meltwater pulse at ~19 cal ka BP. was able to fill these channels with marine mud, the rate of isostatic uplift was suppressed locally, possibly because deglaciation of the basin was extremely rapid and residual ice masses remained a short distance inland. Glaciomarine deposits at Belderg record minimum sea levels of ~20 m IOD and are probably associated with destabilisation of the ice sheet on the margins of Donegal Bay. These data from east central and northwestern ice sheet sectors demonstrate that deep isostatic depression still existed on the periphery of the full ice sheet while most of the northern half of the island lay beneath the remaining ice sheet. A net isostatic uplift of ~170 m is estimated for the Mourne coast and 100 m for Donegal Bay based on comparisons between relative sea levels and relative sea level data from far-field sites (Bard et al., 1993; Siddall et al., 2003; Yokoyama et al., 2000). This difference in the net amount of isostatic depression may demonstrate the general eastward shift in the main centres of ice dispersal and the averaged centre of ice sheet dispersal.

5. Renewed ice sheet growth followed widespread early deglaciation and culminated in at least two regional ice sheet readvances which effected drumlinisation during the Heinrich 1 interval. This period (~18.5 cal ka BP) immediately prior to ice sheet readvance was marked by marine sedimentation with sea levels ~5 m higher than present in east central and northwestern ice sheet sectors. During readvance maximum in these ice sheet sectors isostatic depression increased relative sea levels to ~30 m OD. However, in western Ireland relative sea levels up to ~80 m (Thomas and Chiverrell, 2006) record even greater amounts of isostatic depression exceeding 180 m. The latter may reflect exceptional ice growth in Connemara with ice sheet limits reaching the outer coastal fringe.

6. Data from the north Irish Sea Basin show that ice margins remained at terminal limits for ~1000 years before stagnation zone retreat followed. Dates from Rough Island record marine inundation up to 19 m OD of the drumlin lowlands in northeastern ice sheet sectors ~15 cal ka BP while the area was still isostatically depressed. It is argued that the absence of any recessional moraines in eastern Ulster is a signature for very rapid ice wastage and this is consistent with the extensive marine transgression recorded by marine clays draping subglacial bedforms. This phase of high relative sea level could well reflect a eustatic sea level component immediately prior to MWP 1a because renewed isostatic uplift must have been well underway at this time. Raised gravel barriers and emergent facies sequences in northeastern Ireland (~12 m OD) record falling sea level at 14.5 cal ca BP eventually leading to a marked lowstand (-20 m) at 13.5 cal ka BP in Belfast Lough.

The ice sheet model for Ireland depicting persistence of high relative sea levels around the decaying ice sheet is based largely on direct and age-constrained field evidence. The wealth of data recording deep isostatic depression for 20,000 years is at variance with the predictions of geophysical models which only partially capture the much later (~10 ka) postglacial sea levels (Lambeck, 1996; Lambeck and Purcel, 2001). The models do not have the sensitivity or sophistication to predict (i.e. sea levels) which is one of the essential features of any general model. McCabe et al., (2005) stressed that these standard models are oversimplified and unrelated to field evidence because they are based on static ice sheet concepts, minimum ice sheet thickness, incorrect ice sheet limits, an amalgam of generalised LGM flowlines of mixed ages, assumed monotonic ice retreat and a slowly responding lithosphere. It is also significant that the geophysical models tend to use postglacial sea level data retrogressively back into the deglacial period. It is now known that deglaciation was characterised by very rapid environmental change when lithospheric and sea level changes oscillated on sub-millennial timescales. It is argued therefore that the key to the combined ice sheet and sea level model for Ireland was ice sheet mobility during the course of glaciation and rapid changes in ice sheet configuration. The presence of discrete isostatic provenances around the island is strongly supportive of a dynamic coupling between ice and crust.

BIBLIOGRAPHY

Anderson, E., Harrison, S., Passmore, D. G. and Mighall, T. M. 1998. Geomorphic evidence of Younger Dryas glaciation in the Macgillycuddy's Reeks, south west Ireland. *In:* Owen, L. A. (Ed.), *Mountain Glaciation*, Quaternary Proceedings No. 6, Chichester, Wiley, 75–90.

Andrews, J. T. 1978. Sea level history of Arctic coasts during the upper Quaternary: dating, sedimentary sequences and history. *Progress in Physical Geography*, **2**, 185–199.

Andrews, J. T., King, C. A. M. and Stuiver, M. 1973. Holocene sea level changes, Cumberland coast, northwest England: eustatic and glacioisostatic movements. *Geologie en Mijnbouw*, **52**, 1–12.

Apjohn, J. 1841. Address delivered at the tenth annual meeting of the Geological Society. *Journal of the Geological Society*, Dublin, 1839–1842, 131–165.

Atkinson, T. C., Briffa, K. R., Coope, G. R., Joachim, M. J. and Perry, D. W. 1986. Climatic calibration of Coleopteran data. *In:* Berglund, B. E. (Ed.), *Handbook of Holocene Palaeoecology and Palaeohydrology*, J. Wiley and Son, London, 815–858.

Austin, W. E. N. and McCarroll, D. 1992. Foraminifera from the Irish Sea glacigenic deposits at Aberdaron, western Lleyn, North Wales: palaeoenvironmental implications. *Journal of Quaternary Science*, **7**, 311–317.

Aylsworth, J. M. and Shilts, W. W. 1989. Bedforms of the Keewatin ice sheet. *Sedimentary Geology*, **62**, 407–428.

Ballantyne, C. K., McCarroll, D. and Stone, J. O. 2006. Vertical dimensions and age of the Wicklow Mountains ice dome, eastern Ireland and implications for the extent of the last Irish ice sheet. *Quaternary Science Reviews*, **25**, 2048–2058.

Ballantyne, C. K., McCarroll, D. and Stone, J. O. 2007. Ice over Ireland. QRA annual discussion meeting, 2007, St. Andrews Handbook, p.7.

Bard, E., Hamelin, B., Fairbanks, R. G. and Zindler, A. 1990. Calibration of the ^{14}C timescale over the past 30,000 years using mass spectrometric U–Th ages from Barbados corals. *Nature*, **345**, 405–410.

Bard, E., Arnold, M., Fairbanks, R. G. and Hamelin, B. 1993. ^{230}Th–^{234}U and ^{14}C ages obtained by mass spectrometry on corals. *Radiocarbon*, **35**, 191–199.

Bard, E., et al., 1994. The North Atlantic atmosphere-sea surface ^{14}C gradient during the younger Dryas climatic event. *Earth and Planetary Science Letters*, **126**, 275–287.

Bard, E., Rostek, F., Thuron, J. L. and Gendreau, S. 2000. Hydrological impacts of Heinrich events in the subtropical Northeast Atlantic. *Science*, **289**, 1321–1323.

Blundell, D. J., Davey, F. J. and Graves, L. J. 1971. Geophysical surveys over the south Irish Sea and Nymphe Bank. *Journal of the Geological Society, London*, **127**, 339–375.

Bond, G. and Lotti, R. 1995. Iceberg discharges into the North Atlantic on millennial timescales during the last deglaciation. *Science*, **267**, 1005–1010.

Bond, G., Broecker, W. S., Johnsen, S., McManus, J., Labeyrie, L., Jouzel, J. and Bonani, G. 1993. Correlations between climate records from North Atlantic sediments and Greenland ice. *Nature*, **365,** 143–147.

Bond, G., et al., 1997. A pervasive millennial-scale cycle in North Atlantic Holocene and glacial climates. *Science*, **278**, 1257–1266.

Bond, G., Showers, W., Elliot, M., Evans, M., Lotti, R., Haslda, I., Bonani, G. and Johnson, G. 1999. The North Atlantic's 1–2 kyr climate rhythm: relation to Heinrich events, Dansgaard/Oescher Cycles and the Little Ice Age. *In:* Clark, P. U., Webb, R. S. and Keigwin, L. D. (Eds.), *Mechanisms of Global Climate Change at Millennial Timescales*, Geophysical Monograph 112, American Geophysical Union, Washington, DC, 35–58.

Bondevik, S., Mangerud, J., Birks, H., Guilliksen, S. and Reimer, P. 2006. Changes in north Atlantic radiocarbon reservoir ages during the Allerod and younger Dryas. *Science*, **312**, 1515–1517.

Bouchard, M. A. 1989. Subglacial landforms and deposits in central and northern Québec, Canada, with emphasis on Rogen moraines. *Sedimentary Geology*, **62,** 293–308.

Boulter, M. and Mitchell, I. 1977. Middle Pleistocene (Gortian) deposits from Bemburb, Northern Ireland. *Irish Naturalist's Journal*, **19**, 2–3.

Boulton, G. S., Jones, A. S., Clayton, K. M. and Kenning, M. J. 1977. A British ice sheet model and patterns of glacial erosion and deposition in Britain. *In:* Shotton, F. W. (Ed.), *British Quaternary Studies-Recent Advances*. Clarendon Press, Oxford, 231–246.

Bowen, D. Q. 1978. *Quaternary Geology: A Stratigraphic Framework for Multidisciplinary Work*. Oxford University Press, 221pp.

Bowen, D. Q. 1991. Time and space in the glacial sediment systems of the British Isles. *In:* Ehlers, J., Gibbard, P. L. and Rose, J. (Eds.), *Glacial Deposits in Great Britain and Ireland*, Balkema, Rotterdam, 3–13.

Bowen, D. Q. 1994. The Pleistocene history of northwest Europe. *Science Progress*, **76**, 209–223.

Bowen, D. Q. (Ed.), 1999. *A revised correlation of Quaternary deposits in the British Isles*. Geological Society Special Report 23, Geological Society, London.

Bowen, D. Q., Philipps, F. M., McCabe, A. M., Knutz, P. C. and Sykes, G. A. 2002. New data for the last glacial maximum in Great Britain and Ireland. *Quaternary Science Reviews*, **21**, 89–101.

Bowen, D. Q., Richmond, G. M., Fullerton, D. S., Sibrava, V., Fulton, R. J. and Velichko, A A. 1986a. Correlation of Quaternary glaciations in the northern hemisphere. *Quaternary Science reviews*, 5, 509–510.

Bowen, D. Q., Rose, J., McCabe, A. M. and Sutherland, D. G. 1986b. Correlation of Quaternary glaciations in England, Ireland, Scotland and Wales. *Quaternary Science Reviews*, 5, 299–340.

Brenchley, P. J. 1985. Storm influenced sandstone beds. *Modern Geology*, **9**, 369–396.

Brennand, T. A. 1994. Macroformss, large bedforms and rhythmic sedimentary sequences in subglacial eskers, south-central Ontario: implications for esker genesis and meltwater regime.

Sedimentary Geology, **91**, 9–55.

Carter, R. W. G. 1993. Age, origin and significance of the raised gravel barrier at Church Bay, Rathlin Island, County Antrim. *Irish Geography*, **26**, 141–146.

Chappel, J., Omura, A., Esat, T., McCulloch, M., Pandolfi, J., Ota, Y. and Pillans, B. 1996. Reconstruction of late Quaternary sea levels derived from coral terraces at Huon Peninsula with deep sea oxygen isotope records. *Earth and Planetary Science Letters*, **141**, 227–236.

Charlesworth, J. K. 1924. The glacial geology of the north-west of Ireland. *Proceedings of the Royal Irish Academy*, **36B**, 174–314.

Charlesworth, J. K. 1926. The Evishanoran esker, Co. Tyrone. *Geological Magazine*, **63**, 223–225.

Charlesworth, J. K. 1928a. The glacial geology of north Mayo and west Sligo. *Proceedings of the Royal Irish Academy*, **38B**, 100–115.

Charlesworth, J. K. 1928b. The glacial retreat from central and southern Ireland. *Quarterly Journal of the Geological Society of London*, **84**, 293–344.

Charlesworth, J. K. 1929. The glacial retreat from Iar Connacht. *Proceedings of the Royal Irish Academy*, **39B**, 95–106.

Charlesworth, J. K. 1937. A map of the glacier-lakes and local glaciers of the Wicklow hills. Proceedings of the Royal Irish Academy, **44B**, 29–36.

Charlesworth, J. K. 1939. Some observations on the glaciation of north-east Ireland. *Proceedings of the Royal Irish Academy*, **45B**, 255–295.

Charlesworth, J. K. 1955. The Carlingford readvance between Dundalk, County Louth and Kingscourt and Lough Ramor, County Cavan. *Irish Naturalist's Journal*, **11**, 299–302.

Charlesworth, J. K. 1963. Historical Geology of Ireland, Oliver and Boyd, Edinburgh, 565pp.

Clark, C. D. and Meehan, R. T. 2001. Subglacial bedform geomorphology of the Irish ice sheet reveals major configuration changes during growth and decay. *Journal of Quaternary Science*, **16**, 483–496.

Clark, J., McCabe, A. M., Schnabel, C., Freeman, S., Maden, C. and Xu, S. 2006. New constraints on the deglaciation of the western margin of the British-Irish ice sheet, Ireland, from [10]Be dating. *Geophysical Abstracts*, 8, 10272.

Clark, J., McCabe, A. M., Schnabel, C., Clark, P. U., Freeman, S., Maden, C. and Xu, S. 2007a. A cosmogenic chronology of the last deglaciation of western Ireland (in prep).

Clark, J., McCabe, A. M., Schnabel, C., Clark, P. U., Freeman, S., Maden, C. and Xu, S 2007b. [10]Be chronology of the last deglaciation of County Donegal, northwestern Ireland (in prep).

Clark, P. U. and Hansel, A. K. 1989. Till lodgement, clast ploughing and glacier sliding over a deformable glacier bed. *Boreas*, **18**, 201–207.

Clark, P. U. and Pisias, N. G. 2000. Interpreting iceberg deposits in the deep sea. *Science*, **290**, 51.

Clark, P. U. and Walder, J. S. 1994. Subglacial drainage, eskers and deforming beds beneath the Laurentide and Eurasian ice sheets. *Geological Society of America Bull*etin, **106**, 304–314.

Clark, P. U., McCabe, A. M., Mix, A. C. and Weaver, A. S. 2004. Rapid rise of sea level 19,000 years ago and its global implications. *Science*, **304**, 1141–1144.

Clark, P. U., Hostetler, S. W., Pisias, N. G., Schmittner, A. and Meissner, K. J. 2007. Mechanisms for a ~7–kyr climate and sea-level oscillation during marine isotope stage 3 (in prep).

Close, M. H. 1866. Notes on the general glaciation of Ireland. *Journal of the Royal Geological Society of Ireland*, **1**, 1864–1867, 207–242.

Cohen, J. M. 1979. Deltaic sedimentation into Glacial Lake Blessington, Co. Wicklow, Ireland. *In:* Schlüchter, Ch. (Ed.), *Moraines and Varves*. Balkema, Rotterdam, 357–367.

Colhoun, E. A. 1970. On the nature of the glaciations and final deglaciation of the Sperrin Mountains and adjacent areas in the north of Ireland. *Irish Geography*, **6**, 162–185.

Colhoun, E. A. 1971a. Late Weichselian periglacial phenomena of the Sperrin mountains, Northern Ireland. *Proceedings of the Royal Irish Academy*, **71B**, 53–71.

Colhoun, E. A. 1971b. The glacial stratigraphy of the Sperrin Mountains and its relation to the glacial stratigraphy of north-west Ireland. *Proceedings of the Royal Irish Academy*, **71b**, 37–52.

Colhoun, E. A. 1972. The deglaciation of the Sperrin mountains and adjacent areas in Counties Tyrone, Londonderry and Donegal, Northern Ireland. *Proceedings of the Royal Irish Academy*, **72B**, 91–137.

Colhoun, E. A. 1973. Two Pleistocene sections in south-western Donegal and their relation to the last glaciation of the Glengesh plateau. *Irish Geography*, **6**, 594–609.

Colhoun, E. A. and McCabe, A. M. 1973. Pleistocene glacial, glaciomarine and associated deposits of Mell and Tullyallen townlands, near Drogheda, eastern Ireland. *Proceedings of the Royal Irish Academy*, **73**, 165–206.

Colhoun, E. A. and Mitchell, F. 1971. Interglacial marine formation and lateglacial freshwater formation in Shortalstown townland, County Wexford. *Proceedings of the Royal Irish Academy*, **71**, 211–245.

Colhoun, E. A. and Synge, F. M. 1980. The cirque moraines at Lough Nahanagan, County Wicklow, Ireland. *Proceedings of the Royal Irish Academy*, **80B**, 25–45.

Colhoun, E. A., Dickson, J. H., McCabe, A. M. and Shotton, F. W. 1972. A Middle Midlandian freshwater series at Derryvree, Maguiresbridge, County Fermanagh, Northern Ireland. *Proceedings of the Royal Society of London, series B*, **180**, 273–292.

Coope, G. R. 1968. An insect fauna from Mid-Weichselian deposits at Brandon, Warwickshire. *Philosophical Transactions of the Royal Society of London*, **244B**, 389–412.

Coxon, P. 1993. Irish Pleistocene biostratigraphy. *Irish Journal of Earth Sciences*, **12**, 83–105.

Coxon, P. 1996. The Gortian temperate stage. *Quaternary Science reviews*, **15**, 425–436.

Coxon, P. (Ed.), 2005. The Quaternary of Central Western Ireland, 220pp. Irish Association for Quaternary Studies, Dublin.

Coxon, P. and Browne, P. 1991. Glacial deposits and landforms of central and western Ireland. *In:* Ehlers, J., Gibbard, P. and Rose, J. (Eds.), *Glacial Deposits of Great Britain and Ireland*, A. A. Balkema, Rotterdam, 355–365.

Coxon, P. and Coxon, C. E. 1997. A pre-Pliocene or Pliocene land surface in County Galway, Ireland. *In:* Widdowaon, M. (Ed.), *Palaeosurfaces: Recognition, Reconstruction and Palaeoenvironmental Interpretation*. Geological Society Special Publication No. 120, 37–55.

Coxon, P. and Flegg, A. M. 1985. A middle Pleistocene interglacial deposit from Ballyline, Co. Kilkenny. *Proceedings of the Royal Irish Academy*, **85B**, 107–120.

Coxon, P. and Flegg, A. M. 1987. A Late-Pliocene/Early Pleistocene deposit at Pollnahallia, near Headford, Co. Galway. *Proceedings of the Royal Irish Academy*, **87B**, 15–42.

Coxon, P. and Hannon, G. 1991. The interglacial deposits at Derrynadivva and Burren townland. *In:* Coxon, P. (Ed.), *Fieldguide to the Quaternary of North Mayo*, Irish Association for Quaternary Studies (IQUA), Dublin.

Coxon, P. and Waldren, S. 1995. The floristic record of Ireland's Pleistocene temperate stages. *In:* Preece, R. C. (Ed.), *Island Britain: A Quaternary Perspective*, Geological Society Special Publication No. 96, 243–268.

Coxon, P., Hannon, G. and Foss, P. 1994. Climatic deterioration and the end of the Gortian interglacial in sediments from Derrynadivva and Burren townland, near castlebar, County Mayo, Ireland. *Journal of Quaternary Science*, 9, 33–46.

Craig, A. J. 1973. Studies on the ecological history of south-east Ireland, using pollen influx analysis and other methods. Unpublished Ph. D. thesis, Trinity College, Dublin.

Craig, A. J. 1978. Pollen percentage and influx analysis in south-east Ireland: a contribution to the ecological history of the late Glacial Period. *Journal of Ecology*, 66, 297–324.

Creighton J. R. 1974. A study of the late Pleistocene geomorphology of north-central Ulster, Northern Ireland. Unpublished Ph. D. thesis, The Queen's University, Belfast.

Cunningham, A. and Wilson, P. 2004. Relict periglacial boulder sheets and lobes on Slieve Donard, Mountains of Mourne, Northern Ireland. *Irish Geography*, 37 (2), 187–201.

Cutler, K. B., Edwards, R. L., Taylor, F. W., Cheng, H., Adkins, J., Gallup, C. D., Cutler, P. M., Burr, G. S. and Bloom, A. L. 2003. Rapid sea level fall and deep-ocean temperature change since the last interglacial period. *Earth and Planetary Science Letters*, 206, 253–271.

Cwynar, L. C. and Watts, W. A. 1989. Accelerator-mass spectrometer ages for late-glacial events at Ballybetagh, Ireland. *Quaternary Research*, 31, 377–380.

Dacombe, R. V. and Thomas, G. S. P. 1985. Field guide to the Quaternary of the Isle of Man. Quaternary Research Association, Cambridge, 122pp.

Dardis, G. F. 1980. The Quaternary sediments of central Ulster. *In:* Edwards, K. J. (Ed.), Field Guide No. 3, County Tyrone, Northern Ireland, 1–30. Irish Association for Quaternary Research.

Dardis, G. F. 1981. Stagnant ice topography and its relation to drumlin genesis, with reference to south-central Ulster. *Annals of Glaciology*, 2, 183.

Dardis, G. F. 1982. *Sedimentological aspects of the Quaternary geology of south-central Ulster, Northern Ireland.* Unpublished PhD Thesis, Ulster Polytechnic, Jordanstown, 422p.

Dardis, G. F. 1985a. Till facies associations in drumlins and some implications for their mode of origin. *Geografiska Annaler*, 67A, 13–22.

Dardis, G. F. 1985b. Genesis of late Pleistocene cross-valley moraine ridges, south-central Ulster, Northern Ireland. *Earth Surface Processes and Landforms*, 10, 483–495

Dardis, G. F. 1986. Late Pleistocene glacial lakes in south-central Ulster, Northern Ireland. Irish Journal of Earth Sciences, 7, 133–144.

Dardis, G. F. 1990. Glacial history. *In:* Wilson, P. (Ed.), *IQUA North Antrim Quaternary Field Guide*, University of Ulster, Coleraine.

Dardis, G. F. and McCabe, A. M. 1983. Facies of subglacial channel sedimentation in Late-Pleistocene drumlins, Northern Ireland. *Boreas*, 12, 263–278.

Dardis, G. F. and McCabe, A. M. 1987. Subglacial sheetwash and debris flow deposits in late-Pleistocene drumlins, Northern Ireland. *In:* Rose, J. and Menzies, J. (Eds.), *Drumlins*, Proceedings of first conference on Geomorphology, Manchester.

Dardis, G. F., McCabe, A. M. and Mitchell, W. I. 1984. Characteristics and origins of lee-side stratification sequences in late-Pleistocene drumlins, Northern Ireland. *Earth Surface Processes and Landforms*, 9, 409–424.

Dardis, G. F., Mitchell, W. I. and Hirons, K. R. 1985. Middle Midlandian interstadial deposits at Greenagho, near Belcoo. County Fermanagh, Northern Ireland. *Irish Journal of Earth Sciences*, 7, 1–6.

Davies, G. L. 1964. From flood and fire to rivers and ice. *Irish Geography*, 5, 1–16.

Davies, G. L. 1970. Richard Prior's description of an Irish esker. *Journal of Glaciology*, 9, 147–148.

Davies, G. L. H. and Stephens, N. 1978. *Ireland.* Methuen & Co Ltd., London. 1–250.

Delaney, C. 1997. The Ballymahon esker. *In:* Mitchell, F. and Delaney, C. (Eds.), *The Quaternary of the Irish Midlands*, Irish Association for Quaternary Studies Field guide Bo. 21, 71–75.

Delaney, C. 2001. Morphology and sedimentology of the Rooskagh esker, County Roscommon. *Irish Journal of Earth Science*, 19, 5–22.

Delaney, C. 2002. Sedimentology of a glaciofluvial landsystem, Lough Ree area, central Ireland: implications for ice margin characterisatics during Devensian glaciation. *Sedimentary Geology*, 149, 111–126.

Delantley, L. J. and Whittington, R. J. 1977. A re-assessment of the Neogene deposits of the south Irish Sea and Nymphe Bank. *Marine Geology*, M23–M30.

Dionne, J.-C. 1981. A boulder-strewn tidal flat, north shore of the Gulf of St. Lawrence, Quebec. *Géographie physique et Quaternaire*, 35, 261–267.

Domack, E. W. 1983. Facies of Late-Pleistocene glacial-marine sediments on Whidbey Island, Washington: An isostatic glacial-marine sequence. *In* Molnia B. F., (Ed.), *Glacial-marine Sedimentation*, 535–570, Plenum Press, New York.

Dott, R. H. and Bourgeois, J. 1982. Hummocky stratification: Significance of its variable bedding sequences. *Geological Society of America Bulletin*, 93, 633–680.

Dowling, L. A. and Coxon, P. 2001. Current understanding of Pleistocene stages in Ireland. *Quaternary Science Reviews*, 20, 1631–1642.

Dowling. L. A., Sejrup, H. P., Coxon, P. and Heijnis, H. 1998. Palynology, aminostratigraphy and U-series dating of marine Gortian interglacial sediments in Cork harbour, southern Ireland. *Quaternary Science Reviews*, 17, 945–962.

Dunlop, P. 2004. The Characteristics of Ribbed Moraine and Assessment of Theories for their Genesis. Unpublished Ph.D. Thesis, Dept. of Geography, University of Sheffield.

Dunlop, P. and Clark, C. 2006. The morphological characteristics of ribbed moraine. *Quaternary Science Reviews*, 25, 1668–1691.

Dury, G. H. 1957. A glacially breached watershed in Donegal. *Irish Geography*, 3, 171–180.

Dury, G. H. 1958. Glacial morphology of the Blue Stack area, Donegal. *Irish Geography*, 3, 242–253.

Dury, G. H. 1959. A contribution to the geomorphology of central Donegal. *Proceedings of the Geologists Association*, 70, 1–27.

Dwerryhouse, A. R. 1923. The glaciation of north-eastern Ireland. *Quarterly Journal of the Geological Society of London*, 79, 352–422.

Dyke. A. S., Andrews, J T., Clark, P. U., England, J. H., Miller, G. H., Shaw, J. and Veillette, J. 2002. The Laurentide and Innuitian ice sheets during the last glacial maximum. *Quaternary Science Reviews*, 21, 9–32.

Elliot, M., Labeyrie, T., Dokken, S. and Manthe, S. 2001. Coherent patterns of ice-rafted debris deposits in the Nordic regions during the last glacial. *Earth Planet Science Letters*, 194, 151–163.

Eyles, C. H. 1988. A model for striated boulder pavement formation on glaciated, shallow-marine shelves – an example from the Yakataga Formation, Alaska. *Journal of Sedimentary*

Petrology, **58**, 62–71.

Eyles, C. H. 1994. Intertidal boulder pavements in the northeastern Gulf of Alaska and their geologic significance. *Sedimentary Geology*, **88**, 161–173.

Eyles, N. and McCabe, A. M. 1989a. The Late Devensian (<22,000 BP) Irish Sea Basin: the sedimentary record of a collapsed ice sheet margin. *Quaternary Science Reviews*, **8**, 307–351.

Eyles, N. and McCabe, A. M. 1989b. Glaciomarine facies within subglacial tunnel valleys: the sedimentary record of glacioisostatic downwarping in the Irish Sea basin. *Sedimentology*, **36**, 431–448.

Eyles, N. and McCabe, A. M. 1991. Glaciomarine deposits of the Irish Sea Basin: the role of glacioisostatic disequilibrium. *In:* Ehlers, J., Gibbard, P. and Rose, J. (Eds.), *Glacial Deposits in Great Britain and Ireland*. Balkema, Rotterdam, 311–332.

Eyles, N., Eyles, C. H. and Miall, A. D. 1983. Lithofacies types and vertical profile models: an alternative approach to the description and environmental interpretation of glacial diamict and diamictic sequences. *Sedimentology*, **30**, 393–410.

Farrington, A. 1934. The glaciation of the Wicklow mountains. *Proceedings of the Royal Irish Academy*, **42B**, 173–209.

Farrington, A. 1936. The glaciation of the Bantry Bay district. *Scientific Proceedings of the Royal Dublin Society*, **21**, 345–361.

Farrington A. 1938. The local glaciers of Mount Leinster and Blackstairs Mountain. *Proceedings of the Royal Irish Academy*, **45**, 65–71.

Farrington, A. 1942. The granite drift near Brittas, on the border between county Dublin and county Wicklow. *Proceedings of the Royal Irish Academy*, **47B**, 279–291.

Farrington, A. 1944. The glacial drifts of the district around Enniskerry, County Wicklow. *Proceedings of the Royal Irish Academy*, **50B**, 133–157.

Farrington, A. 1949. The glacial drifts of the Leinster mountains. *Journal of Glaciology*, **1**, 220–225.

Farrington, A. 1953. Local Pleistocene glaciation and the level of the snow line of Croaghaun Mountain in Achill Island, County Mayo. *Journal of Glaciology*, **2**, 262–267.

Farrington, A. 1954. A note on the correlation of the Kerry-Cork glaciations with those of the rest of Ireland. *Irish Geography*, **3**, 47–53.

Farrington, A. 1957. Glacial lake Blessington. *Irish Geography*, **3**, 216–222.

Farrington, A. 1965. The last glaciation in the Burren, County Clare. *Proceedings of the Royal Irish Academy*, **64B**, 33–39.

Farrington, A. 1966. The early-glacial raised beach in County Cork. *Scientific Proceedings of the Royal Dublin Society*, **13**, 197–219.

Farrington, A. and Mitchell, F. M. 1973. Some glacial features between Pollapuca and Baltinglass, Co. Wicklow. *Irish Geography*, **6**, 543–561.

Farrington, A. with Synge, F. M. 1970. The eskers of the Tullamore district. *In:* Stephens, N. and Glasscock, R. E. (Eds.), *Irish Geographical Studies in Honour of E. Estyn Evans*, Belfast, 49–52.

Finch, T. F. and Synge, F. M. 1966. The drifts and soils of west Clare and the adjoining parts of counties Kerry and Limerick. *Irish Geography*, **5**, 161–172.

Finch, T. F. and Walsh, M. 1973. Drumlins of county Clare. *Proceedings of the Royal Irish Academy*, **73B**, 405–413.

Fisher, T. G. and Shaw, J. 1992. A depositional model for Rogen moraine, with examples from the Avalon Peninsula, Newfoundland. *Canadian Journal of Earth Sciences*, **29**, 669–686.

Fronval, T., Jansen, E., Blomendal, J. and Johnsen, S. 1995. Oceanic evidence for coherent fluctuations in Fennoscandian and Laurentide ice sheets on millennial timescales. *Nature*, **374**, 443–446.

Gallagher, C. 1998. A reconstruction of Pleistocene ice limits in Slieve Bloom using heavy minerals. *Irish Geography*, **31**, 100–110.

Gallagher, C., Thorp, M and Steenson, P. 1996. Glacier dynamics around Slieve Bloom, central Ireland. *Irish Geography*, **29**, 67–82.

Garrard, R. A. 1977. The sediments of the south Irish Sea and Nymphe Bank area of the Celtic Sea. *In:* Kidson, C. and Tooley, M. J. (Eds.), *The Quaternary History of the Irish* Sea, Seel House Press, Liverpool, 69–92.

Gennard, D. E. 1984. A palaeoecological study of the interglacial deposits at Benburb, County Tyrone. *Proceedings of the Royal Irish Academy*, **84B**, 43–56.

Gorrell, G. and Shaw, J. 1991. Deposition in an esker, bead and fan complex, Lanark, Ontario, Canada. *Sedimentary Geology*, **72**, 285–314.

Gray, J. M. and Coxon, P. 1991. The Lough Lomond Stadial glaciation in Britain and Ireland. *In:* Ehlers, J., Gibbard, P. L. and Rose, J. (Eds.), *Glacial Deposits in Great Britain and Ireland*. Balkema, Rotterdam, 89–105.

Gregory, J. W. 1925. The Evishanoran esker, Co. Tyrone. *Geological Magazine*, **62**, 451–458.

Grootes, P. M., Stuiver, M., White, J. W. C., Johnston, S. and Jouzel, J. 1993. Comparison of oxygen isotope records from the GISP2 and Grip Greenland ice cores. *Nature*, **366**, 552–554.

Gustavson, T. C. and Boothroyd, J. C. 1982. Subglacial fluvial erosion: a major source of stratified sediment, Malaspina glacier, Alaska. *In:* Davidson, R., Nickling, W. and Fahey, B D. (Eds.), *Research in glacial, Fluvioglacial and Glaciolacustrine Systems*. Guelph Symposium on Geomorphology. Geobooks, Norwich, 93–116.

Gustavson, T. C. and Boothroud, J. C. 1987. A depositional model for outwash, sediment sources and hydrologic characteristicsa, Malaspina glacier, Alaska: a modern analog of the southeastern margin of the Laurentide ice sheet. *Geological Society of America Bulletin*, **19**, 187–200.

Gustavson, T. C., Ashley, G. M. and Boothroyd, J. C. 1975. Depositional sequences in glaciolacustrine deltas. *In:* Jopling, A. V. McDonald, B. C. (Eds.), *Glaciofluvial and Glaciolacustrine Sedimentation*. SEPM Special Publication No.23. Tulsa, Oklahoma. 264–280.

Hald, M., Steinsund, P. J., Dokken, T., Korsun, S., Polyak, L and Aspeli, R. 1994. Recent and Late Quaternary distribution of *Elphidium exclavatum f. clavatum* in Arctic seas. *Cushman Foundation Special Publication*, **32**, 141–153.

Hallissy, T. 1914. Explanatory memoir to sheet 58, illustrating parts of Counties Armagh, Fermanagh and Monaghan. Memoirs of the Geological Survey of Ireland, Department of Agriculture and Technical Instruction in Ireland. LCHMT, Dublin, 1–26.

Hanebuth, T., Stattegger, K. and Grootes, P. M. 2000. Rapid flooding of the Sunda Shelf: a Late-glacial sea-level record. *Science*, **288**, 1033–1035.

Hansom, J. D. 1983. Ice-formed intertidal boulder pavements in the subantarctic. *Journal of Sedimentary Petrology*, **53**, 135–145.

Hansom, J. D. 1986. Intertidal forms produced by floating ice in Vestfirdir, Iceland. *Marine Geology*, **71**, 289–298.

Hanvey, P. 1987. Sedimentology of lee-side stratification sequences in late-Pleistocene drumlins, north-west Ireland. *In:* Menzies, J. and Rose, J. (Eds.), *Drumlin Symposium*, Balkema, Rotterdam, 241–253.

Hanvey, P. M. 1988. *The Sedimentology and Genesis of Late-*

Pleistocene Drumlins in Counties Mayio and Donegal, Western Ireland. Unpublished Ph. D. Thesis, Ulster Polytechnic, Jordanstown, Northern Ireland, 614pp.

Hanvey, P. M. 1989. Stratified flow deposits in a late Pleistocene drumlin in northwest Ireland. *Sedimentary Geology,* **62**, 211–221. .

Hättestrand, C. and Kleman, J. 1999 Ribbed moraine formation. *Quaternary Science Reviews,* **18**, 43–61.

Hay. A. E., Murray, J. W. and Burling, R. W. 1983. Submarine channels in Rupert Inlet, British Columbia: I. Morphology. *Sedimentary Geology,* **36**, 289–315.

Haynes, J. R. 1964. Live and dead foraminifera between the Sarns, Cardigan Bay. *Nature,* 204, (4960), 774.

Haynes, J. R., McCabe, A. M. and Eyles, N. 1995. Microfaunas from late Devensian glaciomarine deposits in the Irish Sea Basin. *Irish Journal of Earth Sciences,* **14**, 81–103.

Heijnis, H., Ruddock, J. and Coxon, P. 1993. A uranium–thorium dated late Eemian or early Midlandian organic deposit from near Kilfenora between Spa and Fenit, Co. Kerry, Ireland. *Journal of Quaternary Science,* **8**, 31–43.

Hein, F. J. 1982. Depositional mechanisms of deep-sea coarse clastic sediments, Cap Etrangé Formation, Québec. *Canadian Journal of Earth Sciences,* **19**, 267–287.

Hill, A. R. 1973. The distribution of drumlins in county Down. *Annals Association of American Geographers,* **63**, 226–240.

Hill, A. R. and Prior, D. B. 1968. Directions of ice movement in north-east Ireland. *Proceedings of the Royal Irish Academy,* **66B**, 71–84.

Hinch J. de W. 1913. The shelly drift of Glenulra and Belderrig, Co. Mayo. *Irish Naturalists' Journal,* **22**, 1–6.

Hoare, P. G. 1975. The pattern of deglaciation in county Dublin. *Proceedings of the Royal Irish Academy,* 75B, 207–224.

Hoare, P. G. 1976. Glacial meltwater channels in county Dublin. *Proceedings of the Royal Irish Academy,* **76B**, 173–185.

Hoare, P. G. 1977. The glacial stratigraphy in Shangannagh and adjoining townlands, south-east county Dublin. *Proceedings of the Royal Irish Academy,* **77B**, 295–305.

Hughes, T. 1987. Ice dynamics and deglaciation models when ice sheets collapsed. *In:* Ruddiman, W. F. and Wright, H. E. (Eds.), *North America and Adjacent Oceans During The Last Deglaciation,* The Geology of North America. Geological Society of America, Boulder, CO. K–3, 183–220.

Hunter, L. E., Powell, R. D. and Smith, G. W. 1996. Facies architecture and grounding-line fan processes of morainal banks during the deglaciation of coastal Maine. *Bulletin of the Geological Society of America,* **108**, 1022–1038.

Jessen, K. 1949. Studies in late Quaternary deposits and flora-history of Ireland. *Proceedings of the Royal Irish Academy,* **52B**, 85–290.

Jessen K. and Farrington A. 1938. The bogs at Ballybetagh, near Dublin, with remarks on lateglacial conditions in Ireland. *Proceedings of the Royal Irish Academy,* **44B**, 205–260.

Jessen, K., Anderson, S. T. and Farrington, A. 1959. The interglacial deposit near Gort, Co. Galway, Ireland. Proceedings of the Royal Irish Academy, **60B**, 1–77.

Keigwin, L. D., Jones, J. A., Lehman, S. J. and Boyle, E. A. 1991. Deglacial meltwater discharge, North Atlantic deep-water circulation. *Journal of Geophysical Research,* **96**, 16811–16826.

Kelley, J., Cooper, J. A. G., Jackson, D. W. T., Belnap, D. F. & Quinn, R. J. 2006. Sea-level change and inner shelf stratigraphy off Northern Ireland. *Marine Geology,* **232**, 1–15.

Kershaw, P. J. 1986. Radiocarbon dating of Irish Sea sediments. *Estuarine, Coastal and Shelf Science,* **23**, 295–303.

Kilroe, J. R. 1888. Directions of ice flow in the north of Ireland.

Quarterly Journal of the Geological Society of London, **44**, 827–833.

Kinahan, G. H. 1864. The eskers of the central plain of Ireland. *Journal of the Geological Society, Dublin,* **10**, 1862–1864, 109–112.

Kinahan, G. H. 1865. Explanation to accompany sheets 115 and 116, Geological Survey of Ireland, Dublin.

Kinahan, G. H. 1878. *Manual of the Geology of Ireland,* Dublin.

King, E. L., Haffidason, H., Sejrup, H. P., Austin, W. E. N., Duffy, M., Helland, E., Klitgaard-Kristensen, D. and Scourse, J. D. 1998. End moraines on the northwest Irish continental shelf. Third ENAM II Workshop. Edinburgh, 1998 (Abstract volume).

Kleman, J. and Hättestrand, C. 1999. Frozen-bed Fennoscandian and Laurentide ice sheets during the Last Glacial Maximum. *Nature,* **402**, 63–66.

Knight, J. 1999. Geological evidence for neotectonic activity during deglaciation of the southern Sperrin mountains, Northern Ireland. *Journal of Quaternary Science,* **14**, 45–57.

Knight, J. 2004. Sedimentary evidence for the formation mechanism of the Armoy moraine and Late Devensian events in the north of Ireland. *Geological Journal,* **39**, 403–417.

Knight, J. 2006. Geomorphic evidence for active and inactive phases of late Devensian ice in north-central Ireland. *Geomorphology,* **75**, 4–19.

Knight, J. and McCabe, A. M. 1997a. Identification and significance of ice flow-transverse subglacial ridges (Rogen moraines) in north central Ireland. *Journal of Quaternary Science,* **12**, 219–224.

Knight, J. and McCabe, A. M. 1997b. Drumlin evolution and ice sheet oscillations along the NE Atlantic margin, Donegal Bay, western Ireland. *Sedimentary Geology,* **111**, 57–72.

Knight, J., McCarron, S. G. and McCabe, A. M. 1999. Landform modification by palaeo-ice streams in east central Ireland. *Annals of Glaciology,* **28**, 161–167.

Knight, J., McCarron, S. G., McCabe, A. M. and Sutton, B. 1999. Sand and gravel aggregate resource management and conservation in Northern Ireland. *Journal of Environmental Management,* **56**, 195–207.

Knutz, P. C., Austin, W. E. N. and Jones, E. J. W. 2001. Millennial scale depositional cycles related to the British ice sheet variability and North Atlantic circulation since 45 kyr B.P., Barra Fan, UK margin. *Palaeoceanography,* **16**, 53–64.

Knutz, P. C., Zahn, R. and Hall, I. R. 2007. Centennial-scale variability of the British ice sheet: implications for climate forcing and Atlantic meridional overturning circulation during the last deglaciation. *Paleoceanography,* **22**, 1–14.

Kor, P. S. G., Shaw, J. and Sharpe, D. R. 1991. Erosion of bedrock by subglacial meltwater, Georgian Bay, Ontario: a regional view. *Canadian Journal of Earth Sciences,* **28**, 623–642.

Lambeck, K. 1995. Late Devensian and Holocene shorelines of the British Isles and North Sea from models of glacio-hydro-isostatic rebound. *Journal of the Geological Society, London,* **152**, 437–448.

Lambeck, K. 1996. Glaciation and sea-level change for Ireland and the Irish Sea since Late Devensian/Midlandian times. *Journal of the Geological Society,* **153**, 853–872.

Lambeck, K. and Purcel, A. P. 2001. Sea-level change in the Irish Sea since the last glacial maximum: constraints from isostatic modelling. *Journal of Quaternary Science,* **16**, 497–505.

Lauriol, B. V. and Gray, J. T. 1980. Processes responsible for the concentration of boulders in the intertidal zone in Leaf Basin, Ungava. *In:* McCann, S. B. (Ed.), *The Coastline of Canada.* *Geological Association of Canada,* Paper 80–10, 281–292.

Leithold, E. L. and Bourgeois, J. 1984. Characteristics of coarse-grained sequences deposited in nearshore, wave-dominated environments – examples from the Miocene of south-west Oregon. *Sedimentology*, **31**, 749–775.

Lewis, H. C. 1894. Papers and Notes on the Glacial Geology of Great Britain and Ireland. Longman, London.

Loeblich, A. R. and Tappan, H. 1953. Studies of Arctic foraminifera. *Smithsonian Miscellaneous collection*. **121**, 1–142.

Lowe, D. R. 1979. Sediment gravity flows: their classification and some problems of application to natural flows and deposits. *In:* Doyle, L. J. and Pilkey, O. H. (jr.) (Eds.), *Geology of continental Slopes*, Special Publication Society of Economic Paleontologists and mineralogists, 27, 75–82.

Lowe, D. R. 1982. Sediment gravity flows. II. Depositional models with special reference to the deposits of high-density turbidity currents. *Journal of Sedimentary Petrology*, **52**, 279–297.

Lowe, D. R. and Lopiccolo, D. R. 1974. The characteristics and origins of dish and pillar structures. *Journal of Sedimentary Petrology*, **44**, 484–501.

Lowell, T. V. 1995. Interhemispheric correlation of late Pleistocene glacial events. Science, 269, 1541–1549.

Lundqvist, J. 1969. Problems of the so-called Rogen moraine. *Sveriges Geologiska Undersökning, Series C*, **648**, 1–32.

Lundqvist, J. 1989. Rogen (ribbed) moraine – identification and possible origin. *Sedimentary Geology*, **62**, 281–292.

Mackiewicz, N. E., Powell, R. D., Carlson, P. R. and Molnia, B. F. 1984. Interlaminated ice-proximal glaciomarine sediments in Muir Inlet, Alaska. *Marine Geology*, **57**, 113–147.

Markgren, M. and Lassila, M. 1980. Problems of Rogen moraine morphology: Rogen moraine are Blattnick moraine. *Boreas*, **9**, 271–274.

McCabe, A. M. 1969. The glacial deposits of the Maguiresbridge area, County Fermanagh, Northern Ireland. *Irish Geography*, **6**, 63–77.

McCabe, A. M. 1972. Directions of Late-Pleistocene ice flows in eastern Counties Meath and Louth, Ireland. *Irish Geography*, **6**, 443–461.

McCabe, A. M. 1973. The glacial stratigraphy of eastern Counties Meath and Louth. *Proceedings of the Royal Irish Academy*, **73B**, 355–382.

McCabe, A. M. 1979. Field guide to east central Ireland. Quaternary Research Association, 63pp.

McCabe, A. M. 1985. Glacial Geomorphology. *In:* Edwards, K. J. and Warren, W. P. (Eds.), *The Quaternary History of Ireland*, Academic Press, London, 67–93.

McCabe, A. M. 1986. Glaciomarine facies deposited by retreating tidewater glaciers – an example from the Late Pleistocene of Northern Ireland. *Journal of Sedimentary Petrology*, **56**, 880–894.

McCabe, A. M. 1987. Quaternary deposits and glacial stratigraphy in Ireland. *Quaternary Science Reviews*, **6**, 259–299.

McCabe, A. M. 1993. The 1992 Farrington lecture: Drumlin bedforms and related ice-marginal depositional systems in Ireland. *Irish Geography*, **26**, 22–44.

McCabe, A. M. 1994. Sand and gravel landscapes in Northern Ireland. *Earth Heritage*, **2**, 18–22.

McCabe, A. M. 1995a. Quaternary geology of Donegal. *In:* Wilson, P. (Ed.), *Northwest Donegal Field Guide*, Irish Association for Quaternary Studies, 15–20.

McCabe, A. M. 1995b. Marine molluscan shell dates from two glaciomarine jet efflux deposits, eastern Ireland. *Irish Journal of Earth Sciences*, **14**, 37–45.

McCabe, A. M. 1996. Dating and rhythmicity from the last deglacial cycle in the British Isles. *Journal of the Geological Society, London*, **153**, 499–502.

McCabe, A. M. 1997. Geological constraints on geophysical models of relative sea-level change during deglaciation of the western Irish Sea Basin. *Journal of the Geological Society, London*, **154**, 601–604.

McCabe, A. M. and Clark, P. U. 1998. Ice-sheet variability around the North Atlantic Ocean during the last deglaciation. *Nature*, **392**, 373–377.

McCabe, A. M. and Clark, P. U. 2003. Deglacial chronology from County Donegal, Ireland: implications for the British-Irish ice sheet. *Journal of the Geological Society, London*, **160**, 847–855.

McCabe, A. M. and Coxon, P. 1993. A resedimented interglacial peat ball containing Carpinus pollen within a glacial efflux sequence, Blackwater, Co. Wexford: evidence for part of the last interglacial cycle in Ireland? Proceedings of the Geologist's association, **104**, 201–207.

McCabe, A. M. and Dardis, G. F. 1989a. Sedimentology and depositional setting of Late Pleistocene drumlins, Galway Bay, western Ireland. *Journal of Sedimentary Petrology*, **59**, 544–559.

McCabe, A. M. and Dardis, G. F. 1989b. A geological view of drumlins in Ireland. *Quaternary Science Reviews*, **8**, 169–177.

McCabe, A. M. and Dardis, G. F. 1992. Tullywee delta. *In:* McCabe, A. M., Dardis, G. F. and Hanvey, P. M. (Eds.), *Glacial Sedimentology in Northern and Western Ireland*. Pre- and post-Symposium Field Excursion Guide book, University of Ulster, 191–194.

McCabe, A. M. and Dardis, G. F. 1994. Glaciotectonically induced water-throughflow structures in a late Pleistocene drumlin, Kanrawer, county Galway, western Ireland. *Sedimentary Geology*, **91**, 173–190.

McCabe, A. M. and Dunlop, P. 2006. *The Last glacial Termination in Northern Ireland*. Geological Survey of Northern Ireland, Belfast, 93pp.

McCabe, A. M. and Eyles, N. 1988. Sedimentology of an ice-contact glaciomarine delta, Carey valley, Northern Ireland. *Sedimentary Geology*, **59**, 1–14.

McCabe, A. M. and Haynes, J. R. 1996. A late Pleistocene intertidal boulder pavement from an isostatically emergent coast, Dundalk Bay, eastern Ireland. *Earth Surface Processes and Landforms*, **21**, 555–572.

McCabe, A. M. and Hirons, K. R. 1986. Field guide to the Quaternary of South-East Ulster. Quaternary Research Association, Cambridge.

McCabe, A. M. and Hoare, P. G. 1978. The Late Quaternary history of east central Ireland. *Geological Magazine*, **115**, 397–413.

McCabe, A. M. and O'Cofaigh, C. 1994. Sedimentation in a subglacial lake, Enniskerry, eastern Ireland. *Sedimentary Geology*, **91**, 57–95.

McCabe, A. M. and O'Cofaigh, C. 1995. Late Pleistocene morainal bank facies at Greystones, eastern Ireland: an example of sedimentation during ice marginal re-equilibration in an isostatically depressed basin. *Sedimentology*, **421**, 647–663.

McCabe A. M. and O'Cofaigh, C. 1996. Upper Pleistocene facies sequences and relative sea-level trends along the south coast of Ireland. *Journal of Sedimentary Petrology*, **66**, 376–390.

McCabe, A. M., Carter, R. W. G. and Haynes, J. R. 1994. A shallow marine emergent sequence from the northwestern sector of the last British ice sheet, Portballintrae, Northern Ireland. *Marine Geology*, **117**, 19–34.

McCabe, A. M., Clark, P. U. and Clark, J. 2005. AMS ^{14}C dating of deglacial events in the Irish Sea Basin and other sectors of the British-Irish ice sheet. *Quaternary Science Reviews*, **24**, 1673–1690.

McCabe, A. M., Clark, P. U. and Clark, J. 2007a. Radiocarbon constraints on the history of the western Irish ice sheet prior to the last glacial maximum. *Geology*, **35**, 147–150.

McCabe, A. M., Cooper, J. A. G. and Kelley, J. 2007c. Relative sea level changes in northeastern Ireland during the last glacial termination. *Geological Society of London*, in press.

McCabe, A. M., Dardis, G. F. and Hanvey, P. M. 1984. Sedimentology of a late Pleistocene submarine-moraine complex, County Down, Northern Ireland. *Journal of Sedimentary Petrology*, **54**, 716–730.

McCabe, A. M., Dardis, G. F. and Hanvey, P. M. 1987. Sedimentation at the margins of a late Pleistocene ice-lobe terminating in shallow marine environments, Dundalk Bay, eastern Ireland. *Sedimentology*, **34**, 473–493.

McCabe, A. M., Dardis, G. F. and Hanvey, P. M. 1992. *Glacial Sedimentology in Northern and Western Ireland*. Pre- and post-Symposium Field Excursion Guide book, Symposium on subglacial processes, sediments and landforms, University of Ulster, 1–236.

McCabe, A. M., Haynes, J. R. and MacMillan, N. F. 1986. Late-Pleistocene tidewater glaciers and glaciomarine sequences from north County Mayo, Republic of Ireland. *Journal of Quaternary Science*, **1**, 73–84.

McCabe, A. M., Knight, J. and McCarron, S. G. 1998. Evidence for Heinrich event 1 in the British Isles. *Journal of Quaternary Science*, **13**, 549–568.

McCabe, A. M., Knight, J. and McCarron, S. G. 1999. Ice flow stages and glacial bedforms in north central Ireland: a record of rapid environmental change during the last glacial termination. *Journal of the Geological Society, London*, **156**, 63–72.

McCabe, A. M., Mitchell, G. F. and Shotton, F. W. 1978. An inter-till freshwater deposit at Hollymount, Maguiresbridge, Co. Fermanagh. *Proceedings of the Royal Irish Academy*, **78B**, 77–89.

McCabe, A. M., Penney, D. N. and Bowen D. Q. 1993. Glaciomarine facies from the western sector of the last British ice sheet, Malin Beg, County Donegal, Ireland. *Quaternary Science reviews*, **12**, 35–45.

McCabe, A. M., Clark, P. U., Clark, J. and Dunlop, P. 2007b. Radiocarbon constraints on readvances of the British-Irish ice sheet in the northern Irish Sea Basin during the last deglaciation. *Quaternary Science Reviews*, **26**, 1204–1211.

McCabe, A. M., Coope, R. G., Gennard, D. E. and Doughty, P. 1987. Freshwater organic deposits and stratified sediments between early and Late Midlandian (Devensian) till sheets, at Aghnadarragh, County Antrim. *Journal of Quaternary Science*, **2**, 11–33.

McCabe, A. M., Eyles, N., Haynes, J. R. and Bowen. D. Q. 1990. Biofacies and sediments in an emergent Late Pleistocene glaciomarine sequence, Skerries, east central Ireland. *Marine Geology*, **94**, 23–36.

McCann, S. B., Dale, J. E. and Hale, P. B. 1981. Subarctic tidal flats in areas of large tidal range, southern Baffin Island, eastern Canada. *Géographie physique et Quaternaire*, **35**, 183–204.

McCarron, S. G., Knight, J., Sutton, B. and McCabe, A. M. 1998. The scientific and aesthetic attributes of glaciofluvial landscapes in Northern Ireland. Closed report to the Countryside and Wildlife Branch of the Department of the Environment for Northern Ireland.

McDonald, B. C. and Vincent, J. S. 1972. Fluvial sedimentary structures formed experimentally in a pipe and their implications for interpretation of subglacial sedimentary environments. Geological Survey of Canada Paper, 72/27, 33pp.

McKenna, J. 2002. Basalt cliffs and shore platforms between Portstewart (Co. Londonderry) and Portballintrae (County Antrim). *In:* Knight, J. (Ed.), *Field guide to the Coastal Environments of Northern Ireland*, International Coastal Symposium, (ICS02), 157–164.

McManaus, J. F., Francois, R., Gherardi, J.-M., Keigwin, L. D. and Brown-Ledger, S. 2004. Collapse and rapid resumption of Atlantic meridional circulation linked to deglacial climate changes. *Nature*, **428**, 834–837.

McMillan, N. F. 1964. The mollusca of the Wexford gravels (Pleistocene) southeast Ireland. *Proceedings of the Royal Irish Academy*, **63B**, 265–290.

Merritt, J. W. and Auton, C. A. 2000. An outline of the lithostratigraphy and depositional history of Quaternary deposits in the Sellafield district, west Cumbria. *Proceedings of the Yorkshire Geological Society*, **53**, 129–154.

Miall, A. D. 1977. A review of the braided-river depositional environment. *Earth-Science Reviews*, **13**, 1–62.

Middleton, G. F. and Hampton, M. A. 1976. Subaqueous sediment transport and deposition by sediment gravity flows. *In:* Stanley, D. G. and Swift, D. J. P (Eds.), *Marine Sediment Transport and Environmental Management*. Wiley, New York, 197–218.

Mitchell, G. F. 1960. The Pleistocene history of the Irish Sea. *Advancement of Science London*, **17**, 313–325.

Mitchell, G. F. 1965. The Quaternary deposits of the Ballaugh and Kirk Michael districts of the Isle of Man. *Quarterly Journal of the Geological Society of London*, **121**, 359–381.

Mitchell, G. F. 1970. The Quaternary deposits between Fenit and Spa on the north shore of Tralee Bay, County Kerry. *Proceedings of the Royal Irish Academy*, **70B**, 141–162.

Mitchell, G.F. 1972. The Pleistocene history of the Irish Sea: second approximation. *Scientific Proceedings of the Royal Dublin Society Series A*, **4**, 181–199.

Mitchell, G. F. 1976. *The Irish Landscape*. Collins, London, 240pp.

Mitchell, G. F. 1981. The Quaternary-until 10,000BP. *In:* Holland, C. H. (Ed.), *A Geology of Ireland*, Scottish Academic Press, Edinburgh, 235–258.

Mitchell, G. F., Penny, L. F., Shotton, F. W. and West, R. G. 1973. A correlation of Quaternary deposits in the British Isles. *Geological Society of London Special Publication* **4, 99pp.**

Mitchell, G. F. and Watts, W. E. 1993. Notes on an interglacial deposit in Ballykeerogemore townland and an interstadial deposit in Battlestown townland, both in County Wexford. *Irish Journal of Earth Sciences*, **12**, 107–117.

Mix, A. C., Bard, E. and Schneider, R. 2001. Environmental processes of the ice age: land, oceans, glaciers (EPILOG). *Quaternary Science Reviews*, **20**, 627–657.

Morrison, M. E. S. and Stephens, N. 1965. A submerged Late Quaternary deposit at Roddans Port on the northeast coast of Ireland. *Philosophical Transactions of the Royal Society of London*, **249B**, 221–255.

Munthe, H. 1897. On the interglacial submergence of Great Britain. *Bulletin Institute of Uppsala*, **3**, 371–411.

Nagy, J. 1965. Foraminifera in some bottom samples from shallow waters in Vestspitzbergen. *Norsk Polarinstitutt Arbok*, **1963**, 109–129.

Nelson, C. H. and Kulm, V. 1973. Submarine fans and channels. *In:* Middleton, G. V. and Bouma, A. H. (Chairmen) *Turbidites and Deep-Water Sedimentation*. Lecture Notes for a Short Course, SEPM Pacific Section, 38–78.

Nemec, W. and Steel, R. J. 1984. Alluvial and coastal conglomerates: their significant features and some comments on gravelly mass flow deposits. *In:* Koster, E. H. and Steel, R. J. (Eds.), *Sedimentology of Gravels and Conglomerates*, Memoir Canada Society Petroleum Geologists, 10, 1–31.

Oldham, T. 1846. On the supposed existence of moraines in Glenmalur, County of Wicklow. *Journal of the Geological Society of Dublin*, **3**, 1843–1849, 197–199.

O'Cofaigh, C. and Evans, D. J. A. 2001a. Deforming bed conditions associated with a major ice stream of the last British ice sheet. *Geology*, **29**, 795–798.

O'Cofaigh, C. and Evans, D. J. A. 2001b. Sedimentary evidence for deforming bed conditions associated with a grounded Irish Sea Glacier, southern Ireland. *Journal of Quaternary Science*, **16**, 435–454.

Peck, V. L. et al., 2006. High resolution evidence for linkages between NW European ice sheet instability and Atlantic Meridional overturning circulation. *Earth and Planetary Science Letters*, **234**, 476–488.

Philcox, M. E. 2000. The glacio-lacustrine delta complex at Blessington, Co. Wicklow and related outflow features. *In:* Graham, J. R. and Ryan, A. (Eds.), IAS Dublin 2000, field trip guidebook, 129–152. Geology Department, Trinity College Dublin.

Portlock, J. E. 1843. Report on the geology of Londonderry and parts of Tyrone and Fermanagh, 784pp.

Postma, G. 1984. Mass-flow conglomerates in a submarine canyon Abrioja fan-delta, Pliocene, southeast Spain. *In:* Koster, R. J. and Steel, R. J. (Eds.), *Sedimentology of Gravels and Conglomerates*. Canadian Society of Petroleum Geologists Memoir 10, 237–258.

Postma, G. 1985. Resedimented conglomerates in the bottomsets of Gilbert-type gravel deltas. *Journal of Sedimentary Petrology*, **55**, 874–885.

Postma, G., Menec, W., Kleinspehn, K. L., 1988. Large floating clasts in turbidites: a mechanism for their emplacement. *Sedimentary Geology*, **58**, 47–61.

Powell, R. D. 1981. A model for sedimentation by tidewater glaciers. *Annals of Glaciology*, **2**, 129–134.

Powell, R. D. 1984. Glaciomarine processes and inductive litho-facies modellings of ice shelf and tidewater glacial sediments based on Quaternary examples. *Marine Geology*, **57**, 1–52.

Powell, R. D. 1990. Glacimarine processes at grounding-line fans and their growth to ice-contact deltas. *In:* Dowdeswell, J. A. and Scourse, J. D. (Eds.) Glacimarine Environments: Processes and Sediments. *Geological Society Special Publication*, **53**, 53–73.

Praeger, R. L. 1892/3. Report of the subcommittee appointed to investigate the gravels of Ballyrudder, County Antrim. *Proceedings of the Belfast Naturalists' Field Club*, 198–209.

Prior, D. B. 1966. Late and Post-glacial shorelines in north-east Antrim. *Irish Geography*, **5**, 173–187.

Prior, D. B., Holland, S. M and Cruickshank, M. M. 1981. A preliminary report on Late Devensian and Early Flandrian deposits on the coast at Carnlough, Country Antrim. *Irish Geography*, **14**, 75–84.

Prior, R. 1699. A description of the ridge of Mary Burrow in the Queen's County of Ireland. pp. 437–8 *in* Waller, R. 'The Posthumous Works of Robert Hooke', London, 1705.

Reineck. H. E. and Singh, I. B. 1972. Genesis of laminated sand and graded rhythmites in storm-sand layers of shelf mud. *Sedimentology*, **18**, 123–128.

Rosen, P. 1979. Boulder barricades in central Labrador. *Journal of Sedimentary Petrology*, **49**, 1113–1124.

Rothlisberger, H. and Lang, H. 1987. Glacial hydrology. *In:* Gurnell, A. M. and Clark, M. J. (Eds.), *Glaciofluvial Sediment Transfer-an Alpine Perspective*. Wiley and Sons, Chichester, 207–274.

Ruddiman, W. F., McIntyre, A., Neibler-Hunt, V. and Durazzi, J. T. 1980. Oceanic evidence for the mechanism of rapid northern hemisphere glaciation. *Quaternary Research*, **13**, 33–64.

Rust, B. R. 1977. Mass flow deposits in a Quaternary succession near Ottawa, Canada: diagnostic criteria for subaqueous outwash. *Canadian Journal of Earth Sciences*, **14**, 175–184.

Rust, B. R. and Romanelli, R. 1975. Late Quaternary subaqueous outwash deposits near Ottawa, Canada. *In:* Jopling, A. V. and McDonald, B. C. (Eds.), *Glaciofluvial and Glaciolacustrine Sedimentation. SEPM Special Publication*, **23**, 177–192.

Schultz, A. W. 1984. Subaerial debris flow deposition in the upper Palaeozoic Cutler Formation, western Colorado. *Journal of Sedimentary Petrology*, **54**, 759–772.

Scouler, J. 1838. Account of certain elevated hills of gravel, containing marine shells which occur in the County of Dublin. *Journal of the Geological Society of Dublin*, **1**, part 4, (1838), 266–276.

Scourse, J. D., Hall, I. R., McCave, N., Young, J. R and Sugdon, C. 2000. The origin of Heinrich layers: evidence from H2 for European precursor events. *Earth and Planetary Science Letters*, **182**, 187–195.

Scourse, J. D., Allen, J. R. M., Austin, W. E. N., Devoy, R. J. N., Coxon, P. and Sejrup, H. P. 1992. New evidence on the age and significance of the Gortian temperate stage: a preliminary report from Cork Harbour site. *Proceedings of the Royal Irish Academy*, **92B**, 21–43.

Severinghaus, J. P. and Brook, E. R. 1999. Abrupt climate change at the end of the last interglacial period inferred from trapped air in polar ice. *Science*, **288**, 930–934.

Shackleton, N. J. 1969. The last interglacial in the marine and terrestrial records. *Proceedings of the Royal Society of London*, **B174**, 135–154.

Shackleton, N. J. and Opdyke, N. D. 1976. Oxygen isotope and palaeomagnetic stratigraphy of Equatorial Pacific core V28-239, Late Pliocene to Latest Pleistocene. *Geological Society of America Memoir*, **145**, 449–464.

Shaw, J. and Carter, R. W. G. 1980. Late-Midlandian sedimentation and glaciotectonics of the North Antrim End Moraine. *Irish Naturalists' Journal*, **20**, 67–69.

Shennan, I., Hamilton, S., Hillier, C. & Woodfoffe, S. 2005. A 16000-year record of near-field relative sea-level changes, northwest Scotland, United Kingdom. *Quaternary International*, **133–134**, 95–106.

Shennan, I., Bradley, S., Milne, G., Brooks, A., Bassett, S. & Hamilton, S. 2006a. Relative sea-level changes, glacial isostatic modelling and ice-sheet reconstructions from the British Isles since the last glacial maximum. *Journal of Quaternary Science*, **21**, 585–599.

Shennan, I., Hamilton, S., Hillier, C., Hunter, A., Woodall, R., Bradley, S., Milne, G., Brooks, A. & Bassett, S. 2006b. Relative sea-level observations in western Scotland since the last glacial maximum for testing models of glacial isostatic land movements and ice-sheet reconstructions. *Journal of Quaternary Science*, **21**, 601–613.

Shotton, F.W. 1962. The physical background of Britain in the Pleistocene. *Advancement of Science*, **19**, 1–14.

Shotton, F. W. and Williams, R. E. G. 1971. Birmingham University Radiocarbon dates VII. *Radiocarbon*, **12**, 385–399.

Siddall, M., Rohling, E. J., Almogi-Laban, A., Hemleben, Ch., Meischner, D., Schmelzer, L. and Smeed, D. A. 2003. Sea-level fluctuations during the last glacial cycle. *Nature*, **423**, 853–858.

Singh, G. 1970. Late-glacial vegetational history of Lecale, Co. Down. *Proceedings of the Royal Irish Academy*, **69B**, 189–216.

Sissons, J. B. 1979. The Loch Lomond Stadial in the British Isles. *Nature*, **280**, 199–203.

Sollas, W. J. 1893–96. A map to show the distribution of eskers in Ireland. *Scientific Transactions of the Royal Dublin Society*,

Series 2, **5**, 785–822.

Smyth, M. 1997. Clara esker. *In:* Mitchell, F. and Delaney, C. (Eds.), *The Quaternary of the Irish Midlands*. Irish Association for Quaternary Studies Field Guide No. 21, 22–24.

Stea, R. W., Piper, D. J. W., Fader, D. B. J. and Boyd, R. 1998. Wisconsin glacial and sea-level history of maritime Canada and the adjoining shelf: a correlation of land and sea events. *Geological Society of America Bulletin*, **110**, 821–845.

Stephens, N. 1958. The evolution of the coastline of north-east Ireland. *Advancement of Science*, **56**, 389–391.

Stephens, N. 1963. Late-glacial sea levels in northeast Ireland. *Irish Geography*, **4**, 345–359.

Stephens, N., Creighton, J. R. and Hannon, M. A. 1975. The late-Pleistocene period in north-eastern Ireland: an assessment 1975. *Irish Geography*, **8**, 1–23.

Stephens, N. and McCabe, A. M. 1977. Late-Pleistocene ice movements and patterns of Late- and Post-Glacial shorelines on the coast of Ulster. *In:* Kidson, C. and Tooley, M. J. (Eds.), *The Quaternary History of the Irish Se*a. Seal House Press, Liverpool. 179–198.

Stephens, N. and Synge, F. M. 1958. A Quaternary succession at Sutton, County Dublin. *Proceedings of the Royal Irish Academy*, **59B**, 131–153.

Stephens, N. and Synge, F. M. 1965. Late Pleistocene shorelines and drift limits in north Donegal. *Proceedings of the Royal Irish Academy*, **64**, 131–153.

Stephens, N. and Synge, F. M. 1966a. Late- and post-glacial shorelines and ice limits in Argyll and north-east Ulster. *Transactions of the institute of British Geographers*, **39**, 101–125.

Stephens, N. and Synge, F. M. 1966b. Pleistocene Shorelines. *In:* Drury, G. H. (Ed.), Essays in Geomorphology. 1–51, London.

Stuiver, M. and Grootes, P. M. 2000. GISP oxygen isotope ratios. *Quaternary Research*, **53**, 277–284.

Stuiver, M., Reimer, P. J. and Braziunas, T. F. 1998. High-precision radiocarbon age calibration for terrestrial and marine samples. *Radiocarbon*, **40**, 1127–1151.

Stuiver, M., Reimer, P. J. and Reimer, R. W. 2005. CALIB 5.0 [WWW program and documentation].

Sutherland, D. G. 1984. The Quaternary deposits and landforms of Scotland and the neighbouring shelves: a review. *Quaternary Science Reviews*, **3**, 157–254.

Synge, F. M. 1950. The glacial deposits around Trim, County Meath. *Proceedings of the Royal Irish Academy*, **53B**, 99–110.

Synge, F. M. 1963. A correlation between the drifts of south-east Ireland and those of west Wales. *Irish Geography*, **4**, 360–366.

Synge, F. M. 1964. The glacial succession in west Caernarvonshire. *Proceedings of the Geologists' Association*, **75**, 431–444.

Synge, F. N. 1968. The glaciation of west Mayo. *Irish Geography*, **5**, 372–386.

Synge, F. M. 1969. The Wurm ice limit in the west of Ireland. *In:* Quaternary Geology and Climate, Publ. 1701, 89–92. National Academy of Sciences, Washington, D. C.

Synge, F. M. 1970a. The Irish Quaternary: current views 1969. *In:* Stephens, N. and Glasscock, R. E. (Eds,), *Irish Geographical Studies in honour of E. Estyn Evans*, Belfast, 34–48.

Synge, F. M 1970b. An analysis of the glacial drifts of south-east Limerick. *Geological Survey of Ireland, Bulletin*, **1**, 65–71.

Synge, F. M. 1971. The glacial deposits of Glenasmole, County Dublin, and the neighbouring uplands. Geological survey of Ireland, Bulletin, 1, 87–97.

Synge, F. M. 1973. The glaciation of south Wicklow and the adjoining parts of neighbouring counties. *Irish Geography*, **6**, 561–569.

Synge, F. M. 1977. The coasts of Leinster. *In:* Kidson, C. and Tooley,

M. J. (Eds.), *The Quaternary History of the Irish Sea*, Seal House Press, Liverpool. 199–222.

Synge, F. M. 1979a. Quaternary glaciation in Ireland. *Quaternary Newsletter*, **28**, 1–18.

Synge, F. M. 1979b. Glacial lake Blessington. *In:* McCabe, A. M. (Ed.), *Field Guide to East Central Ireland*. Quaternary Research Association, Annual field Meeting, April 1979, 40–48.

Synge, F. M. 1981. Quaternary glaciation and changes of sea level in the south of Ireland. *Geologie en Mijnbouw*, **60**, 305–315.

Synge, F. M. and Stephens, N. 1960. The Quaternary Period in Ireland – an assessment. *Irish Geography*, **4**, 121–130.

Tanner, V. 1939. Om de blockrika strandgördlana vid Subarktiska oceankuster. *Förekomstätt og Uppkomst Terra*, **60**, 157–165.

Thomas, G. S. P. and Chiverrell, R. C. 2006. A model of subaqueous sedimentation at the margin of the Late Midlandian Irish Ice sheet, Connemara, Ireland, and its implications for regionally high isostatic sea-levels. *Quaternary Science Reviews*, **25**, 2868–2893.

Thomas, G. S. P. and Kerr, P. 1987. The stratigraphy, sedimentology and palaeontology of the Pleistocene Knocknasilloge Member, Co. Wexford, Ireland. *Geological Journal*, **22**, 67–82.

Thomas, G. S. P. and Summers, A. J. 1981. Pleistocene foraminifera from southeast Ireland-a reply. *Quaternary Newsletter*, **34**, 15–18.

Thomas, G. S. P. and Summers, A. J. 1982. Drop-stone and allied structures from Pleistocene waterlain till at Ely House, County Wexford. *Journal of Earth Sciences Royal Dublin Society*, **4**, 109–122.

Thomas, G. S. P. and Summers, A. J. 1983. The Quaternary stratigraphy between Blackwater Harbour and Tinnaberna, County Wexford. *Journal of Earth Sciences Royal Dublin Society*, **5**, 121–134.

Thomas, G. S. P. and Summers, A. J. 1984. Glaciodynamic structures from the Blackwater formation, County Wexford, Ireland. *Boreas*, **13**, 5–12.

Thomas, G. S. P., Chiverrell, R. C. and Huddard, D. 2004. Ice-marginal depositional responses to readvance episodes in the late Devensian deglaciation of the Isle of Man. *Quaternary Science Reviews*, **23**, 85–106.

Tzedakis, P. C. and Bennett, K. D. 1995. Interglacial vegetation succession: a view from southern Europe. *Quaternary Science Reviews*, **14**, 967–982.

Tzedakis, P. C., Hooghiemstra, H. and Palike, H. 2006. The last 1.35 million years at Tenaghi Philippon: a revised chronostratigraphy and long-term vegetation trends. *Quaternary Science Reviews*, **25**, 3416–3430.

Vaughan, A. P. M., Dowling, L. A., Mitchell, F. J. C., Lauritzen, S. E., McCabe, A. M. and Coxon, P. 2004. Depositional and post-depositional history of warm stage deposits at Knocknacran, Co. Monaghan, Ireland. Implications for preservation of Irish last interglacial deposits. *Journal of Quaternary Science*, **19**, 577–590.

Vernon, P. 1965. Implications of foreign or erratic stones on Slieve Donard, Mourne Mountains. *Irish Naturalists' Journal*, **15**, 36–38.

Vernon, P. 1966. Drumlins and Pleistocene ice flow over the Ards Peninsula/Strangford Lough area, county Down, Ireland. *Journal of glaciology*, **6**, 401–409.

Viau, A. E. and Gajewski, K. 2005. Comments on: 'The magnitudes of millennial- and orbital-scale climate change in eastern north America during the Late Quaternary' by Shuman et al. *Quaternary Science Reviews*, **24**, 2194–2206.

Voelker, A. H. L. and workshop participants. 2002. Global distribution of centennial-scale records for marine isotope stage (MIS) 3: a database. *Quaternary Science Reviews*, **21**, 1185–1212.

Waelbroeck, C., Duplissy, J.-C., Michel, C., Labeyrie, L., Paillard. D. and Duprat, J. 2001. Timing of the last deglaciation in North Atlantic climate rcords. *Nature*, **412**, 724–727.

Walker, R. G. 1975. Generalised facies models for resedimented conglomerates of turbidite association. *Bulletin of the Geological Society of America*, **86**, 737–748.

Walker, R. G. 1984. Turbidite and associated coarse clastic deposits. *In:* Walker, R. G. (Ed.), *Facies Models*, Geological Association of Canada, Reprint Series 1, 91–103.

Walker, R. G. 1992. Facies, facies models and modern stratigraphic concepts. *In:* Walker, R. G. and James, N. P. (Eds.), *Facies Models: Response to Sea Level Change*. Geological Association of Canada/L'Association Géologique du Canada, 1–14.

Warren, W. P. 1979. Moraines on the northern slopes and foothills of the MacGillycuddy's Reeks, south-west Ireland. *In:* Ch. Schluchter (Ed.), *Moraines and Varves*, Balkema, Rotterdam. 223–236.

Warren, W. P. 1985. Stratigraphy. *In:* Edwards, K. J. and Warren, W. P. (Eds.), *The Quaternary History of Ireland*. Academic Press, London, 39–65.

Warren, W. P. 1988a. Leenaun. *In:* O'Connell, M. and Warren, W. P. (Eds.), Field Guide No. 11, Connemara. Irish Association of Quaternary Studies, Galway, 16–18.

Warren, W. P. 1988b. Tullywee Bridge, Kylemore. *In:* O'Connell, M. and Warren, W. P. (Eds.), Field Guide No. 11, Connemara. Irish Association for Quaternary Studies, Galway, 13–16.

Warren, W. P. 1992. Drumlin orientation and the pattern of glaciation in Ireland. *Geologiska Undersokning*, **81**, 359–366.

Warren, W. P. 1993. *Wicklow in the Ice Age*. Geological Survey of Ireland, 46pp.

Warren, W. P. and Ashley, G. 1994. Origins of the ice-contact stratified ridges (eskers) of Ireland. *Journal of Sedimentary Research*, **64A**, 433–449.

Watts, W. A. 1959. Interglacial deposits at Kilbeg and Newtown, Co. Waterford. Proceedings of the Royal Irish Academy, **60B**, 79–134.

Watts, W. A. 1964. Interglacial deposits at Baggotstown, near Bruff, Co. Limerick. *Proceedings of the Royal Irish Academy*, **63B**, 167–189.

Watts, W. A. 1967. Interglacial deposits in Kildromin townland, near Herbertstown, Co. Limerick. *Proceedings of the Royal Irish Academy*, **60B**, 79–134.

Watts, W. A. 1970. Tertiary and interglacial floras in Ireland. *In:* Stephens, N. and Glasscock, R.E. (Eds.), *Irish Geographical Studies*, The Queen's University of Belfast, 17–33.

Watts, W. A. 1977. The late Devensian vegetation of Ireland. *Philosophical Transactions of the Royal Society of London*, **B280**, 273–293.

Watts, W. A. 1985. Quaternary vegetation cycles. *In:* Edwards, K.

J. and Warren, W. P. (Eds.), *The Quaternary History of Ireland*, Academic Press, London, 155–185.

Watts, W. A. 1988. Europe. *In:* Huntley, B. and Webb, T. (Eds.), *Vegetation History*, Dordrechr, Kluwer Academic Publishers, 155–192.

Weaver, T. 1819. Memoir on the geological relations of the east of Ireland. *Transactions of the Geological Society of London*, **5**, part 1, 1819, 117–304.

Wells, J. 1997. The 'Errol Beds' and 'Clyde Beds': a note on their equivalents in the Solway Firth. *Quaternary Newsletter*, **83**, 21–26.

West, R. G. 1977. Flora and fauna, Early and Middle Devensian flora and vegetation. *Philosophical Transactions of the Royal Society of London*, **280B**, 229–246.

West, R. G. 1980. Pleistocene forest history in East Anglia. *New Phytologist*, **85**, 571–622.

Whittington, R. J. 1977. A late-glacial drainage pattern in the Kish Bank area and post-glacial sediments in the central Irish Sea. *In:* Kidson, C. and Tooley, M. J. (Eds.), *The Quaternary History of the Irish Sea*, Seel house Press, Liverpool, 55–68.

Wilson, P. 1990a. Characteristics and significance of protalus ramparts and fossil rock glaciers on Errigal Mountain, County Donegal. *Proceedings of the Royal Irish Academy*, **90B**, 1–21.

Wilson, P. 1990b. Morphology, sedimentological characteristics and origin of a fossil rock glacier on Muckish Mountain, northwest Ireland. *Geografiska Annaler*, **72A**, 237–247.

Wilson, P. 1993. Description and origin of some talus-foot debris accumulations, Aghla Mountains, Co. Donegal, Ireland. *Permafrost and Periglacial Processes*, **4**, 231–244.

Wilson, P. 2004. Relict rock glaciers, slope failure deposits, or polygenetic features? A re-assessment of some Donegal debris landforms. *Irish Geography*, **37**, 77–87.

Wingfield, R. T. R. 1989. Glacial incisions indicating Middle and Upper Pleistocene ice limits off Britain. *Terra Nova*, **1**, 31-52.

Wright, W. B. 1912. The drumlin topography of south Donegal. *Geological Magazine*, **9**, 153–159.

Wright, W. B. 1937. *The Quaternary Ice Age*. Second edition. Macmillan, London.

Wright, W. B. 1920. Minor periodicity in glacial retreat. *Proceedings of the Royal Irish Academy*, **35B**, 93–105.

Wright, W. B. and Muff, H. B. 1904. The pre-glacial raised beach of the south coast of Ireland. *Proceedings of the Royal Dublin Society*, **10**, 250–324.

Yokoyama, Y., Lambeck, K., De Deckker, P., Johnston, P. and Fifield, L. K. 2000. Timing of the last glacial maximum from observed sea-level minima. *Nature*, **406**, 713–716.

Zaragosi, S., Eynaud, F., Pujol., C., Auffret, G. A., Turon, J. L. and Garlin, T. 2001. Initiation of the European deglaciation as recorded in the northwestern Bay of Biscay slope environments (Meriadzek terrace and Trevelyan escarpment): a multi-proxy approach. *Earth and Planetary Letters*, **188**, 493–507.

INDEX

Page numbers in *italic* denote figures